PHILIP'S

ASTRONOMY DICTIONARY

PHILIP'S

ASTRONOMY DICTIONARY

EDITORIAL CONSULTANTS
Ian Ridpath and John Woodruff

Published in Great Britain in 1995
by George Philip Limited,
an imprint of Reed Books,
Michelin House, 81 Fulham Road, London SW3 6RB,
and Auckland, Melbourne, Singapore and Toronto

ISBN 0-540-06016-X (Hardback)
ISBN 0-540-06015-1 (Paperback)

A CIP catalogue record for this book is available from the British Library

Printed in Great Britain

Introduction

Alphabetical order is letter by letter in **extra bold type**, ignoring the spaces between words. So, for example, '**G star**' comes between '**Grus**' and '**Gum Nebula**'. The exception is that '**H I region**' and '**H II region**' appear at the beginning of the '**H**' section. Terms beginning with a number are alphabetized with the number spelt out; an example is '**twenty-one centimetre line**'.

Cross-references are used extensively, and appear in **bold type**. A cross-reference usually indicates that the entry referred to provides secondary information on the topic under discussion. Sometimes it means that the term in **bold type** is defined or discussed at greater length in its own entry. An entry which is just a cross-reference means that the term is defined in another entry, where it will be found in *italic type*. A heading or cross-reference followed by a number, for example '**aberration** (2)', is used whenever the same term has more than one entry.

Most entries are fairly concise, but a number of subjects are treated at greater length, for example '**spectral classification**' and, in particular, the entries for the major planets. The term 'planetary body' is used in this dictionary to refer to large satellites and asteroids in addition to the major planets. For the purposes of the values given for the diameters of the giant planets, their 'surfaces' have been taken to be the levels in their atmospheres at which the pressure is 1 bar, i.e. the average pressure at the surface of the Earth.

Particularly in tables, large numbers are sometimes given in exponential form. Thus 10^3 is a thousand, 2×10^6 is two million, and so on. 'Billion' means a thousand million, or 10^9.

Dates in this book have been expressed in the order day, month, year, although it is customary in astronomy to show dates in the order year, month, day.

In biographical entries, if the first forename of a person is not the one by which they were most commonly known, it is enclosed in parentheses.

List of illustrations

altazimuth mounting 11
anomaly 13
Beta Lyrae star 26
binoculars 28
butterfly diagram 31
Cassegrain telescope 36
celestial sphere 39
Cepheid variable star 40
coudé system 52
Dobsonian telescope 58
eclipsing binary 63
electromagnetic radiation 65
ellipse 66
elongation (conjunction, quadrature and opposition) 67
equatorial telescope 70
Galaxy 79
gravitational lens 85
Hertzsprung–Russell diagram 93
Io 102
Kepler's laws 109
Lagrangian points 111
libration 114
lunar eclipse 117
magnetosphere 121
Mira 132
Newtonian telescope 141
nova 143
orbital elements 149
phase 154
precession 161
pulsar 164
radio telescope 169
R Coronae Borealis star 170
refracting telescope 173
retrograde motion 174
RR Lyrae star 178
RV Tauri star 179
Schmidt camera 185
Schmidt–Cassegrain telescope 186
seasons 188
solar cycle 192
solar eclipse 193
spectroscope 197
spectroscopic binary 198
SS433 199
supernova 209
tides 215
trigonometric parallax 217
tuning fork diagram 220
U Geminorum star 221
Van Allen Belts 225

aberration (1) The small apparent displacement of a star's position produced by the combination of the motion of the Earth and the finite velocity of light.

Annual aberration is the displacement resulting from the motion of the Earth in its orbit around the Sun. This is detected by an observer as an apparent motion of a star in an ellipse over the course of the year. The maximum displacement of the star, given by the **semimajor axis** of this ellipse, is just over 20 seconds of arc. This angle (α) is the *constant of annual aberration* and is determined from the relationship

$$\tan \alpha = v/c,$$

where v is the Earth's orbital velocity and c is the velocity of light. For a star at the pole of the ecliptic the ellipse becomes a circle; for a star on the ecliptic the ellipse becomes a straight line.

Diurnal aberration is a very much smaller displacement of a star, to the east, resulting from the rotation of the Earth on its axis. It reaches a maximum of 0.32 seconds of arc, for an observer at the equator, and is zero at the poles. The Earth's rotation carries an observer in a circle with a velocity that is greatest at the equator and zero at the poles.

The aberration of starlight was discovered by **James Bradley**, who published his findings in 1729. In demonstrating that the Earth is not fixed in space – thus furnishing the first observational evidence to support Copernicus's theory – and that light travels at a finite velocity, the discovery is important in the development of cosmological ideas.

aberration (2) A defect, such as blurring, distortion, or false coloration that can occur in the image produced by a lens or curved mirror. The major aberrations are **chromatic aberration**, spherical aberration, **coma** (1), and **astigmatism.** Spherical aberration results from the unequal deflection of light rays by different zones of a spherical lens or mirror. The rays are not brought to the same focus, causing a lack of definition about the edges of the image. Spherical aberration in lenses can be minimized by using a suitable combination of elements, and in reflecting telescopes it is overcome by the use of a mirror in the shape of a **paraboloid.**

ablation Wearing away of the surface of an object. An example is the erosion of a meteoroid during its high-speed passage through the Earth's atmosphere: atmospheric molecules striking the meteoroid vaporize its surface layers.

absolute magnitude (symbol M) The apparent magnitude a star would have if it were at a standard distance of 10 parsecs from the Earth. The absolute magnitude may be derived from the **apparent magnitude** (m) and the **parallax** (π) by the formula

$$M = m + 5 + 5 \log \pi,$$

where π is in seconds of arc. The absolute magnitudes of most stars lie between -5 and $+15$. The Sun has an absolute magnitude of $+4.8$. SEE ALSO **magnitude**

absolute temperature Temperature as measured on a scale whose zero point is *absolute zero*, the point at which all motion at the molecular level ceases. The unit of absolute temperature is the degree kelvin; an interval of one kelvin (K) is equivalent to an interval of one degree Celsius (°C). The relationship between a Kelvin temperature (T) and the same temperature as measured on the Celsius scale (t) is

$$T = t + 273.16,$$

absolute zero being -273.16°C. Absolute temperatures are widely used in astronomy. (SEE ALSO the table of conversion factors in the Appendix on page 240.)

absorption line A dark line or band in a **continuous spectrum**. Absorption lines are produced when particular wavelengths of radiation from a hot source are absorbed by a cooler, intervening medium. They are typical of the spectra of stars – radiation from the hot interior is absorbed by the star's cooler outer layers, producing absorption lines. The absorption lines in the Sun's spectrum are called **Fraunhofer lines**. The wavelengths at which absorption lines occur are characteristic of the chemical elements present in the radiating source. This enables a star's chemical composition to be determined from its spectrum.

accretion The coalescence of small particles in space as a result of collisions, and the gradual building up of larger bodies from smaller ones by gravitational attraction; also, the accumulation of matter by a star or other celestial object. Matter can also be transferred from one component of a close binary system to another by an accretion process, via an **accretion disk**. Accretion is an important factor in the evolution of stars, planets, and comets.

accretion disk A disk of gas in orbit around an object, formed by inflowing matter. In a binary star system in which the two component stars are very close, gas can be transferred from the larger to the smaller star, creating an accretion disk. If the companion has a strong gravitational field, as it will have if it is a neutron star or black hole, friction will heat the orbiting gas to millions of degrees, hot enough to emit X-rays.

Achernar The star Alpha Eridani, magnitude 0.46, distance 70 light years, luminosity 275 times that of the Sun. It is a main-sequence star of spectral type B3 and the ninth-brightest star in the sky.

Achilles **Asteroid** 588, one of the **Trojan asteroids** sharing Jupiter's orbit, and the first to be discovered, by Max Wolf in 1906. It has a diameter of 147 km (91 miles).

achondrite A type of **stony meteorite**, consisting of rock that was once molten, and which therefore was once part of a planetary body that underwent **differentiation**. Like all stony meteorites, achondrites contain no uncombined metallic elements, and they also lack chondrules – millimetre-sized globules of silicate rock. The principal types are *eucrites*, *diogenites* and *howardites*, differing in chemical make-up and structure, but all with properties (such as radiation damage) that suggest they were once part of a planetary surface. The main candidate is the asteroid Vesta. Achondrites include a handful of meteorites known to have originated on the Moon, and the **SNC meteorites**, believed to have originated on Mars. SEE ALSO **chondrite**

achromatic lens (achromat) A lens designed to reduce **chromatic aberration**, cutting down the amount of false colour in the image. This is usually achieved by using two lens elements made from different types of glass, for example one element of crown glass and the other of flint glass. The combination is chosen such that two selected wavelengths of light come to the same focus, and at the same time the residual aberration is reduced to a minimum. SEE ALSO **apochromat**

active optics A technique for maintaining the primary mirror of large telescopes in accurate shape and alignment, despite the distortions caused by temperature changes or tilting the telescope. The mirror rests on numerous supports which are automatically adjusted by computer to keep its figure smooth under all conditions. The technique can also compensate for any slight imperfections left after the manufacturing of the mirror. Mirrors with active supports can be made thinner, and thus lighter and cheaper, than ordinary telescope mirrors. The term is sometimes used to refer to the more complicated technique properly known as **adaptive optics**.

Adams, John Couch (1819–92) English astronomer. While still an undergraduate at Cambridge, he set out to analyse the motion of Uranus. The planet's observed path was not in agreement with its calculated orbit, and Adams believed that the discrepancies could be accounted for by the gravitational influence of an undiscovered planet. He calculated an orbit for the new planet, but no search was mounted. It was **Johann Galle** who located Neptune, as it was subsequently called, near a position predicted by the French astronomer **Urbain Le Verrier**. After a lengthy priority dispute, the honours of predicting the existence of Neptune came to be shared between Adams and Le Verrier. However, it now seems that neither Adams' nor Le Verrier's calculations were that accurate, and that the two predicted orbits and the actual orbit roughly coincided at the point where Neptune happened to be when Galle found it.

adaptive optics A technique to counteract the blurring of images caused by **seeing** conditions in the Earth's atmosphere. Sensors measure the distortions in a telescope's image caused by atmospheric effects. The image falls on to a small, wafer-thin subsidiary mirror, which can be flexed in a controlled fashion hundreds of times a second to cancel out the imperfections in the image. The resulting image is almost as sharp as if the telescope were in space.

Adrastea One of the small inner **satellites** of Jupiter discovered in 1979 during the Voyager missions. It is irregular in shape, measuring about 25 × 20 × 15 km (15 × 12 × 9 miles), and orbits near the outer edge of Jupiter's main ring.

Advanced X-ray Astrophysics Facility (AXAF) A NASA mission to take images of celestial objects at X-ray wavelengths, planned for launch in 1998.

aerolite Name formerly used for a **stony meteorite**.

airglow A faint glow in the Earth's atmosphere. Molecules in the tenuous upper atmosphere, where the solar radiation is strong, are ionized (split up into charged particles) by ultraviolet radiation or by collision with other charged particles. These particles are short-lived and soon recombine to form the original molecules, emitting a faint light as they do so. The recombination of oxygen molecules is a principal source of emission, which continues long after sunset. From a dark site on a clear, moonless night the airglow is visible as a grey luminous background to the stars, but for many observers **light pollution** makes the airglow impossible to see. The **ashen light** observed on Venus's night side may be a similar phenomenon.

Airy, George Biddell (1801–92) English astronomer and geophysicist. He became Astronomer Royal and Director of the Royal Greenwich Observatory, and re-established the Observatory's standards and authority. Airy discovered irregularities in the orbits of Venus and the Earth, and determined the mass of the Earth by taking gravity measurements at the top and bottom of a mineshaft.

Airy disk The central spot in the image of a star formed in a telescope. The effects of **diffraction** make it impossible for the image to be formed as a point. Instead, it is spread out into an Airy disk, surrounded by diffraction rings and spikes. (Named after George Biddell Airy.)

Al-Battānī, Muḥammad ibn Jābīr (Latinized name **Albategnius**) (858–929) Arab astronomer. His 40 years of observation yielded a catalogue of star positions more accurate than those in Ptolemy's ***Almagest***, and introduced trigonometry into Arab astronomy. He made accurate determinations of the precession of the equinoxes and the obliquity of the ecliptic, and showed that the Earth's distance from the Sun varies over the course of the year.

albedo A measure of the reflectivity, or brightness, of a material or body. The albedo scale goes from 1, for a perfectly reflecting, white, surface, to 0, for a perfectly absorbing, black, surface. In astronomy, albedo is used to indicate the fraction of sunlight reflected by Solar System bodies. Rocky bodies such as Mercury and the Moon have low albedoes, while cloud-covered bodies like Saturn or those consisting of a high proportion of water-ice, like many of the satellites of the outer Solar System, and comets, have high albedoes. Albedo can be defined in various ways. The one most used now is *geometric albedo* – the ratio of the reflectance of the body to that of a flat white surface of the same surface area at the same position.

Aldebaran The star Alpha Tauri, magnitude 0.85 (but slightly variable, of irregular type), 68 light years distant, luminosity 150 times that of the Sun. It is a giant of type K5. It appears to lie in the Hyades cluster, but is actually a foreground object at about half the cluster's distance.

Algol The star Beta Persei, the prototype of a class of eclipsing binaries known as Algol-type variables. The first recorded observation of its variability was made by Geminiano Montanari of Bologna in 1669. **John Goodricke** in 1782 was the first to suggest that the variability resulted from two stars in binary motion, the fainter and larger one from time to time passing in front of the brighter one. There are two stars, of spectral types B8 and G, orbiting each other in 2.867 days. The B8 star is a dwarf and is the visible component; the fainter star is a subgiant. During eclipses, Algol's brightness drops from 2.1 to 3.4. There is also a third star, but it does not take part in the eclipses. Algol lies about 100 light years away.

Algonquin Radio Observatory A radio astronomy observatory at Lake Traverse, Ontario, run by the National Research Council of Canada. Its principal instrument is a 46 m (150 ft) fully steerable radio dish, in operation since 1966.

Almagest An encyclopedia of astronomy compiled by **Ptolemy** in about AD 140. It is thought to be based mainly on the work of **Hipparchus**, whose star catalogue it incorporates. It is the most ancient accurate description of the heavens and remained authoritative until the middle of the 16th century, as nothing else was available.

almanac A yearbook containing predicted positions of celestial objects and other astronomical and calendrical data. For astronomical and navigational purposes the leading publication is *The Astronomical Almanac*, produced jointly each year by the Royal Greenwich Observatory and the US Naval Observatory.

Alpha Centauri The brightest star in the constellation Centaurus, and the third-brightest star in the sky, magnitude −0.27; also known as Rigil Kentaurus. It is a visual binary, with G2 and K1 main-sequence components of magnitudes −0.01 and 1.33, orbital period 80 years, distance 4.3 light years. Associated with it is the much fainter star **Proxima Centauri**.

Al-Sūfi, Abu'l-Husain (903–86) Arab astronomer. In his *The Book of the Fixed Stars* he presented a detailed revision of the star positions in Ptolemy's ***Almagest***, based on his own observations. It contains the first recorded reference to the Andromeda Galaxy. He measured the length of the year, and attempted to determine the length of a degree of the meridian.

Altair The star Alpha Aquilae, magnitude 0.77, distance 17 light years, luminosity ten times that of the Sun. It is a main-sequence star of type A7.

altazimuth A mounting for an instrument such as a telescope that can move in both

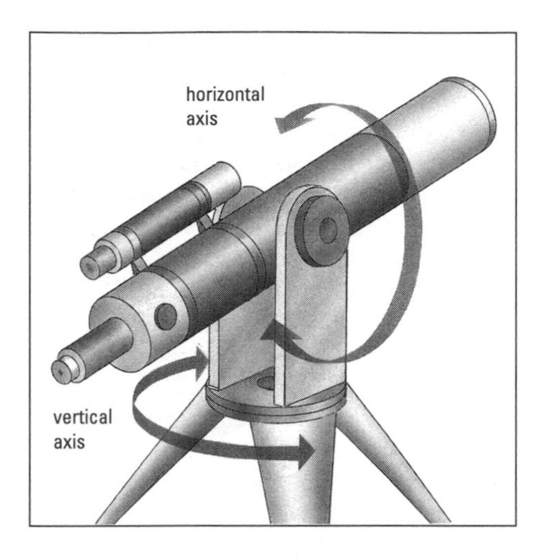

Altazimuth mounting

altitude and azimuth – that is, about both horizontal and vertical axes. The disadvantage of this mounting for an amateur telescope is the problem of moving it simultaneously in azimuth and altitude in order to correct the Earth's diurnal motion, and so keep the object being observed in the field of view. This is made easier with the **equatorial telescope**, and simply ignored with the **Dobsonian telescope**. Since the introduction of computer-controlled drive mechanisms, the altazimuth mounting has been preferred for all large professional instruments.

altitude The angular distance of a celestial body above the observer's horizon. It is measured in degrees from 0 (on the horizon) to 90 (at the zenith) along the **great circle** passing through the body and the zenith. If the object is below the horizon, the altitude is negative. SEE ALSO **coordinates**

Amalthea The largest of Jupiter's small inner **satellites**, discovered in 1892 by Edward Barnard. It is an irregular body, measuring 270 × 165 × 150 km (170 × 100 × 90 miles). Amalthea is dark red in colour, which may indicate that its surface is covered with sulphur from the volcanoes of Io, the innermost of the large satellites.

Amor asteroid One of a class of small **asteroids** whose **perihelion** distances lie inside the orbit of Mars (specifically, those with perihelion between 1.017 and 1.3 AU). They are named after the 1 km (0.6-mile) diameter asteroid 1221 Amor. The first to be discovered, the most studied, and the largest is 588 **Eros**. Some of these small bodies may have been perturbed out of the main asteroid belt by Mars, while others may be the nuclei of extinct comets. Close encounters with Mars or Earth can further perturb Amor asteroids into Earth-crossing **Apollo asteroid** orbits; the two groups are often classified together as *Apollo–Amor objects*. Like all small asteroids in the inner Solar System, Amor asteroids have limited lifetimes, typically 10 million years, and many will eventually impact with Earth or Mars. SEE ALSO **Aten asteroid**

Ananke One of Jupiter's four small outermost **satellites**, discovered in 1951 by Seth Nicholson. It is in a retrograde orbit – SEE **retrograde motion** (1) – and may well be a captured asteroid.

anastigmatic lens (anastigmat) A type of lens designed to compensate for **astigmatism**.

Andromeda A large constellation of the northern hemisphere, adjoining the Square of Pegasus. The main stars lie in a line leading away from Pegasus, and the star Alpha Andromedae (Sirrah or Alpheratz, mag. 2.1) actually forms one corner of the Square. In mythology, Andromeda was the princess rescued by Perseus. Gamma (Almach or Alamak) is a fine double of magnitudes 2.3 and 4.8, spectral types K3 and B9. The open cluster NGC 752 is an easy binocular object, and NGC 7662 is a planetary nebula for small telescopes. The most famous object in the constellation is the **Andromeda Galaxy**.

Andromeda Galaxy A spiral galaxy 2.2 million light years away in the constellation Andromeda, the most distant object visible to the naked eye, also known as M31 or NGC 224. The Andromeda Galaxy has a mass of over 300,000 million Suns. Cepheid variable stars and novae have been observed in it. To the naked eye it appears as a smudge of magnitude 3.5; through binoculars and small telescopes it is seen to have an elongated structure some 1° × 4° in extent, and its plane is tilted at 13° to our line of sight. Its true diameter is about 150,000 light years, somewhat larger than our own Galaxy. The Andromeda Galaxy is accompanied by two elliptical galaxies, M32 and NGC 205, both visible in small telescopes.

Andromedids A **meteor shower** which gave some dramatic displays in the 19th century, following the break-up in the 1840s of its parent comet, **Biela's Comet**. The shower, also known as the *Bielids*, had its radiant in the constellation Andromeda, near the star Gamma. The Andromedid meteor stream no longer intersects the Earth's orbit, but is predicted to do so again early in the 22nd century, when the shower should return.

Anglo-Australian Observatory An observatory at Siding Spring, New South Wales, jointly funded by the UK and Australian governments. It contains the 3.9 m (150-inch) Anglo-Australian Telescope and the 1.2 m (48-inch) UK Schmidt Telescope.

angstrom (angstrom unit) (symbol Å) A unit of length, equal to 10^{-10} m, formerly widely used in spectroscopy for wavelengths of light. It is no longer recommended for use in astronomy, having been superseded by the nanometre, symbol nm (1 nm = 10^{-9} m, 1 Å = 0.1 nm). It is named after Swedish physicist and astronomer Anders Ångström (1814–74), who introduced it in his atlas of the solar spectrum, published in 1868.

angular diameter The apparent diameter of a celestial body expressed in angular measure (usually in **arc minutes** and **seconds**). It is the angle subtended at the observer by the true diameter of the body under observation. If the distance of the celestial body from the observer is known, its true diameter can be calculated.

annual parallax The angle subtended at a celestial object by the radius of the Earth's orbit, which is 1 astronomical unit. It is measured by determining the semimajor axis of the parallactic ellipse traced by the star on the celestial sphere. The reciprocal of the annual parallax in seconds of arc is the distance of the object in **parsecs**. SEE ALSO **parallax**

annular eclipse A type of **solar eclipse** in which the Sun's disk is not completely hidden by the Moon, a thin ring of light (*annulus* is Latin for 'ring') remaining visible around the dark face of the Moon. A solar eclipse is annular if it occurs when the Moon is at **apogee**, at its furthest point from Earth in its elliptical orbit. The tip of its shadow-cone does not reach down to the Earth's surface, and its apparent diameter is a little less than the Sun's.

anomalistic month One revolution of the Moon around the Earth relative to its **perigee**. It is equal to 27.55455 days of **mean solar time**.

anomalistic year One revolution of the Earth around the Sun relative to its **perihelion**. It is equal to 365.25964 days of **mean solar time**. This is about 4m 43.5s longer than a sidereal year, because the perihelion point moves eastwards.

anomaly An angular measurement used in determining the position of a body in an elliptical orbit. It is reckoned from periapsis (SEE **apsides**), in the direction of the object's motion. There are three types of anomaly.

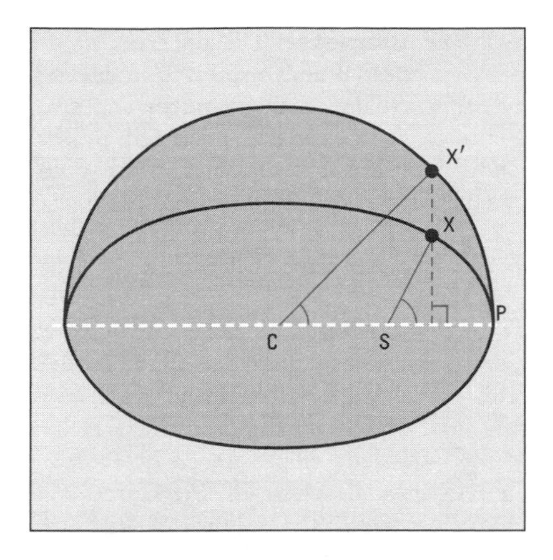

Anomaly: C, centre of ellipse; S, Sun; P, perihelion, X, position of planet

True anomaly, in the case of a planet, is the angle between the planet's radius vector and its perihelion. In the diagram it is the angle PSX. *Mean anomaly* is the same angle, but for an imaginary planet moving along its orbit at constant velocity, rather than at nonuniform velocity as in reality. *Eccentric anomaly* relates to the motion of a point along a circle, the diameter of which is the long axis of the planet's orbit. In the diagram, the eccentric anomaly is the angle PCX′. The point X′ is defined by producing a perpendicular from the long axis through the planet to the circumscribing circle.

ansae (singular **ansa**) The outer parts of Saturn's rings, which through a telescope can appear like handles projecting from the planet's disk. The word is Latin for 'handles'.

antapex SEE **apex**

Antares The star Alpha Scorpii, a supergiant of type M that varies irregularly from about magnitude 0.9 to 1.2. Its luminosity and distance are not well determined. Antares has a main-sequence companion star of magnitude 5.4, type B2, orbital period about 900 years. This star was discovered in 1971 to be a source of weak, variable radio emission.

Antlia A small southern constellation representing an air pump, added to the sky by Lacaille in the 1750s. It has no star brighter than magnitude 4.3.

Antoniadi, Eugène Michael (1870–1944) French astronomer, born in Turkey of Greek parents. He was a highly skilled observer, particularly of the inner planets. His maps of the surface features of Mars and Mercury were not bettered until those planets were visited by spacecraft. The **canals** of Mars were shown by Antoniadi to be illusory. He devised a system known as the **Antoniadi scale**, used to record seeing conditions.

Antoniadi scale The scale devised by Eugène Antoniadi, widely used by amateur astronomers to record the **seeing** conditions (the steadiness of the atmosphere) under which lunar and planetary observations are made. It distinguishes five gradations of seeing:

- I Perfect seeing, without a quiver
- II Slight undulations, with periods of calm lasting several seconds
- III Moderate seeing, with larger air tremors
- IV Poor seeing, with constant troublesome undulations
- V Very bad seeing, scarcely allowing a rough sketch to be made.

ap-, apo- Prefixes referring to the furthest point in an object's orbit around another body, such as **aphelion** and **apogee**.

Apache Point Observatory An observatory in the Sacramento Mountains of New Mexico, containing a 3.5 m (138-inch) reflector. It is operated by a group of US universities called the Astrophysical Research Consortium.

apastron The point in the orbit of one component of a binary system at which it is furthest from the other star.

aperture The clear diameter of an optical telescope's objective lens or primary mirror. More generally, for other types of telescope, the aperture is the size of the principal radiation-collecting element – for example, the antenna of a radio telescope. Aperture is the most significant parameter of any telescope, for the amount of radiation it can collect from distant objects is all-important.

aperture synthesis A technique in radio astronomy in which an array of radio dishes is used to synthesize the resolution of a much larger telescope. The technique involves electronically combining the output from numerous small antennae, which are usually arranged in a line many kilometres long. As the Earth rotates, the array sweeps out an area equivalent to a single large dish with a diameter equal to the maximum length of the line. The technique was developed by the English radio astronomer **Martin Tyle**. SEE also **radio interferometry**

apex The point on the celestial sphere towards which the Sun and the entire Solar System appear to be moving at a velocity of 19–20 km/s relative to the nearby stars; it is also known as the *solar apex*. The apex lies in the constellation Hercules at about RA 18h, dec. +30°. The point on the celestial sphere opposite the apex is the *antapex*, in the constellation Columba.

aphelion The point in the orbit of a planet or a comet at which it is at its furthest distance from the Sun.

apochromat A lens especially corrected to a high degree to eliminate chromatic **aberration**. By using three elements, all of different types of optical glass, a greater amount of correction is possible than with the two-element **achromat**.

apogee The point in the orbit of a body around the Earth at which it is at its furthest distance from Earth.

Apollo asteroid One of a class of small **asteroids** whose **perihelion** distances lie inside the Earth's orbit (specifically, those with perihelion less than 1.017 AU and **semimajor axis** greater than 1 AU). They are named after the 2 km (1.25-mile) diameter object 1862 Apollo. Like the **Amor asteroids**, with which they are often grouped as *Apollo–Amor objects*, some may be extinct cometary nuclei, but others may have been perturbed inwards by Jupiter from one of the **Kirkwood gaps** in the main asteroid belt. Many of them could impact the Earth within a few million years. SEE ALSO **Aten asteroid**

Apollo programme A US space project that landed astronauts on the Moon. The Apollo spacecraft had three main parts: a pressurized *command module* for the three crew members; a *service module* containing oxygen and electrical supplies and the main manoeuvring engine; and the *lunar module* in which two astronauts made the actual Moon landing. The whole affair was launched by a Saturn V rocket. After four preparatory manned flights, Apollo 11 made the first Moon landing in July 1969. Neil Armstrong and Edwin Aldrin became the first men to step on the Moon, remaining on the surface for nearly a day and bringing back 22 kg of lunar rock and dust. Subsequent Apollos extended the exploration of the Moon's surface, bringing back further samples and leaving behind instruments which transmitted data about moonquakes. The last of the series, Apollo 17, was in December 1972.

apparent magnitude (symbol *m*) The brightness of an object as seen from Earth. The apparent magnitude differs from the **absolute magnitude** because of the effect of the object's distance from us. SEE ALSO **magnitude**

apparent solar time Local time, based upon the apparent daily movement of the Sun. It is the time shown on a sundial. Noon occurs when the Sun reaches its maximum altitude, i.e. when it crosses the observer's meridian. Time reckoned in this way is subject to considerable variations because of the Sun's non-uniform motion, a result of both the elliptical shape of the Earth's orbit and the fact that the Sun moves along the ecliptic, not the celestial equator. The time kept by clocks is **mean solar time**, based on the motion of a hypothetical mean Sun.

apparition The appearance in the sky of a comet, or a period during which a planet is best observed.

appulse The apparent close approach of two celestial bodies, such as a planet and a star, whose directions of motion as seen on the celestial sphere converge (though the bodies in question are in reality remote from each other). SEE ALSO **occultation**

apsides (singular **apsis** or **apse**) The two points in an orbit that are respectively closest to (*periapsis*) and furthest from (*apapsis*) the primary body. For example, the apsides of the Earth's orbit are its perihelion and aphelion; in the Moon's orbit they are its perigee and apogee. The *line of apsides* is the line connecting these points.

Apus A far-southern constellation representing the bird of paradise, first shown on Johann Bayer's atlas of 1603. Its brightest star is of magnitude 3.8.

Aquarids Either of two **meteor showers**, the **Delta Aquarids** and the **Eta Aquarids**.

Aquarius A constellation of the zodiac, representing a figure pouring water from a jar. Alpha (Sadalmelik) and Beta (Sadalsuud) are each of magnitude 2.9 and jointly the brightest in the constellation. Zeta is a fine binary of period 850 years, with components of magnitudes 4.3 and 4.5. R Aquarii is a **symbiotic variable**, with a range from 5.8 to 12.4 and a mean period of 387 days. There are two important planetary nebulae: the **Saturn Nebula** and the **Helix Nebula**. Aquarius also contains two globular clusters, M2 and M72.

Aquila A distinctive constellation, led by first-magnitude **Altair**; most of it lies in the northern hemisphere of the sky, although the celestial equator crosses it. Mythologically it represents the eagle sent by Zeus to collect the shepherd-boy Ganymede to become cup-bearer to the gods. Gamma (Tarazed), magnitude 2.7, and Beta (Alshain), 3.7, stand either side of Altair. Eta, a Cepheid variable, has a period of 7.18 days and a range from 3.5 to 4.4; it lies between Theta (3.2) and Delta (3.4), which make good comparison stars. The Mira variable R Aquilae has a range from 5.5 to 12.0 and a period of 284 days. The Milky Way runs through the constellation and is very rich in this region.

Ara A southern constellation between Scorpius and Triangulum Australe, representing the altar of the gods. Beta is its brightest star, magnitude 2.9. The brightest variable in the constellation is the eclipsing binary R Arae (6.0 to 6.9; period 4.43 days, Algol type). There is also a prominent open cluster, NGC 6193, and a scattered globular, NGC 6397, both easy in binoculars.

arc minute, second (symbols ′, ″) Units of angular measure, amounting to 1/60 of a degree and 1/3600 of a degree, respectively. These units are used to express the apparent diameter or separation of astronomical objects and the resolving power of a telescope. Often abbreviated to arcmin, arcsec.

Arcturus The star Alpha Boötis, the fourth-brightest in the sky, magnitude −0.04, distance 36 light years, luminosity 100 times that of the Sun. It is a giant of type K2.

Arecibo Observatory The radio and radar telescope near Arecibo in Puerto Rico, diameter 305 m (1000 ft), the largest single-bowl type of instrument in operation. It is built into a natural hollow in the limestone terrain and cannot be steered. However, an ingenious feed system enables the beam to be swung as far as 20° from the zenith. It was opened in 1963 and is operated by Cornell University.

areo- Prefix referring to the planet Mars, as in *areography*, the description and mapping of the surface features of Mars. (From 'Areos', Greek for Mars.)

Argelander, Friedrich Wilhelm August (1799–1875) German astronomer. In 1852 he began the vast undertaking of preparing an atlas and catalogue of all stars down to magnitude 9.5 in the northern hemisphere. This immense work, covering over 300,000 stars, was published as the ***Bonner Durchmusterung***. He introduced the system of nomenclature for variable stars by which they are assigned one or two capital roman letters, beginning with R and RR, and also the subdivision of magnitudes into tenths.

Ariel One of the five main **satellites** of Uranus, diameter 1158 km (720 miles), discovered in 1851 by William Lassell. It is composed largely of a mixture of rock and water-ice, with a density of 1.6 g/cm^3. Ariel has been geologically active in the past. Its characteristic features are wide, steep-sided troughs like Korrigan Chasma that wind across a landscape made up of cratered terrain, ridged terrain, and plains. **Tectonics** accounts for the ridges, and volcanic flooding for the plains; the troughs formed when the water interior froze and the satellite expanded, splitting the crust.

Aries The first constellation of the zodiac although, since **precession** has now shifted the **vernal equinox** into Pisces, Aries has technically lost this distinction, In mythology, it represents the lamb with the golden fleece. Its brightest stars are Alpha (Hamal), magnitude 2.0, and Beta (Sheratan), magnitude 2.6.

Aristarchus of Samos (3rd century BC) Greek mathematician and astronomer. He tried to calculate the distances of the Sun and Moon from the Earth, as well as their sizes. His method was sound but the results were inaccurate. He explained the immobility of the fixed stars by their great distance from the Earth's orbit. He put the Sun at the centre of the Universe, making him the first to propose a heliocentric theory, but the idea was not taken up because it did not seem to make the calculation of planetary positions any easier.

Aristotle (384–322 BC) Greek philosopher and encyclopedist. He developed a view of the Universe based on the four 'elements' (earth, air, fire and water) and the system of concentric spheres proposed by Eudoxus of Cnidus upon which various celestial bodies were carried. Aristotle added more spheres so as to account for the motion of all celestial bodies. The celestial bodies were, he said, made of a substance called aether, and are perfect and incorruptible, unlike the Earth. He attempted to estimate the size of the Earth, which he maintained was spherical, but unmoving – the centre of the Universe. This 'Aristotelian' view of the world remained almost unchallenged until Copernicus.

Arizona Meteor Crater SEE **Meteor Crater**

array An arrangement of antennae used in radio astronomy. SEE **radio telescope**

artificial satellite A man-made object moving in orbit around the Earth or some other body. The first artificial satellite, Sputnik 1, was successfully put into orbit by the USSR on 4 October 1957.

ascending node The point at which an orbit crosses from south to north of, for example, the celestial equator or the ecliptic. SEE **node**

ashen light A faint luminosity on the night side of Venus when it appears as a thin crescent, close to inferior conjunction. The ashen light is occasionally reported by observers, although it has never been photographed. Its appearance is very similar to **earthshine** on the Moon, although that is produced in a different way. If it is a real phenomenon, and not illusory, it is probably the result of electrical disturbance in the planet's ionosphere, and may therefore be similar in nature to the **airglow** in the Earth's atmosphere.

aspect The position of a planet or the Moon relative to the Sun as seen from Earth. SEE **conjunction**, **elongation**, **opposition**, **quadrature**

association, stellar A group of stars which have formed together, but which are more loosely linked than a star cluster. Many are young and define portions of the spiral arms of our Galaxy. SEE **OB association**, **T association**

A star A star of spectral type A, characterized by the great strength of the hydrogen absorption lines in its spectrum. On the main sequence A stars range in surface temperature from 7500 K at type A9 up to 9900 K at type A0. Their corresponding masses and radii are 1.8 and 1.4 times the Sun's respectively at A9, increasing to 3.2 and 2.5 at A0. Examples are Sirius, Vega, Altair, and Deneb. Roughly 10% of A stars have strong magnetic fields. In these, certain chemical elements become concentrated in their atmospheres and produce unusually strong spectral lines. These are classified as Ap (for peculiar) stars. Another kind of spectral anomaly produces the metallic-line (Am) stars, of which Sirius is one.

asterism A group of easily recognizable stars, part of a constellation. An example is the seven stars that form the shape of the Plough or Big Dipper in Ursa Major.

asteroid (minor planet) A small Solar System body in an independent orbit around the Sun. The majority move between the orbits of Mars and Jupiter. The largest asteroid, and the first to be discovered, on 1 January 1801 by Giuseppe Piazzi, is **Ceres**, with a diameter of 913 km (567 miles). The smallest measured are the so-called **near-Earth asteroids**, which include the object designated 1993 KA_2, estimated to be between 5 m and 11 m (16 ft and 36 ft) across. There are thought to be a million asteroids with a diameter greater than 1 km (0.6 mile); below this, they extend in size down to dust particles. Some of the very small objects have found their way to Earth as **meteorites**.

The second asteroid to be found was **Pallas**, in 1802, followed by **Juno** in 1804 and **Vesta** in 1807. It was not until 1845 that the fifth asteroid, Astraea, was discovered, by Karl Hencke (1791–1866), but from then on the pace of discovery quickened. In 1891 **Max Wolf** made the first asteroid discovery by photography, and went on to find hundreds more. Since then the introduction of a new technique has often increased the rate of discovery; the most recent example is the CCD. Now nearly 6000 asteroids have been catalogued and have had their orbits calculated, and this figure is increasing by several hundred a year. At least 10,000 more have been observed, but not often enough for an orbit to be calculated. Some of the larger asteroids are spherical, but most are irregularly shaped. Since the 1970s, polarization and infrared techniques have allowed improved measurements of asteroids' albedos and sizes.

Most asteroids orbit within the main *asteroid belt*, a torus-shaped (like a ring doughnut) region of space between about 2.2 and 3.3 AU from the Sun. However, within the

ASTEROIDS						
Name	Number	Discovery	Diameter (km)	Av. distance from Sun (million km)	Orbital period (years)	Rotation period (hours)
The first discovered						
Ceres	1	1801, Piazzi	913	413.6	4.60	9.08
Pallas	2	1802, Olbers	525	414.0	4.61	7.81
Juno	3	1804, Harding	244	399.3	4.36	7.21
Vesta	4	1807, Olbers	501	353.2	3.63	5.34
Astraea	5	1845, Hencke	125	385.2	4.13	16.81
Hebe	6	1847, Hencke	192	362.8	3.78	7.27
Iris	7	1847, Hind	203	357.0	3.69	7.14
Flora	8	1847, Hind	141	329.4	3.27	13.6
Metis	9	1848, Graham	146	357.1	3.69	5.08
Hygeia	10	1849, De Gasparis	429	469.2	5.55	17.50
Other large asteroids						
Davida	511	1903, Dugan	337	475.4	5.67	5.13
Interamnia	704	1910, Cerulli	333	458.1	5.36	8.72
Europa	52	1858, Goldschmidt	312	463.3	5.46	5.63
Eunomia	15	1851, De Gasparis	272	395.5	4.30	6.08
Sylvia	87	1866, Pogson	271	521.5	6.52	5.19
Psyche	16	1852, De Gasparis	264	437.1	5.00	4.20
Euphrosyne	31	1854, Ferguson	248	472.1	5.58	5.53
Cybele	65	1861, Tempel	246	513.0	6.37	6.07
Bamberga	324	1892, Palisa	242	401.4	4.41	29.43
Camilla	107	1868, Pogson	237	521.8	6.50	4.84
Some Trojan asteroids						
Hektor	624	1907, Kopff	225	775.3	11.80	6.92
Agamemnon	911	1919, Reinmuth	175	780.3	11.91	—
Diomedes	1437	1937, Reinmuth	171	764.7	11.56	18.0
Patroclus	617	1906, Kopff	149	782.7	11.97	—
Achilles	588	1906, Max Wolf	147	774.4	11.78	—
Some Amor asteroids						
Ganymed	1036	1924, Baade	41	398.7	4.35	—
Eros	433	1898, Witt	21	218.6	1.76	5.27
Betulia	1580	1950, Johnson	6.5	328.3	3.25	6.13
Lick	1951	1949, Wirtanen	5.0	208.0	1.64	—
Amor	1221	1932, Delporte	1	287.2	2.66	—
Some Apollo asteroids						
Toro	1685	1948, Wirtanen	12.2	204.5	1.60	10.20
Phaethon	3200	1983, Green	5.23	190.2	1.43	4
Daedalus	1864	1971, Gehrels	3.3	218.5	1.77	8.57
Apollo	1862	1932, Reinmuth	2.0	220.1	1.78	3.07
Oljato	2201	1947, Giclas	1.9	325.5	3.21	24
Icarus	1566	1949, Baade	1.7	161.3	1.12	2.27

Name	Number	Discovery	Diameter (km)	Av. distance from Sun (million km)	Orbital period (years)	Rotation period (hours)
ASTEROIDS *continued*						
Some Aten asteroids						
Ra-Shalom	2100	1978, Helin	3.5	124.5	0.76	19.79
Aten	2062	1976, Helin	1	144.6	0.95	—
Khufu	3362	1984, Dunbar and Barocci	1	148.0	0.98	—
Hathor	2340	1976, Kowal	0.5	126.2	0.78	—

Diameters and rotation periods of asteroids are often uncertain, and different measurement techniques give different results; most diameters and rotation periods in this table were obtained by the IRAS satellite. For irregular asteroids, the diameter given is an average.

main belt are the **Kirkwood gaps** at distances from the Sun at which **resonances** rule out stable orbits. The typical asteroid orbit is elliptical but more eccentric than those of the major planets, and with a greater inclination to the ecliptic. Some asteroids have orbits that bring them into the inner Solar System (SEE **Amor asteroid**, **Apollo asteroid**, and **Aten asteroid**), while others, such as **Hidalgo** and **Chiron**, have orbits that lie partially or entirely in the outer Solar System.

Asteroids almost certainly originate from **planetesimals** left over from the formation of the Solar System, prevented by Jupiter's gravitational influence from forming into a planet. In the early days of the Solar System Jupiter's gravitation diverted many asteroids into planet-crossing orbits, causing a period of intense cratering of planetary and satellite surfaces, and ejected many other asteroids from the Solar System altogether. The larger of those remaining underwent **differentiation**, acquiring metallic cores and rocky outer layers; many of these were broken up by later collisions. This, and the different distances from the Sun at which the original planetesimals formed, accounts for the wide range of compositional types observed today. The present asteroid population does not represent the remnants of a large planet that disintegrated, as was once thought: put together, they would make up a body only about 1500 km (950 miles) in diameter.

New discoveries are designated by the year of discovery and a pair of letters. A permanent number is given when the orbit has been accurately calculated, and the discoverer may then add a name to the number (as with, for example, 1707 Chantal), subject to the approval of the International Astronomical Union.

Two asteroids have been imaged by the space probe Galileo: **Gaspra** in 1991, and **Ida** in 1993. Other space probes to asteroids are planned, such as the Near Earth Asteroid Rendezvous mission. SEE ALSO **Hirayama family**, **Kuiper Belt**, **Trojan asteroid**

astigmatism An **aberration** (2) in lenses and mirrors that affects light from objects away from the centre of the observer's field of view. Instead of being bought to a point focus, these oblique rays of light form an image which is either an ellipse or a straight line, depending on the distance from the lens at which it is viewed. A lens designed so as to minimize astigmatism is called an *anastigmat*.

Astraea **Asteroid** no. 5, diameter 125 km (78 miles), discovered by Karl Hencke in 1845.

astrograph A telescope used to take wide-angle photographs of star fields, mainly for **astrometry**. In particular, the term is used for the instruments designed for the *Carte du Ciel* photographic survey of the sky – equatorially mounted refractors having an aperture of 330 mm and a focal length of 3.4 m, with objectives corrected for colour and coma. The work of astrographs is now largely carried out with **Schmidt telescopes**, which can be made with larger apertures.

astrolabe An early astronomical instrument for showing the appearance of the celestial sphere at a given moment, and for determining the altitude of celestial bodies. The basic form consisted of two concentric disks – one with a star map and one with a scale of angles around its rim – joined and pivoted at their centres (rather like a modern **planisphere**), with a sighting device attached. Astrolabes were used from the time of the ancient Greeks down to the 17th century for navigation, measuring time, and terrestrial measurement of height and angles.

astrology A pseudo-science professing to assess people's personality traits and to predict events in their lives and future trends in general from aspects of the heavens, in particular the positions of the planets. Astrology is based on ideas which are scientifically unsound and which the great majority of rational people dismiss, but horoscopes are still published in the popular press, though mainly for their entertainment value. In many ancient cultures it was for astrological reasons that observational records first came to be kept. Such records from ancient China have proved of great historical importance in research into past eclipses, novae, and comets. Even in later times, casting horoscopes provided astronomers such as **Johnannes Kepler** with the livelihood to sustain them in their astronomical work.

astrometric binary A star which is recognized as a **binary star** from its wavelike **proper motion**, caused by its orbital motion with an unseen companion.

astrometry The branch of astronomy concerned with the measurement of precise positions of celestial objects, usually determined from photographic plates taken specially for the purpose. **Parallaxes** and **proper motions** of stars are derived from positions obtained at widely differing times. The most accurate positional measurements have been made by the astrometry satellite **Hipparcos**.

Astronomer Royal An additional title, until 1972, of the Director of the Royal Greenwich Observatory, but since then an honorary title bestowed upon a prominent British astronomer. The first Astronomer Royal, John Flamsteed, was appointed by King Charles II when he founded the Greenwich Observatory.

The post of Astronomer Royal for Scotland was created initially to provide a director for the Royal Observatory, Edinburgh. The present holder, appointed in 1995, is John Brown. The last Astronomer at the Cape, working originally at the Royal Observatory at the Cape of Good Hope (now

ASTRONOMERS ROYAL

Name	Held office
John Flamsteed (1646–1719)	1675–1719
Edmond Halley (1656–1742)	1720–42
James Bradley (1693–1762)	1742–62
Nathaniel Bliss (1700–64)	1762–64
Nevil Maskelyne (1732–1811)	1765–1811
John Pond (1767–1836)	1811–35
George Biddell Airy (1801–92)	1835–81
William Christie (1845–1923)	1881–1910
Frank Watson Dyson (1868–1939)	1910–33
Harold Spencer Jones (1890–1960)	1933–55
Richard Woolley (1906–86)	1956–71
Martin Ryle (1918–84)	1972–82
Francis Graham-Smith (1923–)	1982–90
Arnold Wolfendale (1927–)	1991–94
Martin Rees (1942–)	1995–

in South Africa), was Richard Stoy; the post was abolished in 1968.

astronomical twilight The dawn and dusk periods when the Sun is less than 18° below the horizon. SEE ALSO **twilight**

astronomical unit (symbol AU) The mean distance between the Earth and the Sun, used as a fundamental unit of distance, particularly for distances in the Solar System. It is equal to 149,598,000 km (92,956,000 miles).

astrophotography The application of photography to astronomy. Photography makes it possible to obtain a permanent record of the positions and appearance of objects, while long exposures show faint objects such as remote galaxies invisible to the naked eye. In 1850 the first star images, of Vega and Capella, were obtained by W. C. Bond at Harvard College Observatory. Henry Draper obtained a spectrogram of Vega in 1872 and photographed the Orion Nebula in 1880. The 1880s marked photography's breakthrough in astronomy, as techniques improved and plates became more sensitive. The first photographic atlases of the sky were made, and libraries of stellar spectra were built up. In the 20th century astrophotography rapidly supplanted visual observation for many purposes, but has itself now been superseded for certain applications by electronic detectors such as CCDs. These have great sensitivity but limited fields of view, so photography remains better suited for wide-field sky surveys.

Astrophotography differs from normal picture-taking in that exposures may be very long, often an hour or more, and the camera or telescope must accurately track the object as the Earth rotates during the exposure. The objects on an astrophotograph have an enormous range of brightness, perhaps a million to one, whereas objects in terrestrial photography have a brightness range usually less than 100 to 1. There is no 'correct' exposure; the longer it is, the fainter the objects that are recorded, although a practical limit is reached by the build-up of background sky fog.

Large observatories use glass plates coated with special emulsions that are designed to be most efficient for long exposures. The speeds of these emulsions can be improved between 10 and 30 times by *hypersensitizing* the plates before use, which involves baking them in nitrogen or nitrogen/hydrogen mixtures. Astronomical emulsions are available in a variety of spectral sensitivities, chosen to isolate well-defined parts of the spectrum so that measurements of star colours, for example, can be compared between observatories. The astronomer can also choose between emulsions on the basis of contrast, resolution and granularity. Modern emulsions have low photographic speeds and very high contrast. The high contrast ensures that objects that are fainter than the night sky will be visible on the plate, while the fine, uniform grain structure gives good resolution. Various post-processing techniques, such as *contact copying* and *unsharp masking,* can be used to enhance faint features on the plates. Plates taken through colour filters can be combined in the darkroom to give true-colour images.

astrophysics The study of the physical properties of celestial bodies. It is based mainly on the study of radiation from these bodies, and has developed since the 19th century through the application of photography, photometry, and spectroscopy. New observing techniques at wavelengths from radio and infrared to X-rays and gamma rays have enormously increased the scope of astrophysics in the second half of the 20th century. Theoretical astrophysics applies mathematical methods and physical laws to explain or predict the behaviour of the Universe and objects within it.

Aten asteroid One of a class of small **asteroids** whose orbits lie largely within the

Earth's (specifically, those with **semimajor axis** less than 1 AU). They are named after the 1 km (0.6-mile) diameter object Aten. About twenty are known, although it is believed there could be as many as a hundred. SEE ALSO **Amor asteroid, Apollo asteroid**

Atlas One of the small inner **satellites** of Saturn, discovered in 1980 during the Voyager missions. It orbits just outside Saturn's A ring, and seems to act as the **shepherd moon** to the whole bright ring system.

atmosphere The envelope of gases around a star, planet or satellite. The ability of a planet or satellite to retain an atmosphere depends on its outer temperature, determined by its distance from the Sun, and its escape velocity, determined by its mass. Smaller bodies in the inner Solar System, such as Mercury and the Moon, have almost no atmosphere or none at all, while the giant planets have deep, massive atmospheres. Far from the Sun, even small worlds like Pluto and Triton have clung on to a tenuous atmosphere. Venus, Earth, and Mars were too small to hang on to the lighter gases, such as hydrogen, from their original atmospheres, which had the composition of the solar nebula from which the Solar System formed. Subsequent geological processes such as **volcanism** have modified their atmospheres considerably, and they now contain carbon dioxide, oxygen, and nitrogen (although in very different proportions). The giant planets have their original, largely hydrogen, atmospheres. Saturn's largest moon, **Titan**, is the only satellite with a substantial atmosphere, of nitrogen.

The outer layers of a star, which is where the lines in the star's spectrum originate, are referred to as its atmosphere.

atmospheric extinction The reduction in the brightness of light from celestial bodies when it passes through the Earth's atmosphere. The effect is more pronounced for objects at low altitude, for their light has then had to pass through a greater volume of atmosphere. Observations of, for example, a variable star have to be corrected for extinction, which can dim the star's light by half a magnitude when viewed at an altitude of 20°, and by a whole magnitude at 10°. The main cause of atmospheric extinction is the **scattering** of light by molecules in the atmosphere. As scattering affects wavelengths at the blue end of the visible spectrum more than it does the red, the light becomes reddened, as is most obvious at sunrise or sunset.

atmospheric refraction The small apparent change in altitude of a celestial object, seen by an observer at the Earth's surface, caused by its light passing through the Earth's atmosphere. Just as light is refracted when it passes from air into glass, so it is refracted when it passes from the vacuum of space into the atmosphere. Atmospheric refraction makes a body appear to be at a higher altitude than it actually is. The effect is greater the closer the body is to the horizon, and a body can even become visible before it has risen. All astronomical observations have to be corrected for atmospheric refraction in order to obtain the true position from the observed apparent position.

AU ABBREVIATION FOR **astronomical unit**

Auriga A large northern constellation, containing the first-magnitude star **Capella**. Mythologically it represents Erichthonius, a king of Athens, who invented the four-horse chariot. Beta (Menkalinan), magnitude 1.9, is an eclipsing variable of very small range. The stars Eta and Zeta are popularly called the Haedi or Kids. **Epsilon Aurigae** is a remarkable eclipsing binary. Zeta is also an eclipsing binary, consisting of a K5 giant orbited by a B7 main-sequence star with a period of 972 days; the visual range is magnitude 3.7 to 4.0. There are three prominent open clusters: M36, M37, and M38.

aurora (plural **aurorae**) An illumination of the night sky, also known as the *aurora borealis* in the northern hemisphere and as the *aurora australis* in the southern hemisphere. An aurora can take many forms, for example a horizontal arc with rays radiating upwards, or a hanging curtain exhibiting beautiful, flickering colours, usually red and green. Aurorae occur at altitudes between 20 km and 1000 km (12 miles and 600 miles), but most typically at 100 km (60 miles). The phenomenon is produced when fast-moving particles arriving in the **solar wind**, attracted by the Earth's magnetic field, collide with atoms of oxygen and nitrogen, which then emit energy at red and green wavelengths. Auroral displays are commonest at times of high solar activity, when solar **flares** emit powerful bursts of charged particles. Aurorae were observed by the Voyager probes on all the giant planets, which all have strong magnetic fields.

Australia Telescope A large **aperture synthesis** telescope in New South Wales, the most powerful radio astronomy instrument in the southern hemisphere. It has three separate parts. One consists of a 6 km line of six dish aerials, each 22 m (72 ft) in diameter, at Culgoora. These can be linked to another 22 m dish, 100 km to the south at Mopra, and also the large radio dish 200 km further south at Parkes. The Australia Telescope came into operation in 1988.

autumnal equinox SEE **equinox**

averted vision Looking slightly to one side of a faint object in order to see it better. The most sensitive part of the retina in the human eye is not right at the centre, but in a small region surrounding the centre. Using averted vision when observing causes the light from the object being observed to fall in this region, making it easier to see.

AXAF ABBREVIATION FOR **Advanced X-ray Astrophysics Facility**

axis (plural **axes**) The imaginary line, joining the north and south poles of a celestial body, about which it rotates. The magnetic axis of a body possessing a magnetic field is the line joining the north and south poles of the field. A body's rotational and magnetic axes do not necessarily coincide; with **Uranus** and **Neptune**, for example, they are markedly different.

azimuth The angle measured eastwards from north in a horizontal plane to the vertical circle (meridian) that runs through a celestial object. SEE ALSO **coordinates**

B

Baade, (Wilhelm Heinrich) Walter (1893–1960) German–American astronomer. From Mount Wilson Observatory in the 1943 wartime blackout, he was able to observe individual stars in the Andromeda Galaxy and distinguish the younger, bluer **Population I** stars from the older, redder **Population II** stars. He went on to improve the use of **Cepheid variable** stars as distance indicators, and showed that the Universe was older and larger than had been thought.

background radiation SEE **cosmic microwave background**

Baily's beads A very short-lived phenomenon seen during a solar eclipse, just before or after totality, in which the extreme edge of the Sun's disk appears to break into a string of bright lights. It is caused by rays of light from the Sun shining through the valleys on the Moon's limb, while other rays are blocked off by mountains on the limb. It was first described by English astronomer Francis Baily (1774–1844), who observed it in 1836. SEE ALSO **diamond-ring effect**

Balmer series The distinctive series of lines in the **hydrogen spectrum**. The Balmer series is a dominant characteristic of the spectra of A stars. The first Balmer line is called the **hydrogen alpha line** (Hα), and is at 656.3 nm; the second is Hβ, at 486.1 nm; the third is Hγ, at 434.2 nm; and so on. The formula governing the wavelengths of lines in the spectrum of hydrogen was worked out in 1885 by Swiss mathematics teacher Johann Balmer (1825–98).

Barlow lens An extra lens used in conjunction with a telescope's eyepiece in order to increase the magnification, usually by a factor of 2. It can also improve definition. (After English physicist Peter Barlow, 1776–1872.)

Barnard, Edward Emerson (1857–1923) American astronomer. He was a skilled visual observer, and also a pioneer of photographic astronomy. Barnard discovered many comets, the fifth satellite of Jupiter (Amalthea), and **Barnard's Star**. He realized that what appear to be regions containing few stars are in fact dark nebulae, and he published a catalogue of them.

Barnard's Star A red dwarf star 6 light years away in the constellation Ophiuchus. It has the largest known **proper motion**, at 10.3 arc seconds per year, and is the closest star to the Sun after the Alpha Centauri system. It was discovered in 1916 by the American astronomer **Edward Barnard**. Barnard's Star is magnitude 9.5, with a luminosity 0.0004 times the Sun's, and type M5. Irregularities in its proper motion have been attributed to the presence of planets similar to Jupiter and Saturn around it, but also to observational deficiencies.

barred spiral galaxy A spiral **galaxy** in which the arms extend from the ends of a bar through the main plane of the system.

Barringer Crater SEE **Meteor Crater**

barycentre The **centre of mass** of two or more celestial bodies – the point around which they revolve. The term is used in particular for the Earth–Moon system. As the Earth is so much more massive than the Moon, the barycentre lies within the Earth.

Bayer, Johann (1572–1625) German astronomer. In 1603 he published a star atlas, *Uranometria*, in which he introduced twelve new constellations. In it he allocated letters of the Greek alphabet (SEE the table on page 239 in the Appendix) to the stars in each constellation: alpha (α), beta (β), gamma (γ), and so on, generally in order of brightness. This system of what came to be known as *Bayer letters* is still in use.

BD ABBREVIATION FOR ***Bonner Durchmusterung***

Becklin–Neugebauer Object SEE **Orion Nebula**

Belinda One of the small inner **satellites** of Uranus discovered in 1986 during the Voyager 2 mission.

Bell Burnell, (Susan) Jocelyn (1943–) English astronomer. As a graduate student, she ran a survey of radio galaxies which led to the discovery of the first **pulsar**. She is now generally credited, with her then supervisor, **Antony Hewish**, as the co-discoverer of pulsars, but she received no share of the Nobel prize that was subsequently awarded.

Bennett, Comet One of the brightest comets of the 20th century, discovered by South African amateur astronomer John Caister Bennett (1914–90) in 1969, and a first-magnitude object in 1970. Its tail extended for 20°.

Bessel, Friedrich Wilhelm (1784–1846) German mathematician and astronomer. He was a pioneer of exact measurements in positional astronomy. His catalogue of 63,000 star positions, based on refinements of previous observations as well as his own, marks the beginning of modern astrometry. In 1838 he became the first to announce that he had determined the parallax of a star (61 Cygni). From the 'wobbles' in the motions of Sirius and of Procyon he deduced the existence of their companion stars, which were not detected until many years after his death. His name is associated with mathematical functions (*Bessel functions*) which were used in the calculation of planetary orbits.

Beta Lyrae star A type of **eclipsing binary** in which gas escapes from a bloated primary star and falls on to an **accretion disk** that surrounds the secondary star. Beta Lyrae stars are members of a broad class of double stars called *semidetached* binaries; they are said to be semidetached because only one of the two stars fills its **Roche lobe**. In Beta Lyrae itself, the primary star is overflowing its Roche lobe and gases stream on to the disk at the rate of 0.00001 solar masses per year; some gas escapes from the system altogether. The lower half of the diagram (on page 26) shows the light curve of Beta Lyrae; the main minimum (A) is when the secondary passes in front of the brighter primary, and the shallower minimum (B) is when the opposite occurs.

Beta Pictoris SEE **Pictor**

Betelgeuse The star Alpha Orionis, and the second-brightest star in Orion. It is an orange-red supergiant of type M2 Iab and a semiregular variable that ranges from magnitude 0 to 1.3 with a period of 6 years or so. Its luminosity is over 60,000 times that of the Sun. Estimates of its distance range from a few hundred light years to about 1400 light years. Betelgeuse has a diameter about 400 times that of the Sun (about 500 million km/ 300 million miles).

Bethe, Hans Albrecht (1906–) German-American physicist. In 1938 he and **Carl von Weizsäcker** independently proposed a detailed theory to account for the production of energy in the Sun and other stars (SEE **carbon–nitrogen cycle**) in which hydrogen is converted into helium by nuclear fusion. This is essentially still accepted today. SEE ALSO **George Gamow**

Bianca One of the small inner **satellites** of Uranus discovered in 1986 during the Voyager 2 mission.

Biela's Comet A now disintegrated comet of period 6.62 years, discovered by Wilhelm von Biela (1782–1856) in 1826. At its return

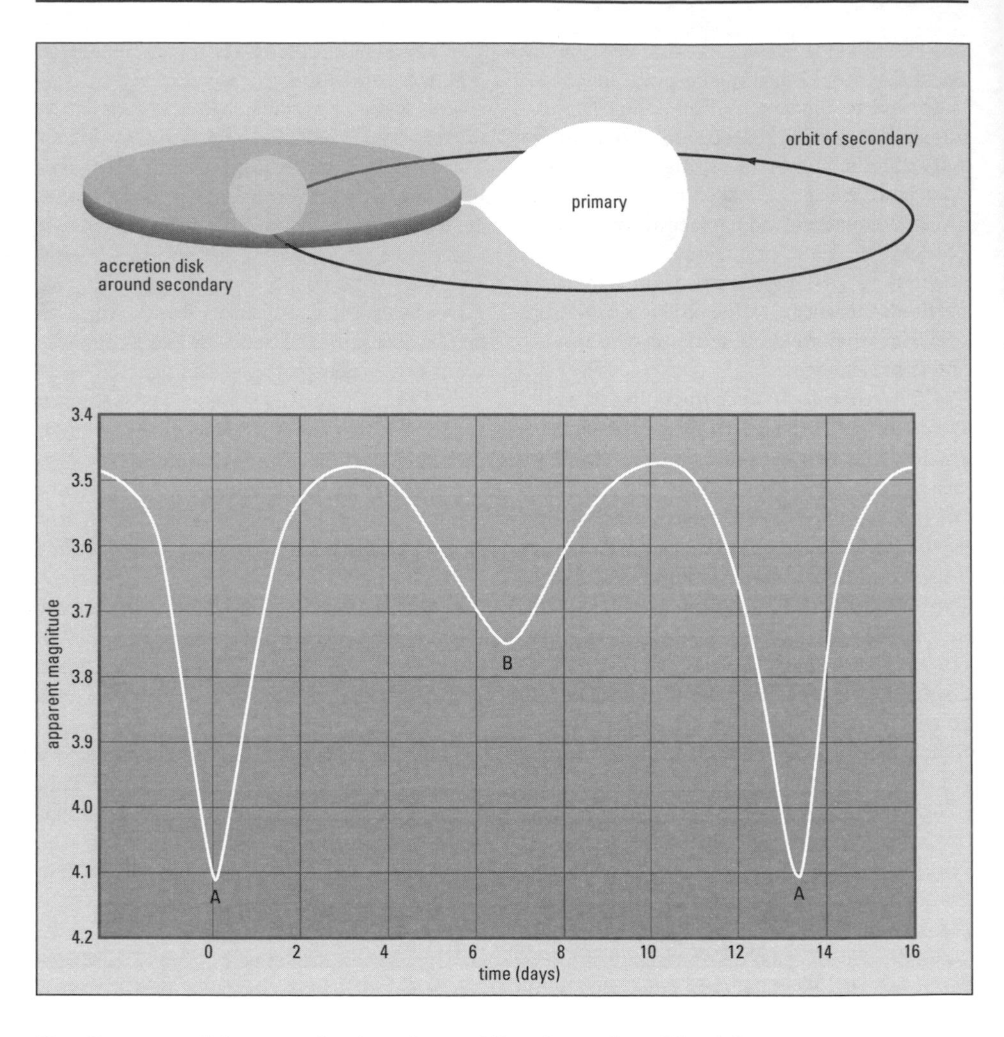

Beta Lyrae star: light curve, showing primary (A) and secondary (B) minima

in 1846 it was observed to have split into two parts, the smaller of which increased in brightness. At the next return in 1852 the two portions were about 2.3 million km (3.7 million miles) apart. The comet has not been seen since, but in 1872 there was a magnificent display of the **Andromedid** meteor shower, at precisely the time when the Earth crossed the orbital path of the vanished comet. This provided dramatic proof that the particles which produce meteors are the debris of disintegrated comets.

Big Bang A theory advanced to explain the origin of the Universe. It was developed in the 1940s by **George Gamow** from the ideas of **Georges Lemaître**. According to the Big Bang theory a giant explosion 10 to 20 thousand million years ago began the expansion of the Universe, which still continues.

Everything in the Universe once constituted an exceedingly hot and compressed gas with a temperature exceeding 10,000 million degrees. When the Universe was only a few minutes old its temperature would have been

1000 million degrees. As it cooled, nuclear reactions would have taken place that would have led to the material emerging from the fireball consisting of about 75% hydrogen and 25% helium by mass, the composition of the Universe as we observe it today. There were fluctuations from place to place in the density or expansion rate. Slightly denser regions of gas whose expansion rate lagged behind the mean value collapsed to form galaxies when the Universe was perhaps 10% of its present age.

The **cosmic microwave background** radiation detected in 1965 is considered to be the residual radiation of the Big Bang explosion. The *ripples* in this radiation discovered by the **Cosmic Background Explorer** satellite in 1992 are evidence for the initial density variations from which galaxies formed.

binary pulsar A **pulsar** orbiting another star, forming a binary. The first to be discovered was PSR 1913+16 in 1974; its companion is also thought to be a neutron star, but one that does not emit pulses. Observations show that the pulsar's orbit is gradually contracting, evidently due to the emission of energy in the form of **gravitational waves** as predicted by Einstein's Special Theory of Relativity. American radio astronomers Joseph Taylor and Russell Hulse were awarded the 1993 Nobel Prize for Physics for the discovery of the first binary pulsar.

binary star Two stars in orbit around a common centre of mass. They are usually classified as visual, spectroscopic or eclipsing binaries, according to the means by which they are observed. Their orbital periods vary enormously, from hours to years for spectroscopic and eclipsing binaries, and from a decade to many centuries for visual binaries. A star with more than two components is called a *multiple star*, **Castor** being an example.

In a *visual binary*, of which about 50,000 are known, the components are sufficiently far apart, with an angular separation greater than about 0.1 arc seconds, for the two to be seen separately through a telescope. The first visual binary discovered was **Mizar** (Zeta Ursae Majoris) in 1650 by Giovanni Battista Riccioli, who saw it had a close companion (as distinct from its wider companion, Alcor).

In an **eclipsing binary**, one star periodically passes in front of the other, so that the total light output appears to fluctuate. Most eclipsing binaries are also spectroscopic binaries. **Algol** (Beta Persei) was the first star of this kind to be identified, by **John Goodricke** in 1782.

A **spectroscopic binary** is a system whose components are too close for their separation to be measured visually. The binary nature is revealed by a Doppler shift in the absorption lines of the stars' spectra as they move around their orbits. The brighter component of Mizar was the first star shown to be a spectroscopic binary, by **Edward Pickering** in 1889.

The study of irregularities in the **proper motions** of stars has in some cases revealed the existence of invisible companions; such a star is known as an *astrometric binary*.

binoculars Two small telescopes joined by a hinge, giving a low-magnification view. The hinge allows the binoculars to be adjusted for the distance between the observer's eyes, and there is usually a means of focusing the eyepieces independently. Compactness is achieved by using prisms to fold the light-paths within the instrument (SEE page 28). Binoculars have designations such as 8 × 40, where 8 is the magnification and 40 is the diameter of each object lens in millimetres.

bipolar flow The flow of material from a star in two streams in opposite directions. Stars that generate bipolar flows are usually surrounded by dusty envelopes oriented perpendicularly to the star's rotation axis. The interplay of stellar rotation and mass outflow is thought to create bipolar flows.

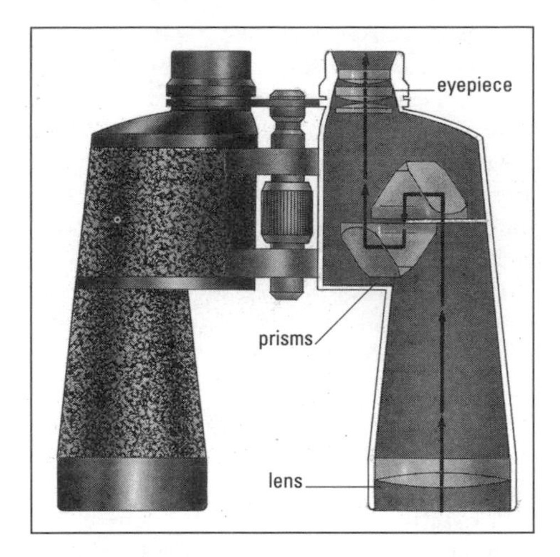

Binoculars: light path through prisms

black-body radiation The radiation that would be emitted by a *black body*, a hypothetical body which absorbs all thermal radiation (heat) falling upon it, and is a perfect emitter of thermal radiation. The spectrum of the radiation emitted by a black body is a **continuous spectrum**, and the intensity of the radiation is greatest at a wavelength which depends only on the body's temperature. Although the black body is only a theoretical object, its importance in astronomy is that the spectra of stars can be interpreted by assuming that they approximate to black bodies. SEE ALSO **effective temperature**

black drop An optical effect, visible at **second contact** or **third contact** during a transit of Mercury or Venus across the Sun's disk, in which a dark ligament – the 'drop' – appears to connect the limb of the planet with the limb of the Sun. The phenomenon has to be allowed for in the timing of transits.

Black Eye Galaxy A spiral galaxy (M64, NGC 4826) in the constellation Coma Berenices that contains a dark cloud of dust near the nucleus, giving the 'black eye' effect. It lies 20 million light years away.

black hole A localized region of space from which neither matter nor radiation can escape – in other words, the escape velocity exceeds the velocity of light. The boundary of this region is called the *event horizon.* Its radius, the **Schwarzschild radius**, depends on the amount of matter that has fallen into the region and it increases linearly as the mass increases.

A black hole of stellar mass is thought to form when a massive star undergoes total gravitational collapse. For stars up to about 1.4 solar masses, gravitational collapse can be physically halted to produce a **white dwarf**. A slightly more massive object will collapse to form a **neutron star**. If, however, the mass exceeds about 3 solar masses, even after a supernova explosion has blasted away the outer layers of the star, the collapse continues beyond even the neutron star stage. As it contracts below its Schwarzschild radius, the object becomes a black hole and effectively disappears. A star of 3 solar masses has a Schwarzschild radius of about 9 km. Inside the event horizon of the black hole, space and time are highly distorted and the stellar matter is increasingly compressed until it forms an infinitely dense **singularity**. Black holes can have an immense range of sizes. *Supermassive black holes* with up to 1000 million solar masses could be the source of energy in **quasars** and other types of active galaxy. At the other end of the scale, some primordial black holes (SEE below) could be truly microscopic.

Since no light or other radiation can escape from black holes, they are extremely difficult to detect. Any matter encountered by the black hole will most likely go into orbit first rather than being drawn directly into it. A rapidly spinning disk of matter, known as an **accretion disk**, forms around the object and heats up through friction to such high temperatures that it emits X-rays. Black holes may therefore appear as X-ray sources in binary stars. The most famous candidate is **Cygnus X-1**. Other promising candidates include A0620-00 in Monoceros, the recurrent nova V404 Cygni, and LMC-3 in the Large Magellanic Cloud.

Not all black holes result from stellar collapse. During the Big Bang, some regions of space might have become so compressed that they formed so-called *primordial black holes*. Such black holes would not be completely black, because radiation could still 'tunnel out' of the event horizon at steady rate, leading to the evaporation of the hole. (Such small-scale effects are not important for larger black holes.) Primordial black holes could thus be very hot.

blazar A term compounded from **BL Lacertae object** and **quasar**, referring to galaxies with exceptionally active nuclei. Blazars show variable optical brightness, strong and variable optical polarization, and strong radio emission. The optical variations can be on timescales as short as days. Much of the activity in active nuclei is related to jets of gas expelled from their central regions at relativistic velocities. One explanation of the exceptional activity in blazars is that in these galaxies we are viewing the jets end-on.

blink microscope SEE **comparator**

BL Lacertae object (BL Lac object) A highly luminous object located in the nuclei of some galaxies, the prototype of which, BL Lacertae, was originally classified as a variable star. BL Lac objects are variable at all wavelengths from radio to X-rays, sometimes over a timescale of just a few hours, and their optical spectra are unusual in being completely featureless, containing neither absorption nor emission lines. In this respect they differ from **quasars**. BL Lac objects are thought to lie in the centres of gas-free galaxies such as giant ellipticals.

blue moon In everyday speech, something rare – as in the phrase 'once in a blue moon'. On rare occasions the Moon does appear blue, when, for example, a large volcanic eruption or an extensive forest fire injects many particles into the atmosphere. The particles scatter light, making it appear bluer.

blue straggler A bright, blue star that remains on the **main sequence** long after stars of similar or lower mass have evolved into red giants. It is thought that they 'straggle behind' because they are close binaries in which mass has been transferred from one to the other, or, in some cases, stars that have merged together.

Bode, Johann Elert (1747–1826) German astronomer. He published the theoretical relationship between the distances of the then known planets previously pointed out by Johann Titius (1729–96). On the basis of this relationship (SEE **Bode's law**), he suggested the existence of an undiscovered planet between Mars and Jupiter. Bode published a catalogue of stars and nebulae.

Bode's law More properly called the *Titius–Bode law* (SEE **Bode, Johann**), a numerical sequence announced by Bode in 1772 which matches the distances from the Sun of the then known planets. It is formed by adding 4 to each number in the sequence 0, 3, 6, 12, 24, 48, 96, and 192, giving 4, 7, 10, 16, 28, 52, 100, and 196. Taking the Earth's distance from the Sun as 10, Mercury, Venus, and Mars fell into place reasonably well (at 3.9, 7.2 and 15.2), as did Jupiter and Saturn (52 and 95). The discovery in 1781 of Uranus at a distance corresponding to 192 prompted a search for a major planet to fill the gap at 28, but led instead to the discovery of the asteroids. The next number in the sequence is 388, but although Neptune (301) does not fit the bill, Pluto (395) does.

Although, in the form in which it was originally proposed, Bode's law has no real basis, it does give an indication of the way in which **resonances** and tidal interactions (SEE **tides**) between bodies in the Solar System can result in them having orbits which are **commensurable**.

Bok, Bartholomeus Jan ('Bart') (1906–83) Dutch–American astronomer. He made extensive studies of the Milky Way, particularly of its star-forming regions (SEE **Bok globule**). He showed that stellar **associations** consist of young stars.

Bok globule A round, dark cloud of gas and dust which is a likely precursor of a **protostar**; named after **Bart Bok**. Small globules have diameters of 0.04 parsecs (8000 AU) and can be seen in front of bright nebulae. Larger globules, up to 1 parsec across, are seen as dark patches against the stellar background of the Milky Way. Estimated globule masses range from about 0.1 solar masses for the smallest to about 2000 solar masses for those in the Rho Ophiuchi dark cloud. Globules undergo gravitational contraction to form firstly protostars, and eventually normal main-sequence stars.

bolide An exceptionally bright meteor – a major **fireball** – that is accompanied by a sonic boom. Objects whose passage through the atmosphere produce bolides almost certainly originate from asteroids rather than comets, and may survive to reach the surface as meteorites.

bolometer An instrument for measuring all the electromagnetic radiation from a source, used in astronomy to measure the total radiation from a star. There are many types of bolometer, but nearly all work by registering the change in electrical resistance that results when radiation from a star falls on a thin conductor.

bolometric magnitude (symbol m_{bol}) A theoretical figure which measures the total amount of energy radiated by a body at all wavelengths. It can be calculated from the visual magnitude.

Bondi, Hermann (1919–) British cosmologist and mathematician, born in Austria. With Thomas Gold and Fred Hoyle he developed the **steady-state theory**, in which the continuous creation of matter drives the expansion of the Universe. He also studied the constitution of stars and general relativity.

Bonner Durchmusterung (BD) A catalogue containing data on 457,857 stars, published by the German astronomers **F. W. A. Argelander** and E. Schönfeld. The stars are numbered in declination zones from +90° to −22°, and are cited in the form 'BD +52° 1638'.

Boötes A prominent northern constellation containing **Arcturus**, the brightest star north of the celestial equator. In mythology Boötes represents a herdsman. Epsilon (Izar or Pulcherrima) is a beautiful double star, with K0 and A2 components of magnitudes 2.7 and 5.1, separation 2.8 arc seconds. Mu (Alkalurops) is a wide double, magnitudes 4.3 and 6.5.

bow shock The boundary around a planet within which the planet's **magnetosphere** predominates over the surrounding **solar wind**. The stream of particles making up the solar wind is deflected around the bow shock, rather like the effect set up in front of a ship's bow as it moves through the water. SEE ALSO **heliosphere**

Bradley, James (1693–1762) English astronomer. In 1742 he succeeded Edmond Halley as Astronomer Royal. He discovered the **aberration** (1) of light – the first observational evidence for Copernicus's heliocentric theory – and the **nutation** of the Earth's axis. His catalogue of 3000 stars was published posthumously, and formed the basis of **Friedrich Bessel's** catalogue.

Brahe, Tycho (1546–1601) Danish astronomer. In 1572 he observed the supernova in Cassiopeia, and his report of it, *De stella nova*, soon made him famous. Under the patronage of King Frederick II of Denmark he built and equipped two observatories – Uraniborg and Stjerneborg – on the island of Hven in the Baltic. He became the most skilled observer of the pre-telescope era, expert in making accurate naked-eye measurements of the positions of stars and planets. He calculated the orbit of the comet seen in 1577, and this, together with his study of the supernova, showed that **Aristotle**'s picture of the unchanging heavens was wrong. He could not, however, accept the world system put forward by **Copernicus**. In his own planetary theory (the *Tychonian system*), the planets move round the Sun, and the Sun itself, like the Moon, moves round the stationary Earth. In 1597 he left Denmark and settled in Prague, where **Johann Kepler** became his assistant. The accurate series of planetary observations made by Tycho Brahe were used by Kepler in deriving his laws of planetary motion, which ironically demonstrated the validity of the Copernican theory.

breccia A type of rock made up of coarse, angular fragments set in a finer-grained material. Impacts produced the fragments, which were later incorporated in a younger rock formation. Most of the lunar rock samples returned by the Apollo missions were breccias, and the surface rocks of many other cratered, geologically inactive bodies in the Solar System are expected to be made up largely of breccias.

bremsstrahlung (German, meaning 'brake radiation') Electromagnetic radiation produced when energetic electrons are decelerated, for example when approaching the nucleus of an atom. It is generated in ionized gas clouds.

brown dwarf A star with mass less than 0.08 solar masses, in which the core temperature does not rise high enough to initiate thermonuclear reactions. Such a star is, however, luminous, for as it slowly shrinks in size it radiates away its gravitational energy. The surface temperature of a brown dwarf is below the 2500 K lower limit of a **red dwarf**.

B star A star of spectral type B, characterized by strong neutral helium absorption lines and the presence of hydrogen lines (although not as strong as in the A stars). On the **main sequence**, the temperatures of B stars range from 10,500 K at B9 to 28,000 K at B0. Their masses and sizes range correspondingly from 3.2 to 17 solar

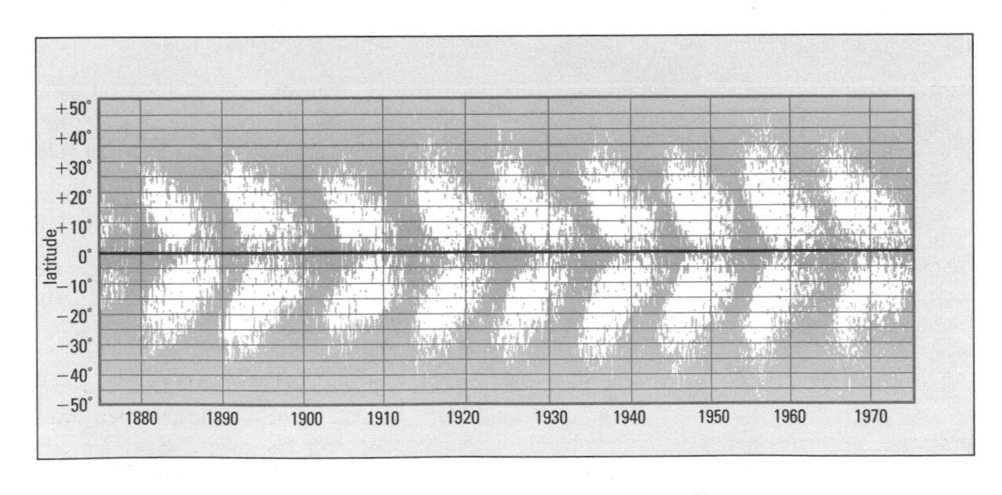

Butterfly diagram: a plot of sunspot latitudes resembles a row of butterflies

masses, and from 2.5 to 10 solar radii. At their hottest, B stars have 20,000 times the luminosity of the Sun. Examples of B stars are Achernar, Regulus, Rigel, and Spica. A B9 star stays on the main sequence for about 500 million years, but a B0 star remains there for only 5 million years. The hottest B stars (and the even hotter **O stars**) form loose groupings known as **OB associations**. O and B stars evolve to become supergiants with luminosities up to 100,000 times that of the Sun.

butterfly diagram A diagram illustrating the changing distribution of sunspots in solar latitude over the course of the 11-year cycle (SEE page 31). At the start of a cycle there are spots at latitudes of up to around 30° north and south, but very few near the equator. As the cycle progresses, spots occur nearer the equator. When all the spots in a cycle are plotted on the diagram, a characteristic pattern in the shape of a butterfly appears. The diagram was devised by Edward Maunder in 1904. SEE ALSO **Maunder minimum**

C

caelum A constellation of the southern sky, adjoining Columba, introduced by **Lacaille,** representing an engraving tool. It has no star above magnitude 4.5. One variable, R Caeli, of Mira type, ranges from magnitude 6.7 to 13.7 with a period of 391 days.

Calar Alto Observatory A German–Spanish observatory near Almeria in southern Spain. Its main telescope is a 3.5 m (138-inch) reflector, opened in 1984.

calendar A list of the days of the year, grouped into months, and based on the apparent motions of the Sun or Moon, or both. The ancient Egyptians used a calendar based on a solar year. The Babylonians (and the modern Hebrews and Muslims) used a lunar year of 12 months, which is 11 days shorter than a solar year, so an extra month is added every third year.

Our present calendar is based on that of the Romans, which originally had only 10 months. In 45 BC Julius Caesar introduced a revised calendar, because the Roman calendar had become badly out of phase with the seasons. In this revised calendar, known as the *Julian calendar,* the year was divided into 12 months. Every fourth year an extra day (leap day) was added to the usual year of 365 days. But the average length of the Julian calendar year, 365.25 days, was longer than a **tropical year** (the interval between two successive passages of the Sun through the vernal equinox) by over 11 minutes. The Julian calendar year thus slowly drifted out of phase with the seasons, by over 18 hours a century.

By the 16th century there was a discrepancy of 10 days in the date of the vernal equinox given by the calendar. In 1582 Pope Gregory XIII introduced a revised calendar, called the *Gregorian calendar.* This revision meant dropping 10 days from the calendar, so 5 October 1582 became 15 October. The average length of the calendar year was reduced by inserting a leap day every four years, except in century years which are not divisible by 400. Thus the years 1600 and 2000 are leap years, but 1700, 1800, and 1900 are not. The Gregorian calendar was adopted in Great Britain and its territories in 1752, when 2 September was followed by 14 September. The Gregorian calendar year of 365.2425 days is accurate to one day in 3300 years.

In about AD 525 Dionysius Exiguus, a chronologist, introduced the method of reckoning the years by reference to the Christian era. He fixed the beginning of the present calendar so that Christ was born in December AD 1. However, it is believed that he made a slight error in his calculations, so that the birth of Christ actually occurred a few years before this. The year that precedes AD 1 is 1 BC; there is no year 0. SEE ALSO **Julian Day**

Callisto The second-largest of Jupiter's **Galilean satellites,** with a diameter of 4800 km (2980 miles), and the outermost of them. It has the lowest average density, at 1.86 g/cm^3, indicating that it contains a high proportion of water-ice. The surface of the satellite is dark and very heavily cratered – in fact it is the most densely cratered object known. As the main period of cratering was early in the Solar System's history, this indicates that no subsequent geological activity has disturbed Callisto's surface. As well as the dense craters, there are several large, multi-ringed **impact features,** the largest of which is Valhalla, with an overall diameter of 4000 km (2500 miles). SEE ALSO **satellite**

Caloris Basin The largest **impact feature** on Mercury. About half of it was photographed by the Mariner 10 probe in 1974–5. It is 1300 km (800 miles) in diameter, and –

like a number of similar features elsewhere in the Solar System – has a multi-ringed structure.

Calypso A small **satellite** of Saturn, discovered by Bradford Smith and others in 1980 between the two Voyager encounters. Calypso is irregular in shape, measuring 34 × 22 km (21 × 14 miles). It is co-orbital with Tethys and Telesto; Calypso and Telesto orbit near the L_4 and L_5 **Lagrangian points** of Tethys's orbit around Saturn (SEE **co-orbital satellite**).

Camelopardalis A barren northern constellation, representing a giraffe. Its brightest star, Beta, is of magnitude 4.0 and has a wide magnitude 8.6 companion. NGC 1502 is an open cluster visible with binoculars.

canals A network of dark linear markings on the surface of Mars, reported to exist by observers from the 1870s until well into the 20th century. They were first described by **Angelo Secchi** and **Giovanni Schiaparelli** as *canali*, the Italian word for 'channels'. In some English reports canali was mistranslated as 'canals' – with the connotation that they were artificial constructions. This idea was seized upon by, in particular, **Percival Lowell**, who was convinced that the canals were artificial waterways built by intelligent beings to irrigate a dying planet. Despite the demonstration by **Eugene Antoniadi** that the canals were illusory, this romantic notion of Mars persisted in many minds until spacecraft visited Mars and sent back pictures that showed no sign of a canal network.

Cancer A dim and largely barren zodiacal constellation, between Gemini and Leo. In legend it represents the crab sent by Juno, queen of the gods, and trodden on by Hercules during his fight with the multi-headed Hydra. Beta, magnitude 3.5, is its brightest star. Zeta is a multiple system; the main components are of magnitudes 5.1 and 6.2, and the brighter star is again a close double. R Cancri is a Mira variable, range 6.1 to 11.8, period 362 days. RS Cancri is a semiregular variable (5.1 to 7.0, 120 days) and so is X Cancri (5.6 to 7.5, period about 195 days). The open cluster **Praesepe**, M44, is flanked by Gamma, 4.7, and Delta, 3.9, the 'asses'. Also in Cancer is M67, a very old open cluster visible in binoculars.

Canes Venatici A constellation introduced by **Hevelius** on his map of 1690, representing the dogs Asterion and Chara, held by the herdsman Boötes. The only star above fourth magnitude is Cor Caroli ('Charles's Heart'), magnitude 2.9. This name was given to it by Edmond Halley, in honour of King Charles I of England. The star is the prototype of the class of *magnetic variable* stars whose spectra contain lines that vary in intensity. It is a double, with a companion of magnitude 5.6, separation 19 arc seconds, excellent for small telescopes. Y is a semiregular variable, ranging in magnitude from 5.2 to 6.6 in a period of 157 days. The famous **Whirlpool Galaxy**, M51, lies in Canes Venatici. Other spirals are M63, M94, and M106. There is also the 6th-magnitude globular cluster M3, which is distinct in binoculars.

Canis Major A constellation representing one of Orion's hunting dogs. It is distinguished by the presence of **Sirius**, the brightest star in the sky. Other bright stars are Epsilon (Adhara), magnitude 1.5; Delta (Wezen), 1.8; and Beta (Mirzam), 2.0. Beta is variable over a very small range and is a prototype of the class of variables known either as *Beta Canis Majoris stars* or *Beta Cephei stars*. There are three fairly bright variables: R (5.7 to 6.3, 1.14 days, Algol type), W (6.4 to 7.9, irregular), and UW (4.8 to 5.3, 4.39 days, Beta Lyrae type). There is one prominent open cluster, M41, of 5th magnitude, south of Sirius.

Canis Minor A constellation representing the smaller of Orion's two dogs. The first-magnitude star **Procyon** makes it easy to locate. Beta (Gomeisa), magnitude 2.9, is a slightly variable **shell star**. There are no other objects of note, although three Mira variables (V, R, and S) can rise to above magnitude 8 at maximum.

cannibalism SEE **cluster of galaxies**

Cannon, Annie Jump (1863–1941) American astronomer. Working at the Harvard College Observatory, initially as an assistant to **Edward Pickering**, she reorganized the classification of stars into spectral types, and classified the 225,000 stars in the ***Henry Draper Catalogue*** of stellar spectra.

Canopus The star Alpha Carinae, the second-brightest star in the sky, magnitude −0.72. Its luminosity and distance are not accurately known, but one estimate classifies it as a bright giant of type A9, 800 times as luminous as the Sun, 74 light years away.

Capella The star Alpha Aurigae, magnitude 0.08 (the sixth-brightest in the sky), distance 41 light years. It is in fact a spectroscopic binary, consisting of two giant stars of types G6 and G2, with an orbital period of 104 days.

Capricornus One of the less prominent of the zodiacal constellations; it has been identified with the Greek god Pan. Its brightest star is Delta (Deneb Algedi), magnitude 2.9. Alpha (Algedi) is a wide optical double, magnitudes 3.6 and 4.2. The semiregular variable RT has a range from 6th to 9th magnitude and a period of 393 days. The most important cluster is M30, a globular of magnitude 7.5.

captured rotation SEE **synchronous rotation**

carbonaceous chondrite A rare type of chondritic meteorite (SEE **chondrite**) with a higher-than-average content of carbon, in the form of organic compounds. Not counting the **volatiles**, the proportions of chemical elements in a carbonaceous chondrite are very similar to the composition of the Sun. These meteorites are therefore believed to have come from bodies that formed very early in the Solar System's history, which makes them important in the study of its origin.

carbon–nitrogen cycle A sequence of nuclear reactions which accounts for the production of energy inside main-sequence stars that are more massive and hence hotter than the Sun. It was first described in 1938 by **Hans Bethe**. The reactions involve the fusion of four hydrogen nuclei into one helium nucleus with the release of a huge amount of energy. Nitrogen and oxygen are formed as intermediate products. The presence of carbon is essential, but it behaves as a catalyst and remains unchanged at the end of the cycle. The process takes place at temperatures above 15 million degrees K. It is also known as the *carbon cycle* or the *carbon–nitrogen–oxygen cycle*.

carbon star (C star) A cool red giant star whose surface composition shows carbon-based molecules such as carbon monoxide (CO), cyanogen (CN), molecular carbon (C_2), and other compounds of carbon, and also the presence of much lithium. Carbon stars are allocated spectral class C in the Morgan–Keenan system of **spectral classification**.

Carina Part of the dismembered constellation Argo Navis, the ship Argo. It is the brightest and richest part of Argo, representing the ship's keel, and contains **Canopus**, the second-brightest star in the sky. Apart from Canopus, its brightest stars are Beta (Miaplacidus), magnitude 1.7, and Epsilon (Avior), magnitude 1.9. Epsilon and Iota (2.2) form part of the **False Cross**. Upsilon is a double, 3.0 and 6.0. Carina's variable stars include the unique **Eta Carinae**, which lies within the nebula NGC 3372. Other variables with maximum above magnitude 6.5 are the Cepheids U (range 5.7 to 7.0, period 38.77 days) and l (3.3 to 4.2, 35.54 days), as well as the Mira-type stars R (3.9 to 10.5, 309 days) and S (4.5 to 9.9, 150 days). There are several open clusters visible to the naked eye: NGC 2516, NGC 3114, and IC 2581. The cluster IC 2602 around Theta is spectacular when viewed with binoculars. There is also a 6th-magnitude globular cluster, NGC 2808.

Carme One of Jupiter's four small outermost **satellites**, discovered in 1938 by Seth Nicholson. It is in a retrograde – SEE **retrograde motion** (1) – orbit, and may well be a captured asteroid.

Cassegrain telescope A reflecting telescope in which the light received by the primary mirror is reflected back from the secondary mirror through a hole in the primary. The light is brought to a focus – the *Cassegrain focus* – on the axis of the telescope, where the image can be observed by the eye or by instruments. The combination of a concave primary (in the form of a paraboloid) and a convex secondary (usually a hyperboloid) tends to cancel **aberrations** (2), such as coma, of the separate mirrors. The arrangement is used for both amateur and professional telescopes. It was devised in 1672 by a Frenchman named Cassegrain whose identity is uncertain. SEE ALSO **Schmidt–Cassegrain telescope**

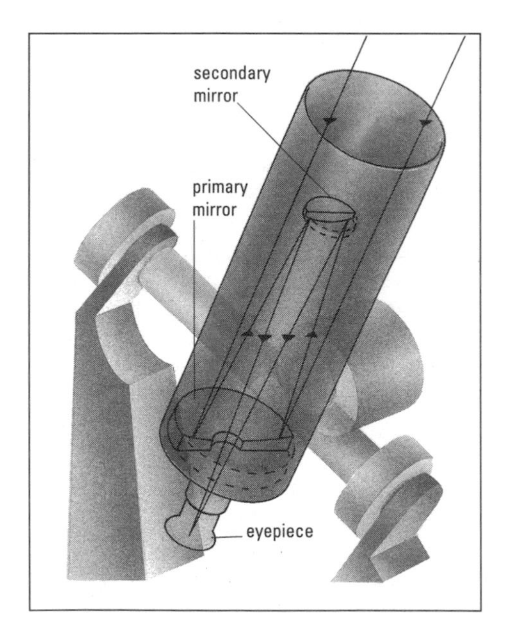

Cassegrain telescope: light path

Cassini French family of Italian origin that produced four generations of astronomers and cartographers, all of whom ran the Paris Observatory.

Cassini, Giovanni Domenico, later also known as **Jean Dominique** (1625–1712), moved to Paris in 1667. He was the first to accurately measure the dimensions of the Solar System. He discovered the division in Saturn's rings that now bears his name, and also four of that planet's satellites. He measured Jupiter's rotation period, and also improved the tables of Jupiter's satellites, which helped Ole Römer in his determination of the velocity of light.

Cassini, Jacques (1677–1756), son of Giovanni. He accurately measured an arc of a meridian running through France. The results helped to show that the Earth is flattened at the poles, although he and his father had believed there to be a polar elongation. He determined the proper motion of the star Arcturus.

Cassini, César François (1714–84), son of Jacques, and **Cassini, Jacques Dominique** (1748–1845), son of César François, produced a topographic map of France.

Cassini A joint space mission between NASA and the European Space Agency to explore Saturn and its moon Titan. Cassini is due to be launched in October 1997 and reach Saturn in June 2004. Cassini will go into orbit around Saturn, while the Huygens sub-probe will land on Titan.

Cassini Division The gap between Rings A and B in the ring system of **Saturn**, discovered by Giovanni Cassini in 1675. The Voyager spacecraft showed that it is not a true gap, but contains several narrow rings.

Cassiopeia A distinctive northern constellation, representing the mother of Andromeda. The five leading stars make up a W or M pattern, on the opposite side of the Pole Star from Ursa Major. Alpha (Schedar) is its brightest star, magnitude 2.2. Gamma, magnitude range 1.6 to 3.0, is the prototype of the variables known as **shell stars**. Rho is a semiregular variable, usually of around magnitude 5 but which on rare occasions falls to below 6. Other variables are R (4.7 to 13.5, 430 days, Mira type), RZ (6.2 to 7.7, 1.2 days, Algol type), and SU (5.7 to 6.2, 1.95 days, Cepheid). Eta is a fine coloured binary for viewing with small telescopes, magnitudes 3.5 and 7.5; the orbital period is just under 500 years. Cassiopeia contains several open clusters resolvable in binoculars: M52, NGC 457, NGC 559, NGC 663 and M103.

Cassiopeia A The strongest radio source in the sky apart from the Sun, and the remnant of a supernova that exploded around 1660 but was not recorded on Earth. It lies about 10,000 light years away and appears optically as a faint nebula.

Castor The star Alpha Geminorum, magnitude 1.57, distance 47 light years. Castor is one of the finest visual binaries in the sky for small telescopes, consisting of main-sequence components of magnitudes 1.9 and 2.9, types A1 and A2, period nearly 500 years. Each of the stars is itself a spectroscopic binary. A distant companion, Castor C, orbits the system. It is an eclipsing binary also known as YY Geminorum, made up of two faint red dwarfs, range 9.2 to 9.6.

cataclysmic variables (eruptive variables) A term given to a diverse group of stars that erupt in brightness, including **supernovae, novae, dwarf novae, flare stars, shell stars**, and some X-ray objects. With the exception of supernovae, these objects are very close binaries whose outbursts are caused by interaction between the two components. A typical system of this type has a low-mass secondary which fills its **Roche lobe** so that material is transferred on to the primary, which is usually a white dwarf. The transferred material has too much angular momentum to fall directly on to the primary, and so forms an accretion disk. A hot spot arises on the outer edge of the **accretion disk** where the infalling material hits it. Outbursts occur when the flow rate varies, which can be at irregular intervals from weeks to many years.

catadioptric A type of telescope in which a lens–mirror combination forms the image. The **Schmidt–Cassegrain telescope** is the most common type of catadioptric design for amateur use; another is the **Maksutov telescope**. A catadioptric telescope is basically a reflector with a full-aperture lens at its top end. The lens corrects for **aberration** (2), and produces a divergent light-beam which is reflected on to the secondary mirror by a short-focus spherical or paraboloidal primary mirror. The design is compact and often used for portable instruments.

CCD Abbreviation for *charge-coupled device*, a solid-state electronic imaging device used in astronomy as an alternative to conventional astrophotography. A CCD is an integrated circuit whose upper layer contains a square array of closely spaced electrodes. A ray of light from a celestial object falling on an electrode is converted into an electronic charge, and the charges from all the electrodes are 'coupled' together and read off as a pattern – an instantaneous image of the object. A series of these images produced over a period of time is stored on a computer and combined to create a picture, just as a photographic emulsion builds up a picture over the course of its exposure. Pictures requiring hours of exposure on photographic emulsion can be recorded in minutes by using a CCD. Fainter objects can be detected, the effects of **light pollution** can be compensated for, and image processing by computer can be used to bring out detail. CCDs are widely used in professional astronomy, and are becoming increasingly cheaper and readily accessible to amateur astronomers.

celestial equator The projection of the Earth's equator on to the celestial sphere, dividing the sky into the northern and southern hemispheres.

celestial latitude, longitude SEE **coordinates**

celestial mechanics The branch of astronomy dealing with the motions of heavenly bodies. It uses the laws of physics to explain and predict the orbits of planets, satellites and other celestial bodies.

celestial meridian The great circle passing through the observer's zenith, the nadir, and the celestial poles, cutting the horizon at the north and south points.

celestial poles SEE **celestial sphere, pole**

celestial sphere The imaginary sphere, of infinite radius and with the Earth at its centre, upon which all the celestial bodies appear to be projected as though they were at a uniform distance. The most obvious behaviour of the celestial sphere is its apparent daily east-to-west rotation, due to the axial spin of the Earth. The celestial sphere appears to rotate about the *celestial poles*, as though on an extension of the Earth's own axis. The Earth's axis points towards the celestial poles. Although the positions of the celestial poles seem to be fixed, they are in fact slowly drifting because of the effect of **precession**.

The celestial sphere is divided into two halves by the *celestial equator*, which is equidistant from the poles. We can thus visualize a set of **great circles** similar to circles of latitude on Earth, called *declination circles*, each parallel to the equator. Declination circles allow us to specify the position of an object in terms of its angular distance north or south of the celestial equator.

Right ascension, the celestial equivalent of longitude, is also an extension of the terrestrial system. The zero of right ascension is chosen using the apparent motion of the Sun around the sky during the year (SEE below). This motion is actually the result of the Earth's orbital motion around the Sun, and the track traced by the Sun on the celestial sphere, called the **ecliptic**, indicates the plane of the Earth's orbit. The ecliptic is not the same as the celestial equator because the Earth's axis is tilted by about 23½°.

For half of each year the Sun is in the northern hemisphere of the sky, and for the other half it is in the southern hemisphere. On two dates each year the Sun crosses the celestial equator; these points are known the **equinoxes**. It is the *spring equinox*, when the Sun crosses into the northern hemisphere, that is chosen as the zero of right ascension. Right ascension is measured eastwards from this point but, unlike terrestrial longitude, it is measured in units of time rather than degrees, minutes and seconds of arc (SEE **sidereal time**).

Centaurus A brilliant southern constellation representing a centaur, with 13 stars above magnitude 3.5. The brightest is **Alpha Centauri**. Beta Centauri (Hadar or Agena) is magnitude 0.61, a giant of type B1. Alpha and Beta indicate the direction of the Southern Cross. Gamma is a close binary with an 85-year period; the components are equal, at magnitude 2.9, and their combined brightness is magnitude 2.2. R is a Mira-type variable, range 5.3 to 11.8, and the period is unusually long, 546 days. T is a semiregular variable (5.5 to 9.0, 90 days). The globular

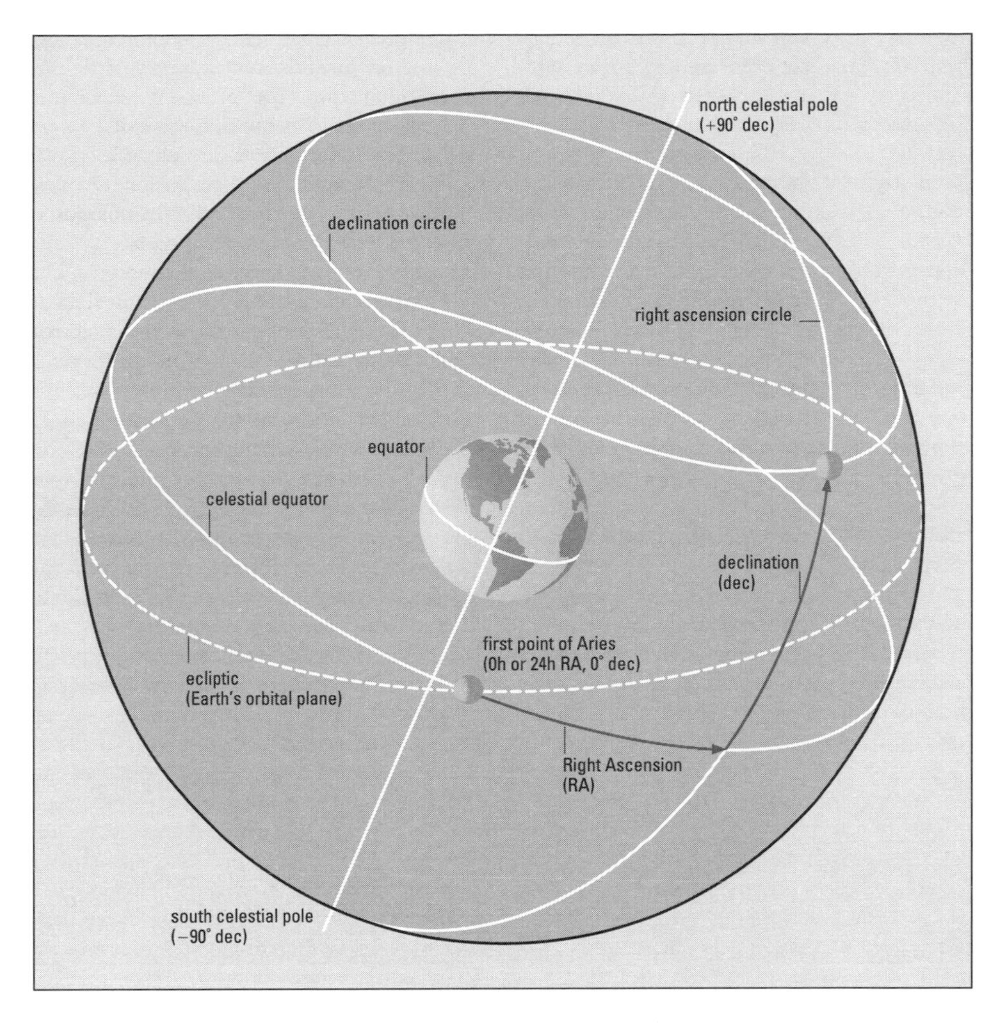

Celestial sphere: the imaginary surface on which astronomical positions are measured

cluster **Omega Centauri** is the finest in the entire sky. There are also two bright open clusters, NGC 5460 and NGC 3766, and a fine planetary nebula, NGC 3918. The elliptical galaxy NGC 5128 is the radio source **Centaurus A**.

Centaurus A The galaxy NGC 5128, a strong radio source about 10 million light years away in the constellation Centaurus. Optically it appears to be an elliptical galaxy crossed by a prominent dust lane. It is thought to result from the encounter of a massive elliptical galaxy with a smaller spiral with a high content of dust.

central meridian The imaginary line joining the north and south poles of a planet or the Sun, bisecting the disk. It is used by observers to estimate the longitude of features as the body rotates.

centre of mass The point in a body, or a system of bodies, at which the total mass of the body or system may be regarded as concentrated. It is the point from which gravitational attraction of the system as a whole appears to act. SEE ALSO **barycentre**

Cepheid variable One of an important class of variable stars that pulsate in a regular manner, accompanied by changes in luminosity. They take their name from the first of the type to be discovered, Delta Cephei, the variability of which was noted by **John Goodricke** in 1784. Cepheids periodically expand and contract, changing in size by as much as 30% in each cycle. The diagram shows the pulsations and light curve of Delta Cephei itself. A typical Cepheid has a surface temperature that varies between 6000 and 7500 K and a spectral type that ranges from F2 at maximum to G2 at minimum. The average luminosity is 10,000 times that of the Sun. Cepheids became important in cosmology in 1912, when **Henrietta Leavitt** discovered a simple relationship between the period of light variation and the absolute magnitude of a Cepheid. This relationship, the **period–luminosity** law, enables the distances of stars to be ascertained, not only in our Galaxy but in other galaxies too.

There are two classes of Cepheid: type I, or *classical Cepheids*, are younger and more massive than type II, or **W Virginis stars**. Both types follow a period–luminosity relationship, but their curves are different. The classical Cepheids, such as Delta Cephei itself, are yellow supergiants of **Population I**, while the second type are older **Population II** stars and are found in globular clusters and the centre of the Galaxy. In using Cepheids to determine distances it is necessary to know which type is being observed. When Cepheids were first used to determine distances it was not known that there were two types. This resulted in applying the value for type II Cepheids to the classical Cepheids in error, which affected the measured distances to external galaxies. When the error was found, by **Walter Baade** in 1952, the result was to double the size of

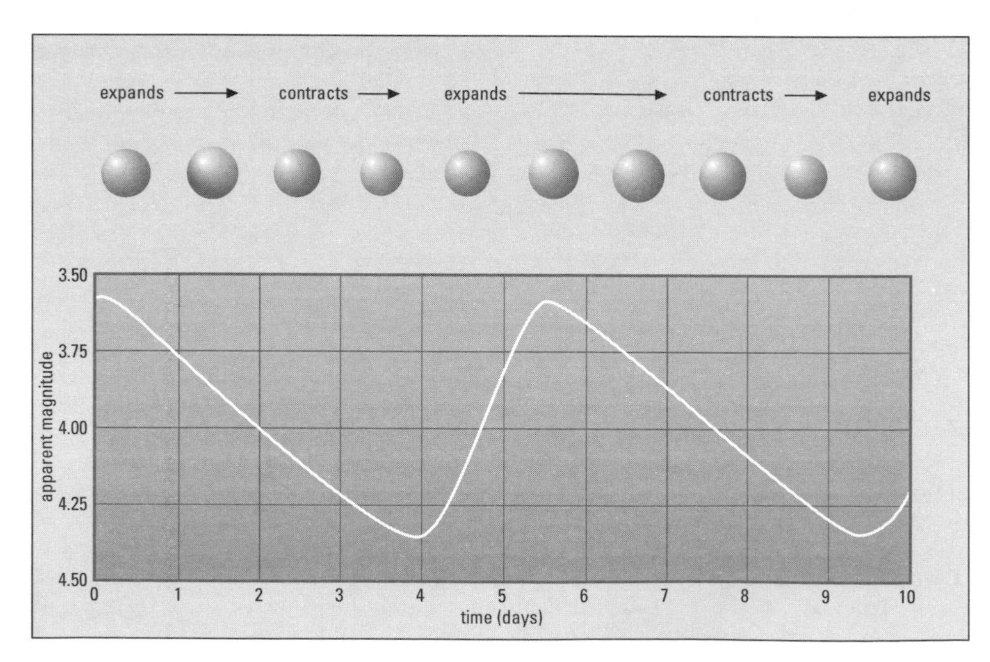

Cepheid variable star: light curve corresponds to pulsations of the star

the Universe. Classical Cepheids are one to two magnitudes brighter than type II Cepheids of the same period. The periods of Cepheids are in the range 1–50 days; classical Cepheids have a somewhat shorter period on average than those of type II.

Cepheus A constellation in the far north, adjoining Ursa Major, but not very distinctive. In mythology, it represents the husband of Cassiopeia and the father of Andromeda. Alpha (Alderamin) is magnitude 2.4. Beta (Alfirk), magnitude 3.2, is variable over a small range, and is a prototype of the class known as either *Beta Cephei stars* or *Beta Canis Majoris stars*. Delta is the prototype **Cepheid variable**, with a range from magnitude 3.5 to 4.4 and a period of 5.37 days; it has a wide optical companion of magnitude 6.3. Mu, which William Herschel named the Garnet Star, is a semiregular red supergiant with a range of 3.4 to 5.1. T Cephei is a Mira variable (5.2 to 11.3, 388 days); VV is a giant eclipsing variable, 4.8 to 5.4, period 7430 days.

Ceres The first **asteroid** to be discovered, by Giuseppe Piazzi on 1 January 1801. Its diameter of 913 km (567 miles) makes it the largest asteroid. It orbits in the main asteroid belt, at an average distance from the Sun of 414 million km (257 million miles), the distance of the 'missing' planet predicted by **Bode's law**.

Cerro Tololo Inter-American Observatory An observatory on Cerro Tololo peak near La Serena, Chile, part of the US National Optical Astronomy Observatories. Its largest telescope is a 4 m (158-inch) reflector, the twin of that at Kitt Peak. Other instruments include the 0.61 m (24-inch) Curtis Schmidt Telescope of the University of Michigan.

Cetus A large constellation, crossed by the celestial equator, but not very conspicuous; it represents the sea monster of the Andromeda legend. Its brightest star is Beta, magnitude 2.0. The most famous star in the constellation is the prototype long-period variable, **Mira**. Tau Ceti, magnitude 3.5, 11.7 light years away, is one of the closest stars to bear any real resemblance to the Sun, being a main-sequence star of type G8. UV Ceti is a **flare star**; it is a pair of 13th-magnitude red dwarfs 8.7 light years away that exhibit large flares every few hours.

Chamaeleon A small constellation near the south pole, representing a chameleon. Its brightest stars are only of magnitude 4.1.

Chandrasekhar, Subrahmanyan (1910–) Indian-American astrophysicist. He has made significant contributions to the theories of stellar evolution, stellar atmospheres, and relativity. His theory of white dwarf stars suggests there is a limit to their mass (SEE **Chandrasekhar limit**).

Chandrasekhar limit The maximum possible mass for a **white dwarf** star, first calculated in 1931 by **Subrahmanyan Chandrasekhar**. The value is about 1.4 solar masses, although more recent computations suggest that a higher value is possible for a rapidly rotating white dwarf. If a star has a mass which exceeds the Chandrasekhar limit it will collapse to become a **neutron star** or possibly a **black hole**.

charge-coupled device SEE **CCD**

Charon The **satellite** of **Pluto**, discovered in 1978 by James Christy (1939–). Its diameter of 1270 km (790 miles) makes it the largest satellite in relation to its primary in the Solar System. Estimates of its mass vary from 8 to 16% of Pluto's, and estimates of its density from 1.3 to 2.0 g/cm^3. Also, it is unique among planetary satellites in having a synchronous orbit with a period (6.4 days) matching the rotation period of its primary. From Charon's spectrum, it seems to have a greyish surface, probably of water-ice. There is almost certainly no permanent atmosphere, but some of the nitrogen outgassed by Pluto when at perihelion could be captured and retained by Charon as a temporary atmosphere.

Chiron A small body in the outer Solar System, originally designated as an asteroid (no. 2060) following its discovery in 1977 by Charles Kowal (1940–). However, it has an eccentric, cometary orbit (period 50.7 years) which takes it from within Saturn's orbit out nearly as far as Uranus, and in 1989, 7 years before reaching perihelion, it was found to have a comet-like coma. Chiron has a diameter of roughly 200 km (125 miles), much larger than any comet, and does not have a bright surface; also, its orbit is unstable. Chiron may well be a **Kuiper Belt** object, perturbed inwards a few hundred million, or even just a few hundred thousand, years ago.

chondrite A type of **stony meteorite**, consisting of an agglomeration of *chondrules* – millimetre-sized globules believed to have remained unchanged since they condensed out of the nebula from which the Sun and the Solar System formed. They are therefore more ancient than **achondrites**. Chondrites are composed mainly of the silicates of iron, calcium, aluminium, magnesium, and sodium. They are classified into various types, including the **carbonaceous chondrites**.

chromatic aberration An **aberration** (2) in lenses that results from the unequal refraction of the different colours that make up white light. Each colour is brought to a focus at a different distance from the lens, causing the image of an object to be fringed with prismatic colours. Chromatic aberration is reduced by the use of an **achromatic lens**.

chromosphere ('sphere of colour') The layer of the Sun's atmosphere between the photosphere and the corona. The chromosphere is normally invisible because of the glare of the photosphere shining through it, but it is briefly visible near the beginning and end of a total solar eclipse as a spiky red rim around the Moon's disk. At other times it can be studied by spectroscopy or with the use of a special filter. The chromosphere is about 10,000 km (6000 miles) thick. At its base the temperature is about 4000 K, rising to 100,000 K at the top, where the chromosphere gives way to the corona. Powerful magnetic fields are believed to cause this rise in temperature.

Circinus A small constellation in the region of Alpha and Beta Centauri, representing a pair of compasses. Its only fairly bright star is Alpha, magnitude 3.2.

circumpolar stars Stars which from a given place of observation never dip below the horizon. Their polar distance (90° minus their declination) is less than the observer's latitude.

cislunar Between the Moon and the Earth.

civil twilight SEE **twilight**

Clementine A US space probe which in April and May 1994 imaged the entire surface of the Moon in several wavelength bands. This made possible not only the first complete mapping of the Moon at visual wavelengths, but also a detailed geological survey. A complete laser altimetry of the surface revealed previously undetected variations in surface elevation, and the 12 km (7½-mile) deep floor of the farside crater Aitken. An intended extension of the mission to rendezvous with asteroid 1620 Geographos in August 1994 was abandoned after a mishap with Clementine's attitude-control fuel supply.

Cluster A European Space Agency mission to study the effect of the solar wind on the Earth's magnetosphere. Cluster consists of a fleet of four identical satellites planned for launch into orbit around the Earth in December 1995.

cluster of galaxies A group of galaxies, which can range in size from a few dozen members, as in our own **Local Group,** to a rich cluster containing thousands of galaxies. Poor clusters are usually irregular in shape, but rich clusters tend to be spherical and condensed at the centre, and with a giant elliptical galaxy often dominating the central region. Such giant galaxies have probably grown by *cannibalism*, their overwhelming gravity dragging in smaller galaxies. The nearest rich clusters of galaxies are the **Virgo Cluster**, 45 million light years away, and the Coma Cluster, 300 million light years away. Clusters themselves are grouped into *superclusters* up to about 100 million light years across. The local supercluster, of which the Local Group is part, is centred on the Virgo Cluster.

Coalsack A prominent dark nebula in the constellation Crux, the Southern Cross. It is seen in silhouette against the background stars of the Milky Way.

COBE ABBREVIATION FOR **Cosmic Background Explorer**

coelostat Two mirrors arranged so that slowly rotating one of them counteracts the apparent rotation of the sky, enabling light from a celestial object to be directed along a fixed axis. The first mirror is mounted equatorially and is clock-driven at half the rate of the Earth's rotation. The beam of light it receives is reflected on to a second (fixed) mirror, which in turn reflects the beam into the instrument being used for the observation. A coelostat is used with heavy, cumbersome, or fixed equipment, for example a spectroheliograph. It is an improved form of **heliostat** and **siderostat**. SEE ALSO **coudé**

collimation The process of aligning the optical components of a telescope so as to bring the rays of light from the object being observed to a focus at the correct position. A refracting telescope requires only occasional adjustment of the object glass in its cell. Collimation is more of a problem with reflectors, as the primary and secondary mirrors both need to be adjusted, and the secondary can become misaligned quite easily. An incorrectly collimated telescope can suffer from **coma** (1).

colour index The difference in brightness of a star at two different wavelengths, used as a measure of the star's colour and hence its temperature. The brightness of the star is measured through coloured filters, usually at blue (B) and yellow (V, for visual) wavelengths, and the difference between the two readings is the B−V colour index. The B−V colour index of cool red stars is positive (up to about +2.0), while that of hot blue stars is negative (about −0.5 maximum). A white star of spectral type A0 such as Vega has a B−V colour index of exactly 0.0. A measurement at U (ultraviolet) wavelengths may also be taken, and a U−B colour index derived. Other measurements in the red and infrared can also be used.

The colour index of light from a star may change as the light passes through dust in interstellar space. Dust scatters blue light but allows red light to pass more freely, thereby reddening the star. This effect must be taken into account when determining the colour index of a star. SEE ALSO **UBV system**

Columba A southern constellation adjoining Canis Major and Carina, representing a dove. Alpha (Phakt) is magnitude 2.6, but the constellation contains little of interest.

colure A great circle on the celestial sphere that passes through both celestial poles. SEE ALSO **equinoctial colure, solstitial colure**

coma (1) A defect in an optical system which results in the image of a point appearing as a blurred pear-shaped patch with a flared appearance resembling a comet. It is caused by incident light striking a lens obliquely. The defect increases with distance from the centre of the field of view.

coma (2) The spherical cloud of gas and dust surrounding the nucleus of an active **comet**.

Coma Berenices A northern constellation adjoining Boötes, representing the hair of Queen Berenice of Egypt. Its brightest stars, of 4th and 5th magnitudes, form a very wide star cluster. It is rich in faint galaxies; some are part of the **Virgo Cluster**, but others are part of the separate and more distant *Coma Cluster*, about 300 million light years away.

comes (plural **comites**) The fainter companion in a **binary star**.

comet A small, icy Solar System body in an independent orbit around the Sun. The solid *nucleus* of a comet is small – that of Halley's Comet measures just 16 × 8 km (10 × 5 miles). The nucleus is made up of ices, largely water-ice, in which are embedded rock and dust particles, and is covered by a thin dark crust. As the comet approaches the Sun and gets warmer, evaporation begins, and jets of gas and dust escape through the crust to form the luminous, spherical *coma*. The coma can be huge, a million kilometres or so across, but extremely tenuous. Later, radiation pressure from the Sun and the solar wind may send dust and gas streaming away as a *tail*, tens of millions of kilometres long. The *gas tail* or *ion tail* is straight and long, up to 10 million km, while the *dust tail* is shorter, broader, and curved.

Comets are thought to originate in the **Oort Cloud**. From there, a comet may be perturbed (by a passing star, for example) and displaced towards the Sun. It may swing past the Sun on a parabolic or even a hyperbolic orbit and return to the Oort Cloud, or enter an elliptical orbit as a *long-period comet*, with a period of thousands or even millions of years (SEE **parabola**). Further perturbation by one of the giant planets may drive it into a smaller orbit as a *short-period comet*, with a period of 200 years or less. Further such close encounters can subsequently modify the orbit.

With their loss of material at every approach to the Sun, periodic comets have a limited lifetime. The debris a comet leaves in its wake spreads around its orbit to form a **meteor stream**. Some comets, once they

SOME PERIODIC COMETS AND ASSOCIATED METEOR SHOWERS

Name	Discovery	Orbital period (years)	Perihelion distance (million km)	Eccentricity	Inclination (°)	Shower(s)
Encke	1786, 1818	3.31	51	0.85	12	Taurids
Pons–Winnecke	1819, 1858	6.16	187	0.64	22	Pons–Winneckids
Biela (lost)	1772, 1826	6.62	129	0.76	13	Andromedids (Bielids)
Giacobini–Zinner	1900, 1913	6.59	154	0.71	32	Draconids (Giacobinids)
Tuttle	1790, 1858	13.8	152	0.82	55	Ursids
Tempel–Tuttle	1865	32.9	147	0.91	163 *R*	Leonids
Halley	240 BC	76	88	0.97	162 *R*	Eta Aquarids, Orionids
Swift–Tuttle	1737, 1862	130*	144	0.96	114 *R*	Perseids
Thatcher (1861 I)	1861	410*	75*	0.98	80	Lyrids

** approximate values   R indicates retrograde orbit*

have expended all their gas and dust, may be strong enough to hold together and continue to orbit the Sun – the **Apollo asteroids**, **Amor asteroids**, and **Aten asteroids** may be extinct comets. Others, like Comets **Biela**, **West** and **Shoemaker–Levy 9**, have been observed to break up. The distinction between comets as very small, predominantly icy bodies, and asteroids as generally rather larger, rocky bodies is no longer clear-cut. Some objects first classed as asteroids have later behaved like comets – **Chiron**, for example, developed a coma.

Nearly 800 comets have had their orbits calculated in detail, and over three-quarters of these are long-period comets. Several new comets are discovered every year. After its orbit has been confirmed, it is customary for a new comet to be given the name of its discoverer. Donati's comet (or, more formally, Comet Donati), for example, was discovered by Giovanni Donati in 1858. A few comets, such as Halley's Comet and Encke's Comet, are named after the people who first calculated their orbit. In the case of co-discoveries, up to three names are allowed, as in Comet IRAS–Araki–Alcock (discovered independently by two amateur astronomers and the IRAS satellite). Formally, periodic comets are also given the prefix 'P', as in Comet P/Halley.

Three comets have been visited by space probes: **Giacobini–Zinner** in 1985, **Halley** in 1986, and **Grigg–Skjellerup**, in 1992.

commensurable Describing two numbers with a common factor. If the orbital periods of two planets around the Sun, or of two satellites round a planet, can be expressed as a ratio of small whole numbers, then the periods are called commensurable. Because of tidal interactions (SEE **tides**) between bodies of similar size, stable orbits tend to form at (nearly) commensurable distances. Examples are the orbital periods of Saturn and Jupiter, which are in a ratio of nearly 5 : 2, and the periods of Jupiter's **Galilean satellites**. With bodies which are very different in size, orbital **resonances** can have the effect of preventing bodies from having commensurable orbital periods, as with the **Kirkwood gaps** in the main asteroid belt and the gaps in Saturn's ring system. **Bode's law** was partly (but only by chance) successful in predicting undiscovered planets because of the commensurability of orbits.

comparator A device with which the images of the same star field on two photographs taken at different times can be rapidly alternated in the field of view. If any object has changed its position or varied in brightness during the interval between the times when the two photographs were taken, the comparator will show it up. Hunters of novae and asteroids use comparators. Variable stars reveal themselves as a pulsation, because they give images of different sizes according to their brightness. The *blink microscope* or *blink comparator* is a professional instrument in which the two images are viewed alternately in an eyepiece. The *stereo comparator* is a simple device for viewing the two images simultaneously with binocular vision; another arrangement uses two slide projectors and a rotating shutter.

Compton Gamma Ray Observatory (GRO) A NASA satellite for studying the sky at gamma-ray wavelengths, launched in April 1991. It carries four instruments: the Burst and Transient Source Experiment (BATSE), to detect short-lived outbursts; the Oriented Scintillation Spectrometer Experiment (OSSE), to study the spectrum of gamma-ray sources; the Imaging Compton Telescope (Comptel), to map the gamma-ray sky at medium energies; and the Energetic Gamma Ray Experiment Telescope (EGRET), to make an all-sky survey of high-energy sources.

conjunction The alignment of the Earth with two other bodies in the Solar System so that from the Earth the other two appear in (nearly) the same position. The term is used mainly for alignments with the major

THE CONSTELLATIONS			
Name	Genitive	Abbreviation	Order of size
Andromeda	Andromedae	And	19
Antlia	Antliae	Ant	62
Apus	Apodis	Aps	67
Aquarius	Aquarii	Aqr	10
Aquila	Aquilae	Aql	22
Ara	Arae	Ara	63
Aries	Arietis	Ari	39
Auriga	Aurigae	Aur	21
Boötes	Boötis	Boo	13
Caelum	Caeli	Cae	81
Camelopardalis	Camelopardalis	Cam	18
Cancer	Cancri	Cnc	31
Canes Venatici	Canum Venaticorum	CVn	38
Canis Major	Canis Majoris	CMa	43
Canis Minor	Canis Minoris	CMi	71
Capricornus	Capricorni	Cap	40
Carina	Carinae	Car	34
Cassiopeia	Cassiopeiae	Cas	25
Centaurus	Centauri	Cen	9
Cepheus	Cephei	Cep	27
Cetus	Ceti	Cet	4
Chamaeleon	Chamaeleontis	Cha	79
Circinus	Circini	Cir	85
Columba	Columbae	Col	54
Coma Berenices	Comae Berenices	Com	42
Corona Australis	Coronae Australis	CrA	80
Corona Borealis	Coronae Borealis	CrB	73
Corvus	Corvi	Crv	70
Crater	Crateris	Crt	53
Crux	Crucis	Cru	88
Cygnus	Cygni	Cyg	16
Delphinus	Delphini	Del	69
Dorado	Doradus	Dor	72
Draco	Draconis	Dra	8
Equuleus	Equulei	Equ	87
Eridanus	Eridani	Eri	6
Fornax	Fornacis	For	41
Gemini	Geminorum	Gem	30
Grus	Gruis	Gru	45
Hercules	Herculis	Her	5
Horologium	Horologii	Hor	58
Hydra	Hydrae	Hya	1
Hydrus	Hydri	Hyi	61
Indus	Indi	Ind	49

THE CONSTELLATIONS *continued*			
Name	Genitive	Abbreviation	Order of size
Lacerta	Lacertae	Lac	68
Leo	Leonis	Leo	12
Leo Minor	Leonis Minoris	LMi	64
Lepus	Leporis	Lep	51
Libra	Librae	Lib	29
Lupus	Lupi	Lup	46
Lynx	Lyncis	Lyn	28
Lyra	Lyrae	Lyr	52
Mensa	Mensae	Men	75
Microscopium	Microscopii	Mic	66
Monoceros	Monocerotis	Mon	35
Musca	Muscae	Mus	77
Norma	Normae	Nor	74
Octans	Octantis	Oct	50
Ophiuchus	Ophiuchi	Oph	11
Orion	Orionis	Ori	26
Pavo	Pavonis	Pav	44
Pegasus	Pegasi	Peg	7
Perseus	Persei	Per	24
Phoenix	Phoenicis	Phe	37
Pictor	Pictoris	Pic	59
Pisces	Piscium	Psc	14
Piscis Austrinus	Piscis Austrini	PsA	60
Puppis	Puppis	Pup	20
Pyxis	Pyxidis	Pyx	65
Reticulum	Reticuli	Ret	82
Sagitta	Sagittae	Sge	86
Sagittarius	Sagittarii	Sgr	15
Scorpius	Scorpii	Sco	33
Sculptor	Sculptoris	Scl	36
Scutum	Scuti	Sct	84
Serpens	Serpentis	Ser	23
Sextans	Sextantis	Sex	47
Taurus	Tauri	Tau	17
Telescopium	Telescopii	Tel	57
Triangulum	Trianguli	Tri	78
Triangulum Australe	Trianguli Australis	TrA	83
Tucana	Tucanae	Tuc	48
Ursa Major	Ursae Majoris	UMa	3
Ursa Minor	Ursae Minoris	UMi	56
Vela	Velorum	Vel	32
Virgo	Virginis	Vir	2
Volans	Volantis	Vol	76
Vulpecula	Vulpeculae	Vul	55

planets (SEE the diagram at **elongation**). A **superior planet** is at conjunction when on the opposite side of the Sun from the Earth. An **inferior planet** is at *inferior conjunction* when it lies between the Earth and the Sun, and at *superior conjunction* when it is on the far side of the Sun. Because planetary orbits are inclined at various angles to the plane of the ecliptic, alignments at conjunction are rarely exact, and conjunction is defined to occur when the two bodies have the same celestial longitude as seen from the Earth.

Two or more planets – including in this case the Moon – are also said to be in conjunction when they are close together in the sky.

constellation A region of the sky containing an arbitrary grouping of stars, forming an imaginary figure traced on the sky. The stars in a constellation lie at very different distances from us, so the groupings have no physical significance. The original number of 48 constellations given in Ptolemy's ***Almagest*** has grown to 88, which were assigned boundaries on the celestial sphere by the International Astronomical Union in 1930. Constellations are very unequal in size and importance; they range from the vast Hydra down to the tiny but brilliant Crux. Many of the names are drawn from ancient mythology, while more recent mapping of the southern-hemisphere skies introduced some modern-sounding names such as the Octant and the Telescope. Noteworthy stars within a constellation are designated by Greek letters followed by the name of the constellation in the genitive (possessive) case, as with Alpha Lyrae; these are often abbreviated in the form α Lyr.

continuous spectrum The unbroken sequence of colours, merging one into the other, produced when light is decomposed by refraction through a prism. An incandescent solid, liquid, or dense gas emits a continuous spectrum. The spectrum of a star consists of a continuous spectrum crossed by **absorption lines**. SEE **emission spectrum**

continuum SEE **spacetime**

co-orbital satellite A satellite orbiting a planet at the same average distance as another satellite. Although the distances – and hence the periods – are the same, the orbital inclinations and eccentricities may differ slightly. An example is **Helene**, a small satellite of Saturn with a very similar orbit to the larger satellite Dione.

Coordinated Universal Time SEE **Universal Time**

coordinates Reference systems that can be used to define the position of a point or body on the **celestial sphere**. Several systems of coordinates are used in astronomy.

The most commonly used is the *equatorial system*, in which the reference plane is that of the celestial equator, and the coordinates are usually **right ascension** and **declination** (although **hour angle** and **polar distance** can be used instead).

In the *horizontal system* the reference plane is that of the observer's horizon, and the coordinates are **azimuth** and **altitude**.

In the *ecliptic system* the reference plane is that of the ecliptic, and the coordinates are celestial latitude and longitude. *Celestial latitude* is the angular distance between the object and the ecliptic, measured at right angles to the ecliptic. It is measured from 0° to 90°, positively towards the north ecliptic pole and negatively towards the south ecliptic pole. *Celestial longitude* is given by the angle between the object and the vernal equinox, measured along the ecliptic from 0° to 360° eastwards from the vernal equinox. Celestial latitude and longitude are not measured with reference to the celestial equator; they are sometimes more correctly called *ecliptic latitude* and *ecliptic longitude*.

When dealing with the Galaxy, the system of *galactic coordinates* is used, in which the plane of reference is the galactic plane (plane of the Milky Way). Its coordinates are galactic latitude and longitude. *Galactic lati-*

tude is measured from 0° to 90° from the galactic equator, positively towards the north galactic pole. *Galactic longitude* is measured from 0° to 360° eastwards along the galactic equator. As defined by the IAU, the position of zero galactic longitude is at RA 17h 46m, dec. −28° 56′ (2000 coordinates).

Copernicus, Nicholas (1473–1543) Polish churchman and astronomer. A canon at Frauenburg Cathedral and local administrator, he pursued astronomy as a side-interest. The accepted view in his day was the geocentric (Earth-centred) universe as explained by Ptolemy nearly 1500 years before, with its complicated system of orbits. Through his study of planetary motions, Copernicus developed a heliocentric (Sun-centred) theory of the universe. In this *Copernican system*, as it is called, the planets' motions in the sky were explained by having them orbit the Sun, which simplified things considerably. The motion of the sky was simply a result of the Earth turning on its axis, and the stars remained fixed as the Earth orbited the Sun because they were so very far away.

An account of his work, *De revolutionibus orbium coelestium*, was published in 1543. Most astronomers (and possibly even Copernicus himself) considered the new system as merely a means of calculating planetary positions, and continued to believe in **Aristotle**'s view of the world. It was not until the discoveries by **Galileo** and **Johann Kepler** that the reality of it began to take hold. SEE ALSO **aberration** (1)

Copernicus A large lunar crater, 93 km (58 miles) in diameter, in the Mare Imbrium. It is one of the most conspicuous craters, and is surrounded by a system of rays stretching for well over 500 km (300 miles).

Cordelia The innermost small **satellite** of Uranus, discovered in 1986 during the Voyager 2 mission. It and Ophelia act as **shepherd moons** to the planet's Epsilon Ring.

Coriolis force A force that appears to act on a body (such as a projectile or a mass of air) moving freely across the Earth so as to deflect it from a straight path. The deflection is to the right in the northern hemisphere, to the left in the southern. In fact the effect is a consequence of us observing the motion from within a rotating frame of reference – the spinning Earth. In 1735 the English meteorologist George Hadley (1685–1768) recognized the effect of the Earth's rotation on the movement of air currents; general deduction of the displacing force of the Earth's rotation was made by Gaspard de Coriolis (1792–1843). The Coriolis force accounts for the circulation of air around cyclones on the Earth, and is a driving force for the meteorology of the Earth and other planets with substantial atmospheres.

corona The outermost layer of the Sun's atmosphere, extending for many millions of kilometres into space. It springs into view as a white halo during a total solar eclipse; at other times it can be observed in visible light only by using a special instrument called a **coronagraph**. The corona emits strongly in the X-ray region, and has been studied by X-ray satellites. Its overall shape changes during the 11-year solar cycle, from regular and symmetrical at solar minimum to uneven with long streamers at solar maximum. The corona has a temperature of 1–2 million K.

The corona may be divided into the inner *K corona*, which shines by light from the photosphere scattered by high-energy electrons, and the outer *F corona*, which shines by light scattered by lower-energy dust particles. The K corona has a **continuous spectrum** (K stands for the German word *Kontinuum*), on which are superimposed the **Fraunhofer lines** of the F corona. There are also emission lines in the spectrum contributed by the *E corona*, lower-energy metal ions such as calcium present out to about two solar radii. SEE ALSO **coronal holes, flare, Sun, Yohkoh**

Corona Australis A small southern constellation representing a crown or wreath. It contains no star above magnitude 4, but its arc of stars makes it easy to identify. It is sometimes called Corona Austrinus.

Corona Borealis A northern constellation, representing the wedding crown of Ariadne, adjoining Boötes. Alpha (Gemma or Alphecca), magnitude 2.2, is the only bright star, but the curve of five stars makes the constellation easy to find. In the 'bowl' of the crown is the famous variable star **R Coronae Borealis**, prototype of a group of stars that fade suddenly and unpredictably. T Coronae Borealis is a recurrent nova, known as the Blaze Star, that erupted in 1866 and 1946.

coronagraph (sometimes spelt **coronograph**) An instrument, used in conjunction with a telescope, which allows the corona and prominences to be viewed by producing an 'artificial eclipse', blotting out the light from the Sun's disk with a circular obstruction. Before the invention of the coronagraph, such observations were possible only during the brief duration of a total eclipse.

coronal holes Relatively cool regions of the solar **corona**. They are associated with weak regions of the Sun's magnetic field, and are sources of high-speed streams of the **solar wind**.

Corvus A constellation adjoining Hydra and Crater, representing a crow. Its four brightest stars, of third magnitude, form a quadrilateral which is easy to find. Otherwise, Corvus contains little of note.

cosmic abundance The relative proportions of the various chemical elements (SEE **element, chemical**) in the Universe as a whole. It has been found from spectroscopic studies of the atmospheres of the Sun and stars that (with the exception of hydrogen and helium) the cosmic abundance is much the same as in the Sun. Stars are roughly two-thirds hydrogen and one-third helium, with other elements – chiefly carbon and oxygen – accounting for only 2%. The chemical make-up of a particular star depends on its age (SEE **stellar evolution**). Some rare types of star, for example **Wolf–Rayet stars**, show unusual abundances of certain elements; others, such as the *technetium stars*, contain elements it is difficult to explain by currently accepted theories of how stars evolve. SEE ALSO **interstellar matter**

Cosmic Background Explorer (COBE) A NASA satellite that studied the **cosmic microwave background** radiation. It determined that the microwave background has a **black body radiation** spectrum with a temperature of 2.73 K. In 1992 it detected slightly warmer and cooler spots in the background radiation, differing by about 30 millionths of a degree from the average; these *ripples* are attributed to density fluctuations in the early Universe that marked the first stage in the formation of galaxies. COBE was launched in November 1989 and operated until the end of 1993.

cosmic microwave background Weak electromagnetic radiation from the Universe, first detected in 1965 by **Arno Penzias** and **Robert Wilson** of the Bell Telephone Laboratories. The microwave background is black-body radiation at a temperature of 2.73 K and has an almost equal intensity in all directions in space. It is considered to be the remnant of the radiation from the hot, early Universe following the **Big Bang**. The background radiation increased in wavelength as the Universe expanded until today it peaks in the millimetre region of the spectrum. It is called the microwave background because it was first detected in the form of microwaves. Penzias and Wilson won the 1978 Nobel Prize for Physics for their discovery. SEE ALSO **Cosmic Background Explorer**

cosmic rays High-energy atomic particles, moving at speeds approaching the speed of light, which enter the Earth's atmosphere from space. Cosmic rays (so called because they were originally thought to be gamma rays) consist mainly of the nuclei of atoms, mostly hydrogen, stripped of their electrons. While travelling through space they are called *primary cosmic rays*. When they enter the atmosphere they cause disintegration of the atoms they encounter and produce various atomic and subatomic particles, called *secondary cosmic rays*, which can be detected at ground level. The highest-energy cosmic rays are the most energetic particles known; their origin is uncertain, and they may come from outside our Galaxy. Cosmic rays of medium energy originate in supernovae; low-energy cosmic rays are ejected from the Sun by solar **flares**.

cosmogony The study of how the Universe began. Originally the term referred to the creation of the Solar System and the formation of stars, but now it is taken to refer to the first moments after the Big Bang. We can trace the Universe back at least to an age of 10^{-5} seconds after the **Big Bang**, when it contained matter, antimatter, and radiation at a temperature of around 10^{12} K. In the standard Big Bang theory we cannot extrapolate back to the moment of creation itself because the theory predicts the existence of a **singularity**, where the laws of physics break down. Suggestions for an origin have arisen from progress in the unification of the fundamental forces of physics, known as **grand unified theories**, or GUTs. These lead to *inflationary cosmology*, in which a finite region, of which the Universe is a visible part, starts off as a small, hot, expanding bubble of matter in a form that can exist only at a temperature above 10^{27} K. The bubble cools and expands rapidly as it changes into normal matter. The inflationary model would take the Universe back to 10^{-35} or 10^{-45} seconds, depending on whether gravity is included in the unification of forces, but it may be more correct to imagine that time itself begins at this point.

cosmological redshift SEE **redshift**

cosmology The study of the structure of the Universe on the largest scale. Contained within it is **cosmogony**, which deals with the origins and evolution of the Universe.

The starting point for modern cosmology was the discovery in the 1920s by **Edwin Hubble** that the Universe is expanding. The rate of expansion is given by the **Hubble constant**. Other observational evidence shows that the Universe appears much the same in all directions (the so-called *isotropy*) and at all distances (*homogeneity*).

Cosmological models have included the **steady-state theory**, in which the Universe is not only the same in all places but also at all times. It therefore had no beginning, will have no end, and never changes at all when viewed on the large scale. This theory required matter to be created as the Universe expanded in order that the overall density of galaxies should not increase. For this reason it is also referred to as the *continuous creation* model. On the other hand, according to the **Big Bang** theory, the Universe was created in a single instant about 15,000 million years ago, from which it has been expanding ever since. In the future it may continue to expand or possibly collapse back on itself, depending on the total amount of matter and energy in the Universe – on whether it is above or below its **critical density**. An important consideration is the question of the **missing mass**: the amount of matter we see in the Universe is less than a tenth of the amount inferred from the motions of the galaxies.

Cosmologists have made several important advances in discriminating between cosmological models. The discovery of the **cosmic microwave background** radiation provided strong evidence against the steady-state theory. This background radiation is very uniform in all directions, which has led

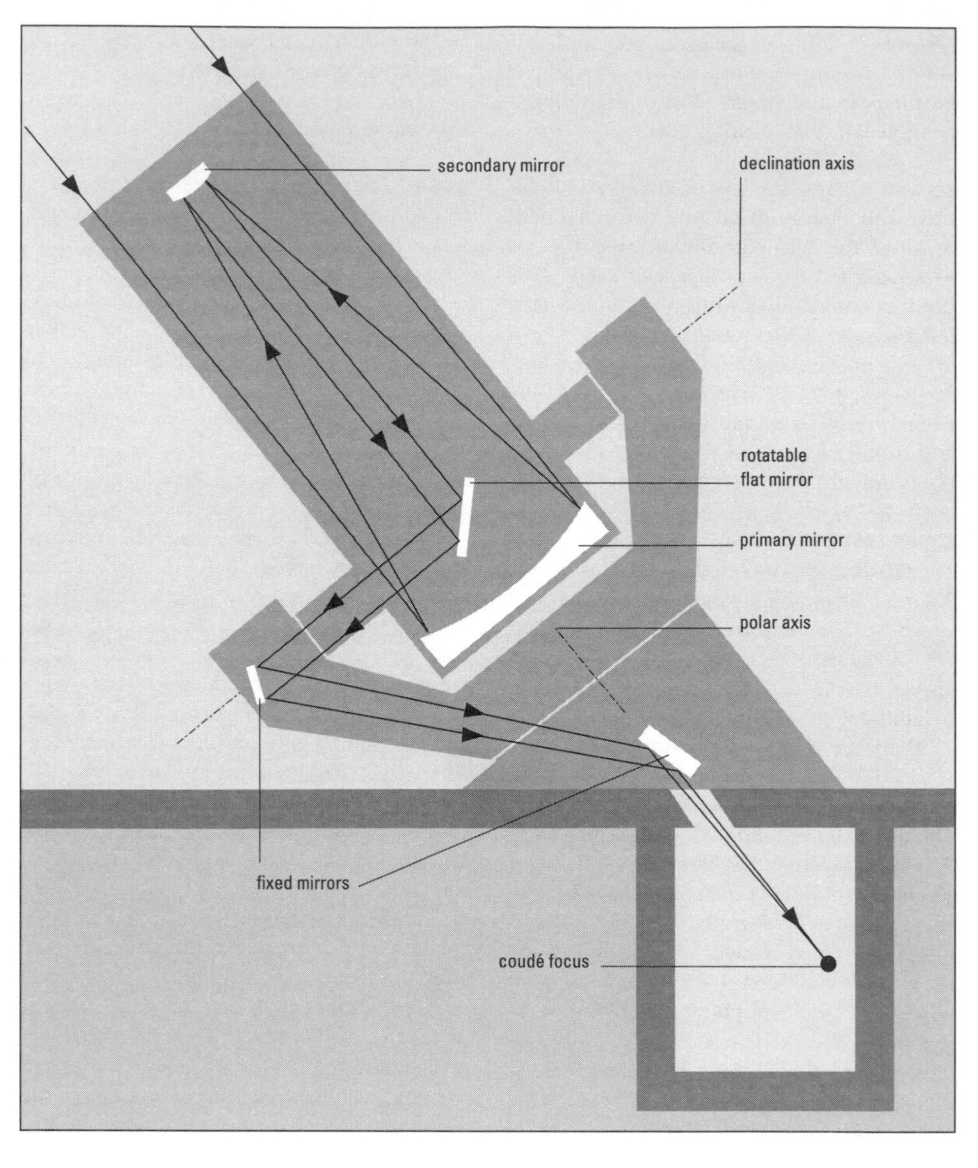

Coudé system: a five-mirror arrangement for directing light to a fixed focus

cosmologists to devise the so-called *inflationary* models (SEE **cosmogony**) in which there was a very rapid period of expansion shortly after the Big Bang. The Hubble constant is known to an accuracy of about 30%. The **deceleration parameter** – which determines whether the Universe will expand for ever or collapse back on itself – has also been estimated, and seems to have a value close to that expected for a Universe which will just continue to expand for ever.

At present, the best observational evidence favours an inflationary, Big Bang model for the Universe with a Hubble constant of around 75 km/s/megaparsec, and a deceleration parameter close to 0.5. Its overall geometry is surprisingly close to the simplest possible flat 'Euclidean' model.

coudé An arrangement of auxiliary mirrors used with a telescope on an equatorial mount to direct the light path down the polar axis to a fixed focal point. The coudé system is used in conjunction with large, immovable instruments, such as a spectrograph, which are too massive to be mounted on the telescope itself. The word comes from the French *coudé*, meaning bent like an elbow, and is not the name of a person, as is sometimes thought. SEE ALSO **Nasmyth focus**

Crab Nebula The nebula M1 (NGC 1952), about 6500 light years away in Taurus, the remnant of a supernova that was noted by Chinese astronomers in July 1054, when it shone as brightly as Venus, being visible even in daylight. It is now about 8th magnitude. The nebula was discovered in 1731 by the English astronomer John Bevis (1695–1771) and independently by **Charles Messier** in 1758. It gained its popular name of Crab Nebula after it was sketched by Lord **Rosse** to resemble that creature.

The Crab Nebula is an intense source of radio emission known as *Taurus A*, and is also a source of X-rays. Gas ejected in the supernova explosion forms filaments that appear red on colour photographs, surrounding an ionized gas that emits a yellow glow. The glow is **synchrotron radiation**, caused by electrons spiralling in an intense magnetic field. The synchrotron radiation extends from the radio domain into the visible, and the Crab Nebula is the only synchrotron nebula that can be seen in a small telescope. Inside the Crab Nebula lies the *Crab Pulsar*, the remaining core of the supernova of 1054. It spins 30 times a second, spraying out electrons to replenish the synchrotron radiation and keep the nebula glowing. The Crab Pulsar also flashes at visible wavelengths. SEE **pulsar**

crater A circular formation, found on many bodies in the Solar System – planets, satellites, and asteroids. Nearly all craters are **impact features**, although on some bodies there are volcanic craters. A typical impact crater has a raised rim, or wall, and its floor is below the level of the surrounding terrain. Some have a central peak. Impact craters range in size from the microscopic (as found on lunar rocks returned by the Apollo missions) to features a third of the diameter of the body on which they are found (as on Saturn's moon **Mimas**). Heavily cratered bodies such as **Callisto** are geologically inactive and preserve a record of the bombardment that took place in the early Solar System, when there were still plenty of free **planetesimals**. Less cratered bodies are usually those on which geological activity has led to some degree of resurfacing, obliterating older impact sites.

Crater A small constellation adjoining Hydra and Corvus, representing a cup or chalice. Its brightest stars are of fourth magnitude and it contains no objects of note.

Crepe Ring A popular name for Ring C of **Saturn**. SEE ALSO **ring, planetary**

Cressida An inner **satellite** of Uranus, discovered in 1986 during the Voyager 2 mission.

critical density The average density of matter in the Universe that would ensure it would just barely continue expanding for ever, without ever collapsing back on itself. The observed density of the Universe is less than a tenth of the critical value; the remainder is thought to be composed of dark matter. The ratio between the observed and critical density is given the symbol Ω. Hence, if the Universe has the critical density, then $\Omega = 1$. If Ω is less than 1, then the Universe will expand for ever. Ω is twice the value of q_0, the **deceleration parameter**. SEE ALSO **missing mass**

crust SEE **differentiation**

Crux The smallest constellation in the sky, popularly known as the Southern Cross. It is almost surrounded by Centaurus. Its brightest star, Alpha (Acrux), is of magnitude 0.76 and lies about 500 light years away. It is actually a visual double, magnitudes 1.3 and 1.7, both type B. With Beta (Mimosa), magnitude 1.2, Gamma (1.6), and Delta (2.8), it makes a pattern that is more like a kite than a cross; the symmetry is disturbed further by Epsilon (3.6), the only other star above magnitude 4. Interesting objects include the **Jewel Box** cluster, and the **Coalsack**, a dark nebula.

C star SEE **carbon star**

culmination The passage of a celestial body across the observer's meridian. At culmination, an object reaches the greatest altitude above, or least altitude below, the observer's horizon. In the case of circumpolar stars, both culminations are observable. *Lower culmination* occurs when a star transits between the pole and horizon; the star's hour angle is then exactly 12h. *Upper culmination* occurs when a star transits between the pole and zenith; the star's hour angle is then exactly 0h.

curvature of space The distortion of **spacetime** in the neighbourhood of matter, one of the consequences of **Einstein**'s General Theory of Relativity. This curvature makes rays of light and particles of matter follow curved paths called *geodesics*. In **cosmology**, if the Universe contains enough mass it will have enough curvature to be *closed*; if not, it will be *open*.

cusp One of the points or horns of the crescent Moon, or of an inferior planet at crescent phase.

cusp caps The bright areas at the **cusps** of the crescent Venus that are sometimes reported by observers.

Cygnus One of the most distinctive constellations, often nicknamed the Northern Cross. Its brightest star is first-magnitude **Deneb**. Beta is Albireo, a beautiful coloured double with a giant primary of type K3, magnitude 3.1, and a main-sequence companion of type B8, magnitude 5.1, separation 34 arc seconds. Chi has the largest visual range of any Mira variable, 3.3 to 14.2, period 408 days. P Cygni is a B-type supergiant variable that has ranged between magnitudes 3 and 6 in the past, because of mass loss, and now appears around 5th magnitude. **61 Cygni** is a binary pair of orange dwarf stars, magnitudes 5.2 and 6.0. The constellation also contains the radio galaxy **Cygnus A**, the black hole candidate **Cygnus X-1**, the **North America Nebula**, and the **Cygnus Loop** supernova remnant.

Cygnus A The strongest source of radio emission outside our Galaxy, situated in the constellation Cygnus. It is thought to be caused by the collision of two galaxies. It is also a source of X-ray emission.

Cygnus Loop The remnant of a supernova, consisting of a vast loop of gas ejected from a star that exploded about 30,000 years ago. It lies about 2500 light years away. The Loop is about 3° in diameter, and is expanding at about 100 km/s. Different parts of it bear the designations NGC 6960, 6979, 6992, and 6995. The brightest part, NGC 6992, is known as the Veil Nebula.

Cygnus X-1 A strong source of X-ray emission in the constellation Cygnus, believed to contain a **black hole**. At the position of the source lies a spectroscopic binary star, HDE 226868, with an orbital period of 5.6 days. The visible star is a blue supergiant. Its invisible companion is calculated to be at least 8 solar masses, far too great for a white dwarf or neutron star. Hence it seems that this massive invisible object may well be a black hole.

Dactyl The satellite of the asteroid **Ida**.

dark adaptation (dark adaption) The process by which the human eye adjusts from vision under conditions of high illumination to night vision. It takes 20 minutes or more to develop fully, during which time the pupil widens and different receptors come into play. Dark adaptation is essential for observers leaving a brightly lit room to begin observing. It is instantly destroyed by any bright light, so observers use dim red lights when they need to consult star charts or record observations.

dark matter Unseen matter, inferred to exist in galactic halos and in the space between galaxies, that is thought to make up at least 90% of the mass of the Universe. It could exist in various forms, such as black holes, brown dwarfs, or unknown atomic particles, all of which are termed *cold dark matter*; or it could be in the form of a 'sea' of fast-moving neutrinos, known as *hot dark matter*; or a mixture of both. SEE ALSO **missing mass**

dark nebula A cloud of gas and dust that is not illuminated, and can thus be seen only in silhouette against stars and bright nebulae beyond. They range in size from minute **Bok globules** to the naked-eye clouds of the **Coalsack** nebula in Crux and the gigantic Rho Ophiuchi Dark Cloud which covers about 1000 square degrees (2%) of the sky. Dust comprises only about 0.1% of the mass of a dark nebula, but is believed to play an important part in the formation of molecules in space, since the surface of the dust particles provides a site for atoms to adhere and combine into molecules. The interiors of these molecular clouds are very cold, typically only 10 K, which allows them to collapse under their own gravity into stars.

David Dunlap Observatory The observatory of the University of Toronto, Canada. Its principal instrument is a 1.88 m (74-inch) reflector, opened in 1935.

Dawes' limit SEE **resolving power**

day The time taken by the Earth to rotate once on its axis. There are various definitions. An *apparent solar day* is the interval between two successive transits of the Sun across the meridian. This interval is variable because the Sun's apparent motion throughout the year is not uniform (SEE **apparent solar time**). For convenience, an imaginary *mean Sun* is defined which moves along the celestial equator at a constant rate, giving the *mean solar day*. A *sidereal day* is defined as the interval between two successive transits of the spring **equinox** across the meridian. This day is about four minutes shorter than the mean solar day because of the Sun's apparent eastward movement of about 1° each day. A mean solar day is 24h 03m 56.555s of mean sidereal time. A mean sidereal day is 23h 56m 04.091s of mean solar time. SEE ALSO **time**

deceleration parameter (symbol q_0) A figure describing the rate at which the expansion of the Universe is slowing down. If q_0 is greater than 0.5, the expansion will eventually stop and reverse. If q_0 is less than 0.5, the Universe will expand for ever. Theoretical considerations suggest that the true value of q_0 should be exactly 0.5. SEE ALSO **critical density, missing mass**

declination (dec., symbol δ) The angular distance of a celestial object north or south of the celestial equator. It is reckoned positively from 0 to 90° from the equator to the north celestial pole, and negatively from 0 to 90° from the equator to the south celestial pole. SEE ALSO **celestial sphere**

decoupling The time, about 300,000 years after the Big Bang, when the Universe had cooled sufficiently for the first atoms to form; also known as *recombination*. The **cosmic microwave background** radiation was released at the time of decoupling.

deep sky The Universe beyond the Solar System. Star clusters, nebulae, and galaxies are known as *deep-sky objects*, but the term is not usually applied to individual stars.

deferent SEE **Ptolemy**

degenerate matter A state of matter existing in stars in the final stage of their evolution (SEE **stellar evolution**), when they have ceased producing energy at their cores. Atomic nuclei and electrons are packed closely together at ultra-high densities, and the laws of classical physics no longer apply. Pressure ceases to be dependent on temperature and is a function only of density.

Deimos The smaller of Mars's two **satellites**, discovered in 1877 by Asaph Hall. It is a dark, irregular body, measuring about 15 × 12 × 11 km (9 × 7.5 × 7 miles), and may well be a captured asteroid.

Delphinus A small but distinctive constellation in the Aquila area, representing a dolphin. Its leading stars, of fourth magnitude, make up what at first sight looks like a widespread cluster.

Delta Aquarids A **meteor shower** which occurs at the end of July and the beginning of August. It has a double radiant in Aquarius, the more active radiant lying near the star Delta Aquarii. The shower is associated with Halley's Comet.

Deneb The star Alpha Cygni, magnitude 1.25, a supergiant of type A2, luminosity 60,000 times the Sun's, distance about 1500 light years. It is the brightest star in the constellation Cygnus.

descending node The point at which an orbit crosses from north to south of a reference plane such as the celestial equator or ecliptic. SEE **node**

Desdemona One of the small inner **satellites** of Uranus discovered in 1986 during the Voyager 2 mission.

de Sitter, Willem (1872–1934) Dutch cosmologist. His hypothetical *de Sitter Universe*, derived from Einstein's general theory of relativity, contained no mass, but provided the first theoretical indication of an expanding Universe.

Despina One of the small inner **satellites** of Neptune discovered in 1989 during the Voyager 2 mission. It appears to be the inner **shepherd moon** to the planet's Le Verrier Ring.

diagonal (1) An optical device, used in conjunction with a telescope's eyepiece, for deflecting the light path through 90° so as to make it easier for the observer to view through the eyepiece when the telescope is at an awkward angle. In this device, also called a *star diagonal*, either a mirror or a right-angled prism is set at 45° to the light path.

diagonal (2) Another name for the flat secondary mirror in a **Newtonian telescope**.

diamond-ring effect A short-lived phenomenon seen during a total solar eclipse, just before or after totality, when just the bright central part of the edge of the Sun's disk is visible as an intense point of light, together with the faint inner corona, giving the appearance of a diamond ring.

dichotomy The phase of Mercury or Venus when it appears exactly half-illuminated by the Sun, and the **terminator** is a straight line. The term is also applied, less commonly, to the half Moon.

differential rotation The rotation of different parts of a non-solid body at different rates. The deep atmospheres of the giant planets rotate slightly faster at the equator than at the poles. The Sun shows a similar differential rotation, but rather more pronounced. The stars and other objects in galaxies are in independent orbits and with periods that increase with distance from the galactic centre, so galaxies may be regarded as showing differential rotation.

differentiation The process by which a planetary body evolves a layered structure. In a body which has grown to a sufficient size by **accretion**, energy supplied by gravitational compression and radioactive decay, supplemented by the kinetic energy of impacting bodies, will cause the interior to melt. Under the action of gravity, denser materials will then sink towards the centre, where they will eventually form the *core*, and less dense materials will rise towards the surface, where they will form the *mantle* and the *crust*. Differentiation has given the terrestrial planets, for example, cores of nickel–iron and outer layers of a predominantly rocky composition. The distinct chemical compositions of some meteorites and asteroids indicate that they come from bodies which were large enough to have undergone differentiation before being broken up by impacts.

diffraction The slight sideways spreading of a beam of light as it passes by a sharp edge or through a narrow slit. Starlight passing through a telescope is diffracted by the edges of various components so that, instead of coming to a sharply focused point, it appears as a small disk of light surrounded by concentric *diffraction rings* of light, and radial *diffraction spikes*. The bright central disk is called the **Airy disk**.

diffraction grating A surface on which a very large number of equidistant parallel lines have been ruled very close together. Light striking a grating is dispersed by **diffraction** into a spectrum. There are metal gratings that function by reflection, and glass gratings that function by transmission. The lines are ruled at, typically, 100 to 1000 per millimetre. Diffraction gratings produce very high-quality spectra, and are used in astronomical **spectrographs**.

diffuse nebula A luminous cloud of gas in space. The term 'diffuse' refers to the fact that diffuse nebulae cannot be resolved into individual stars, unlike star clusters and galaxies. They come in two varieties: **emission nebulae**, also known as **H II regions**, which shine by fluorescence; and **reflection nebulae**, in which the starlight reflects off dust particles.

Dione A medium-sized **satellite** of Saturn, diameter 1120 km (695 miles), discovered in 1684 by Giovanni Cassini. It has the second-highest density, at 1.44 g/cm^3, of Saturn's main satellites. It has a bright icy surface, but one hemisphere is darker than the other. There are three main types of terrain: cratered terrain, cratered plains, and smooth plains. Evidence of past geological activity includes a number of **tectonic** features such as Palatine Chasma – a trough over 600 km (nearly 400 miles) long and up to 8 km (5 miles) wide. Dione shares its orbit with a small **co-orbital satellite, Helene**.

direct motion (prograde motion) (1) Orbital or rotational motion in the same direction as the Earth's motion: anticlockwise as seen from above the Sun's north pole. The orbital motion of all planets and most moons is direct. Orbital or rotational motion is direct if the orbital or axial inclination is less than 90°. COMPARE **retrograde motion** (1)

direct motion (prograde motion) (2) The regular, west-to-east motion of Solar System bodies as seen from Earth relative to background star. COMPARE **retrograde motion** (2)

dispersion The separation of white light into its constituent colours by a lens, prism, or diffraction grating. This happens because the different wavelengths of light are refracted more at the red end than at the violet end of the spectrum. Dispersion has to be countered in telescope lenses in order to avoid **chromatic aberration**, but it is exploited in astronomical instruments such as the **spectrograph** in order to study the spectra of stars.

distance modulus The difference between the apparent magnitude (m) and the absolute magnitude (M) of a star. It is a measure of the star's distance (r) in parsecs:

$$m - M = 5 (\log r) - 5.$$

diurnal motion The apparent daily motion of celestial bodies across the sky from east to west. It is a result of the Earth's axial rotation.

diurnal parallax SEE **parallax**

Dobsonian telescope A Newtonian reflecting telescope on a simple form of **altazimuth** mount. The use of lightweight materials makes it possible to build a low-cost, large-aperture telescope that is portable, making the Dobsonian increasingly popular with amateur observers. A typical Dobsonian is mounted at its base, on a 'rocker-box' assembly which rotates on a horizontal baseplate. The instrument is named after the American amateur John Dobson (1915–).

Dollond, John (1706–61) English optician. Formerly a silkweaver, in 1752 he joined the optician's business founded by his son, **Peter Dollond** (1730–1820). He found that the deviation of light rays could be achieved without **dispersion**, and combined crown and flint glass to make **achromatic lenses**. These he used in commercially successful refracting telescopes, which were further refined by Peter. John also invented

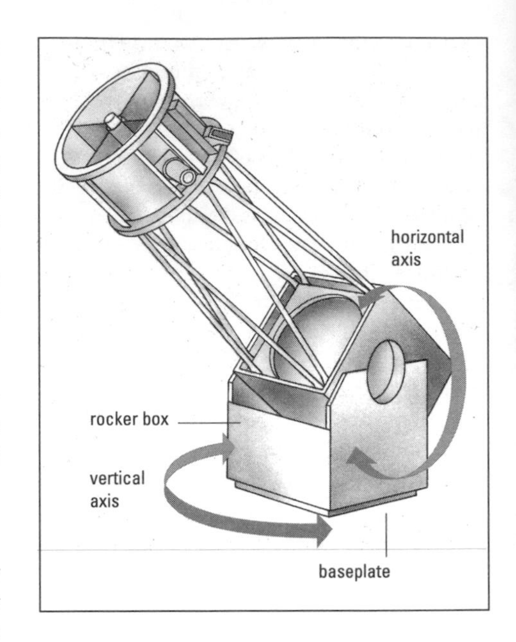

Dobsonian telescope

the **heliometer**, a refractor modified so as measure small angular distances.

Dominion Astrophysical Observatory An observatory near Victoria, BC, Canada, opened in 1917. It has 1.83 m (72-inch) and 1.22 m (48-inch) reflectors.

Doppler effect The apparent increase in frequency (and decrease in wavelength) of radiation from a source moving towards the observer, or the similar decrease in frequency (increase in wavelength) of radiation from a source moving away. In everyday life, the Doppler effect is familiar as, for example, the change in pitch (i.e. frequency) of a siren as an ambulance speeds past.

In astronomy, its importance is that it explains the displacement of spectral lines. If a celestial object is moving towards the Earth, its light will be shifted to shorter wavelengths, towards the violet end of the spectrum. But if is moving away, then the shift will be towards longer wavelengths at the red end of the spectrum – a **redshift**.

This *Doppler shift*, as it is called, makes it possible to detect and measure relative motion in the line of sight, in other words **radial velocity**, and rotation of celestial objects. The redshifts of distant galaxies provided the first evidence of the **expanding Universe**. The effect is named after Austrian physicist Christian Doppler (1803–53), who first described it.

Dorado A southern constellation, representing a goldfish. It contains the main part of the Large **Magellanic Cloud**, including the **Tarantula Nebula** and the highly luminous irregular variable star S Doradus. Alpha, its brightest star, is magnitude 3.3.

Double Cluster Two large star clusters in the constellation Perseus, visible to the naked eye. They bear the designations NGC 869 and NGC 884, and are also known as h and χ (chi) Persei. They both lie about 7400 light years away.

double star Two stars which appear close together in the sky. There are two types of double star. If the stars are genuinely close together in space and are connected by gravity they are known as a **binary star**. Two stars which are quite distant from each other, but appear close together as a result of chance alignment, are known as an **optical double**.

doublet A lens with two components, which are either cemented together or separated by an air-gap. This construction is used to reduce **chromatic aberration**.

Draco A long, winding northern constellation representing the dragon slain by Hercules, extending between Ursa Major and Ursa Minor, with the dragon's head near the star Vega. Gamma (Eltanin), magnitude 2.2, is its brightest star. Alpha (Thuban), magnitude 3.7, was the pole star in ancient times. On the whole Draco is a barren group but one of the stars in the head, Nu, is a wide double easily separable with binoculars; each component is of magnitude 4.9.

draconic month One revolution of the Moon around the Earth relative to its ascending **node**. It is equal to 27.21222 days of **mean solar time**.

Dreyer, Johan Ludvig Emil (1852–1926) Danish astronomer, who from 1874 lived and worked in Ireland, first at Lord **Rosse**'s observatory. He compiled the *New General Catalogue of Nebulae and Clusters of Stars* (SEE **NGC**) and its two supplementary *Index Catalogues*.

Dumbbell Nebula A large planetary nebula in the constellation Vulpecula, and designated M27. It was so named by Lord **Rosse** from its hourglass shape. The Dumbbell is visible in binoculars.

dwarf galaxy A galaxy that is much smaller than a normal galaxy and is of low luminosity. Such galaxies are usually elliptical or irregular. There are several dwarf ellipticals in the **Local Group** of galaxies.

dwarf nova A member of a class of irregular variable stars whose light-curves resemble those of novae. Their luminosity stays the same for long periods, then rapidly increases, and finally slowly returns to normal. **U Geminorum stars** and **Z Camelopardalis stars** are the main types. Dwarf novae are thought to be close binary stars in which one component is a white dwarf.

dwarf star The most common type of star in the Galaxy, constituting 90% of its stars and 60% of its mass. Dwarfs are also known as **main-sequence** stars, from their position on the **Hertzsprung–Russell diagram**. The term 'dwarf' refers to luminosity rather than size, so dwarfs should be thought of as normal rather than diminutive. The Sun is a typical dwarf, roughly midway in the range of properties of dwarf stars.

Dwarfs are common because stars spend most of their lives on the main sequence, converting hydrogen to helium in their cores. In stars heavier than the Sun the principal nuclear reaction is the **carbon–nitrogen cycle**, while in lower-mass stars it is the **proton–proton reaction**. An upper limit to the masses of dwarf stars is determined by radiation pressure in their interiors: above about 60 solar masses, the outward force exerted by radiation exceeds the force of gravity, preventing such a star from being formed, or making it unstable with a very short lifetime if it does form.

The hottest and most massive dwarfs (spectral types O and B) have relatively short lifetimes, a few hundred million years or less, so they are rare in the Galaxy. The lifetimes of the lowest-mass dwarfs (type K and M) are so long that none has evolved from the main sequence since the Galaxy formed. For dwarf stars below about 0.08 solar masses, the core temperatures cannot become high enough to initiate hydrogen-burning. As a result the main sequence terminates at stars with surface temperatures of about 2500 K. Lower-mass stars are known as **brown dwarfs**. SEE ALSO **red dwarf, white dwarf**

Dwingeloo 1 A nearby **galaxy** in Cassiopeia, so close to the plane of the Milky Way that at optical wavelengths it is almost completely obscured by galactic dust and gas. It is a barred spiral galaxy with a mass of about 100,000 million Suns, and lies about 10 million light years away. It was discovered in 1994, initially as a radio source by a team using the radio telescope at Dwingeloo in the Netherlands. There is a small companion galaxy, named Dwingeloo 2.

dynamical parallax The distance of a **binary star** determined from a relationship between the known masses of the components, the size of the orbit, and the period of revolution.

E

Eagle Nebula A bright nebula surrounding a cluster of new-born stars, M16 or NGC 6611, in the constellation Serpens. The nebula and its associated stars are about 8000 light years away.

early-type star A hot star of spectral type O, B, or A. The name was given when it was thought that the sequence of spectral types was an evolutionary sequence, hot stars being the youngest and cool stars the oldest. This is now recognized not to be the case, but the term has remained in use. COMPARE **late-type star**

Earth The third major planet from the Sun, and the largest of the four inner, or terrestrial planets. From space the Earth is predominantly the blue of the oceans, plus the browns and greens of its land masses, the white polar caps, and a continually changing pattern of white cloud. Seventy per cent of the surface is covered by water, and it is this and the Earth's average surface temperature of 13°C (286 K) that make it suitable for life. The continental land masses make up the other 30%. Our planet has one natural satellite, the **Moon.** The main data for the Earth are given in the first table.

Monitoring the propagation of seismic waves from earthquakes has revealed the Earth's internal structure. Like all the terrestrial planets, there is a dense *core* rich in iron and nickel, surrounded by a *mantle* consisting of silicate rocks. The thin, outermost layer of lighter rock is called the *crust*. Continental crust can be as much as 50 km (30 miles) deep, but oceanic crust has an average depth of only 10 km (6 miles). The boundary between the crust and the mantle is called the *Mohorovičić discontinuity* (usually shortened to *Moho*) after its Croatian discoverer, Andrija Mohorovičić (1857–1936). The mantle extends to a depth of 2890 km (1795 miles). The outer part of the core, down to 5150 km (3200 miles), is molten, but the inner core, 2460 km (1525 miles) in diameter, is solid because of the greater pressure there. The temperature and pressure at the Earth's centre are estimated to be 4000°C and 4 million bars (i.e. 4 million times the pressure at the surface). The inner core rotates at a different rate from the solid outer layers, and this, together with currents in the molten outer core, gives rise to the Earth's magnetic field.

The crust and the uppermost layer of the mantle together form the *lithosphere*, which consists of a number of tightly fitting slabs called *plates* – eight major ones and over twenty minor ones. The plates are all moving slowly with respect to one another, and are supported on a semi-molten layer of mantle called the *asthenosphere*. Plates grow at *mid-oceanic ridges*, boundaries between adjoining plates where molten rock rises and solidifies

EARTH: DATA	
Globe	
Diameter (equatorial)	12,756 km
Diameter (polar)	12,714 km
Density	5.52 g/cm^3
Mass	5.976×10^{24} kg
Volume	1.083×10^{12} km^3
Sidereal period of axial rotation	23h 56m 04s
Escape velocity	11.2 km/s
Albedo	0.37
Inclination of equator to orbit	23° 27′
Surface temperature (average)	290 K
Orbit	
Semimajor axis	1 AU = 149.6×10^6 km
Eccentricity	0.0167
Inclination to ecliptic	0 (by definition)
Sidereal period of revolution	365.256d
Mean orbital velocity	29.8 km/s
Satellites	1

EARTH: LAYERS OF THE EARTH'S ATMOSPHERE				
Layer	Upper boundary	Altitude (km)	Temperature (°C)	Pressure (mbar)
Troposphere	Tropopause	20*	−80*	250
Stratosphere	Stratopause	50	0	0.9
Mesosphere	Mesopause	80	−90	0.007
Thermosphere	—	500	1500	10^{-9}

The altitude, temperature and pressure are values at the upper boundary.
** Above the equator. At the poles the altitude of the tropopause is 10 km.*

into new oceanic crust, pushing the plates apart; an example is the Mid-Atlantic Ridge. Where two plates meet, one may be forced below the other, descending into the mantle and melting, causing earthquakes and volcanoes at the edge of the plate above. This is called *subduction*, and occurs principally in the so-called 'ring of fire' around the Pacific Ocean's rim. Where plates collide and no subduction occurs, they crumple up to form folds that develop into mountain ranges like the Himalayas. All these interactions are known collectively as *plate tectonics*.

Because of tectonics, most of the Earth's surface is young compared with the planet's age of 4.6 billion years. The oldest rocks belong to the *Precambrian shields*, areas over 600 million years old, as in parts of Canada; some rocks have been dated to over 4 billion years ago. Most impact craters in these areas have been weathered away, but some surface features (such as **Meteor Crater**) are identifiable as impact structures.

The Earth's atmosphere consists of 78% (by volume) nitrogen, 21% oxygen, 0.9% argon, and 0.03% carbon dioxide, other gases such as neon making up the remaining 0.07%. In addition there can be up to 3% water vapour, depending on geographical location and weather conditions. There are higher concentrations of carbon dioxide and other gases, including sulphur dioxide, over industrial areas. The atmosphere produces a **greenhouse effect**, raising the surface temperature by 35 K.

The atmosphere is divided into a number of layers according to the way in which its temperature varies with altitude (SEE the table above). The *troposphere* contains most of the atmosphere, and is where lifeforms are found and weather systems operate. The *stratosphere* contains the *ozone layer*, which absorbs the high-energy ultraviolet radiation from the Sun that is harmful to life. **Meteors** occur in the mesosphere, and **aurorae** in the thermosphere. The thermosphere is extremely rarefied, and its high temperature indicates the high kinetic energy of its molecules, rather than its heat content. Ionized atoms and molecules in the mesosphere and thermosphere constitute the **ionosphere**. Above the thermosphere is the exosphere, which contains the Earth's **magnetosphere** and **Van Allen Belts**, and merges into interplanetary space.

earthgrazer An alternative name for a **near-Earth asteroid**.

earthshine Sunlight reflected from the Earth on to the part of the Moon's nearside which is in the Sun's shadow. It is easiest to see when the Moon is a thin crescent, when it appears as a pale greyish light over the remainder of its disk. Earthshine is known popularly as 'the old Moon in the new Moon's arms'.

eccentric (1) Describing an orbit with a high **eccentricity**.

eccentric (2) SEE **Ptolemy**

eccentric anomaly SEE **anomaly**

eccentricity (symbol *e*) One of the elements of an **orbit**. It indicates how much an elliptical orbit departs from a circle. The eccentricity is found by dividing the distance between the two foci of the ellipse by the length of the major axis. A circle has an eccentricity of 0, and a parabola has an eccentricity of 1.

echelle grating A **diffraction grating** designed to yield spectra of high **dispersion** and detail. Its lines are ruled further apart than on an ordinary grating, and are shaped so as to produce high resolution over a narrow band of wavelengths at angles of illumination greater than 45°. Such a grating is used in an *echelle spectrograph*.

eclipse The partial or total obscuration of the light from a celestial body as it passes through the shadow cast by another body. A body may be eclipsed by the passage of another body between it and the observer, as in a **solar eclipse**, or by the intervention of another body between it and the source of the light it reflects, as in a **lunar eclipse**. The eclipse of a star by the Moon or by a planet or other Solar System body is called an **occultation**. SEE ALSO **annular eclipse, transit**

eclipse year The interval between successive passages of the Sun through a given node of the Moon's orbit, equal to 346.62003 days of mean solar time. It differs significantly from other types of year because the nodes of the Moon's orbit move westwards by over 19° per year (SEE **regression of the nodes**). Nineteen eclipse years are equal to 6585.78 days, almost exactly the same length of time as a **saros**.

eclipsing binary A type of **binary star** in which the orbital plane of the two stars is viewed almost edge-on, leading to mutual eclipses and consequent variations in combined light output. Two eclipses may be expected during an orbital cycle, one usually causing a bigger drop in light than the other since the two stars are rarely equal in size or brightness (SEE diagram below). The deeper minimum (A) is called the primary

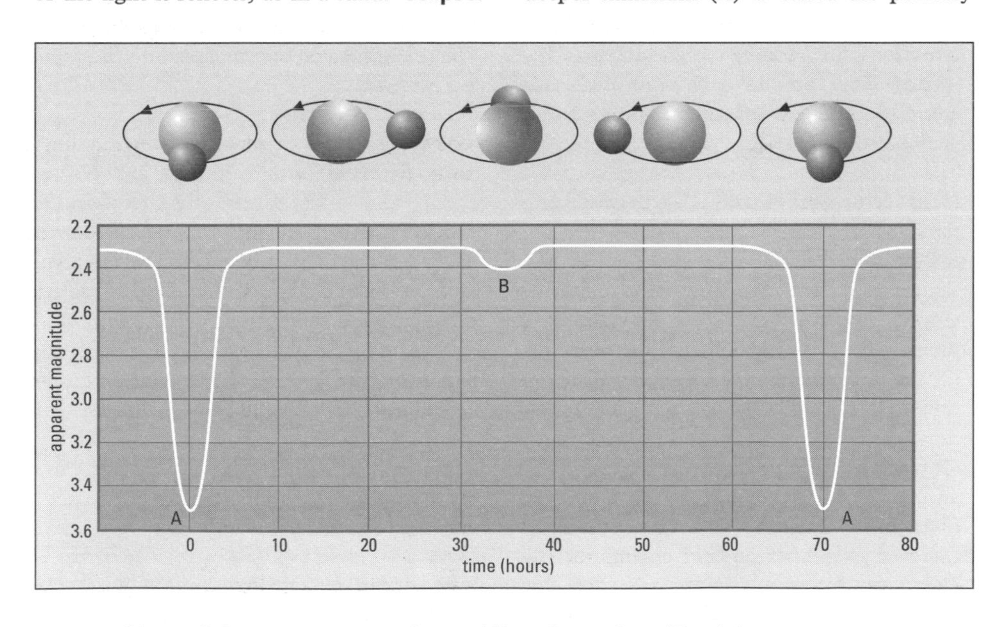

Eclipsing binary: light curve, showing primary (A) and secondary (B) minima

minimum, and the shallower minimum (B) is the secondary minimum. If the stars are very unequal in brightness the secondary minimum may hardly be noticeable.

Eclipsing binaries are classified into three types on the basis of their light curves: **Algol** type (EA), **Beta Lyrae stars** (EB), and **W Ursae Majoris stars** (EW). The EA types have a constant light level outside eclipses, but EB and EW types show continuous variability. This is because one or both stars in the binary are elliptical in shape, having expanded to fill their **Roche lobe**. The EW stars show more or less equal depths of primary and secondary minima. In such systems, both stars are thought to be filling their Roche lobes and are thus touching each other; they are known as *contact binaries*.

ecliptic The apparent yearly path of the Sun against the background stars. It intersects the celestial equator at the two equinoxes. The ecliptic is really the projection of the Earth's orbit around the Sun on to the celestial sphere. Because of the tilt of the Earth's axis the ecliptic is inclined to the celestial equator by about 23½°, a figure known as the *obliquity of the ecliptic*. The ecliptic poles are the points 90° from the ecliptic, and lie in the constellations Draco and Dorado.

ecliptic coordinates SEE **coordinates**

ecliptic limits The greatest angular distance the Sun or Moon can be from the Moon's nodes for which a **solar eclipse** or **lunar eclipse** can take place. The limits are 37° for the Sun at a solar eclipse, and 24° for the Moon at a lunar eclipse.

E corona SEE **corona**

ecosphere The shell-shaped region of space surrounding the Sun or another star within which the temperature is such that any suitable planet would be capable of sustaining life. The dimensions of the ecosphere vary greatly with the type of star. The ecosphere of a small dwarf star would be narrow and lie close in, whereas that of a luminous giant would be broad and a considerable distance from the star.

Eddington, Arthur Stanley (1882–1944) English astronomer and physicist. He pioneered the use of atomic theory to study the internal constitution of stars, and explained the role of radiation pressure in preventing stars from collapsing under gravity. Among his many discoveries were the mass–luminosity relationship and the fact that white dwarfs contain degenerate matter. He supported and popularized Einstein's theory of relativity, and in 1919 he obtained experimental proof of the theory's prediction that gravity bends light by measuring star positions close to the Sun during a solar eclipse.

Eddington limit The maximum ratio of luminosity to mass that a star can have before radiation pressure overcomes the gravitational force holding it together, and the star consequently blows apart. It gives a theoretical limit to stellar mass of between about 50 and 100 solar masses.

effective temperature (symbol T_{eff}) The temperature a star would have if its output were **black-body radiation** at the same energy and at the same wavelengths as the star. As stellar spectra approximate closely to black-body spectra, a star's effective temperature is a good approximation to its actual surface temperature.

Effelsberg Radio Observatory The radio observatory of the Max-Planck-Institut für Radioastronomie near Bonn, Germany, equipped with a 100 m (328 ft) steerable radio telescope, completed in 1970.

Einstein, Albert (1879–1955) German–Swiss–American theoretical physicist. His work has had an enormous impact on 20th-century science. In astronomy, and in particular in cosmology, his theories of relativity have had profound effects. The Special Theory of Relativity (1905) gave the relation $E = mc^2$ between mass and energy; the General Theory of Relativity (1916) extended the theory to encompass gravitation. The General Theory has been borne out, for example, by its explanation of the advance of the perihelion of **Mercury**, the curvature of light by a gravitational field (SEE **Eddington**), and the **gravitational redshift** of spectral lines. In the 1920s relativity made Einstein world-famous, but he never again matched the scientific work of his early years. He spent much time unsuccessfully seeking a unified field theory which would link relativity with electromagnetic forces. SEE ALSO **grand unified theory**

ejecta SEE **rays**

Elara One of the four small **satellites** of Jupiter's intermediate group, discovered by Charles Perrine in 1905.

electromagnetic radiation (em radiation) The emission and propagation of energy in the form of periodic waves that can travel through a vacuum. It originates when charged atomic particles are accelerated, and it consists of oscillating electric and magnetic fields. There is a continuum of electromagnetic radiation – from long-wavelength radio waves of low frequency and low energy, through visible light waves, to short-wavelength X-rays and gamma-rays of high frequency and high energy – known as the *electromagnetic spectrum* (SEE the diagram below). The speed at which em radiation travels in a vacuum (c) is given by its wavelength (λ) times its frequency (ν):

$$c = \lambda\nu.$$

This speed is constant, and is commonly known as the speed of light (SEE **light, speed of**). In the *quantum theory*, an em wave (particularly if its wavelength is that of visible light or shorter) can also be regarded as a stream of particles called *photons*, each having an energy (E) related to the frequency (or wavelength) of the radiation:

$$E = h\nu = hc/\lambda.$$

With X-rays and gamma-rays, this energy, expressed in **electron-volts**, is a more important property than their wavelength. Photons also possess momentum, which means they can exert a pressure, called **radiation pressure**.

Almost all of our information about the Universe and its contents has been learned from studies of the electromagnetic radiation

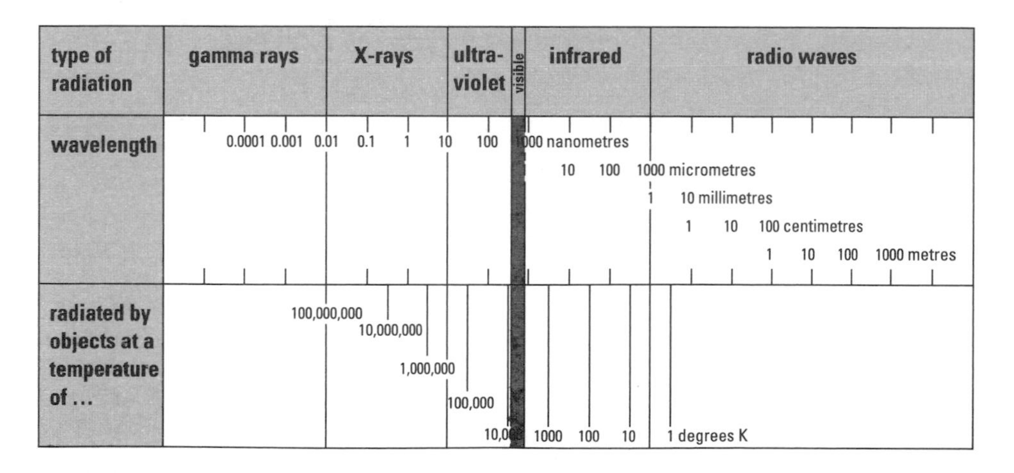

Electromagnetic radiation: the type of radiation a body emits depends on its temperature

that reaches us from celestial objects. Much of the radiation that arrives at the Earth is absorbed by its atmosphere, but there are some bands of the electromagnetic spectrum that can reach the surface of the Earth and can be studied by Earthbound instruments. The two main ones are known as the *optical window* and the *radio window*. Other types of cosmic radiation are detected and analysed by instruments carried above the atmosphere by satellites or rockets.

electron-volt (symbol eV) A unit of energy used mainly in atomic physics for the energies of atomic particles. One electron-volt is the kinetic energy acquired by a particle carrying a charge equal to that on one electron when it passes through a potential difference of 1 volt. 1 eV is equivalent to 1.6021×10^{-19} joules, and, from the equation $E = h\nu$ (SEE **electromagnetic radiation**), corresponds to a frequency of 2.42×10^{14} hertz, or a wavelength of 1240 nm. An electron with a kinetic energy of 1 eV has a velocity of about 580 km/s.

element, chemical One of the basic materials of which everything in the Universe is composed. Elements are chemically homogeneous substances – that is, they cannot be decomposed by chemical means. The atoms of a particular element all have the same number of protons in their nucleus, but the number of neutrons can vary, giving rise to different *isotopes* of the same element. The number of known elements is now over 100. Of these, 90 occur in nature, and the remainder have been prepared artificially, in laboratories. SEE ALSO **cosmic abundance**, **nucleosynthesis**

element, orbital SEE **orbit**

ellipse A closed curve, such that the sum of the distances of any point on it from two points within it – the *foci* (singular *focus*) of the ellipse – is a constant. In mathematics the ellipse is a type of conic section, so

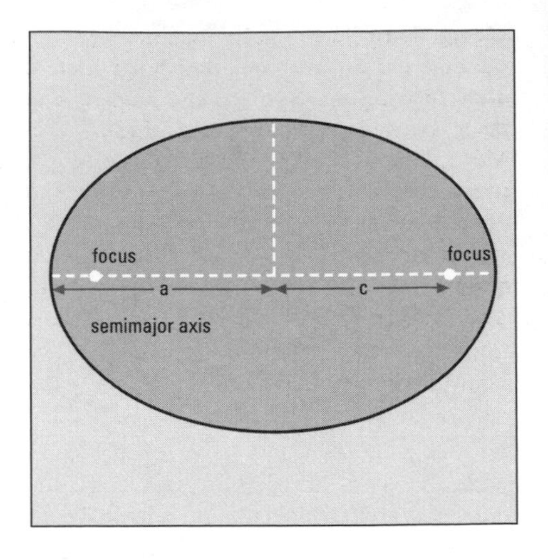

Ellipse

called because it is one of the intersections of a plane with a cone. The ellipse is important in astronomy because the closed **orbits** of planets, satellites, companion stars, comets, and so on are all close approximations to ellipses. (**Perturbations** make orbits depart from true ellipses.) It is also basic to **Kepler's laws**. Important parameters of the ellipse, and therefore of orbits, are the **semimajor axis** (a) and the **eccentricity** (e). The shape of the ellipse is governed by its eccentricity, which is given by $e = c/a$, where c is the distance from the centre of the ellipse to one of its foci. For an ellipse, e is always less than 1.

elliptical galaxy A **galaxy** that is ellipsoidal in shape. Elliptical galaxies are composed of old stars and contain little gas or dust. They are the central dominant galaxies in very rich clusters, and many are powerful radio sources. Ellipticals are classified according to their degree of ellipticity (flattening), and given a designation from E0 to E7. They range widely in size, from 100 million stars for *dwarf ellipticals* to 10 million million stars for *supergiant ellipticals*.

ellipticity Another word for **oblateness**.

elongation The angular distance between the Sun and a planet or other Solar System body orbiting the Sun (or the angular distance between the Sun and the Moon). More accurately, it is the difference in the two bodies' celestial longitudes measured in degrees. A planet with an elongation of 0° is at **conjunction**; at 90° or 270° it is at **quadrature**; and at 180° it is at **opposition**. The inferior planets, Mercury and Venus, when at their maximum angular distance from the Sun, are said to be at *greatest elongation east* when following the Sun (setting at its latest time after it) and at *greatest elongation west* when preceding the Sun (rising at its earliest time before it).

emersion The reappearance of a celestial body from the shadow of another after it has been eclipsed or occulted.

emission nebula A cloud of gas in space that emits light. Ultraviolet radiation from nearby stars heats the gas of the nebula,

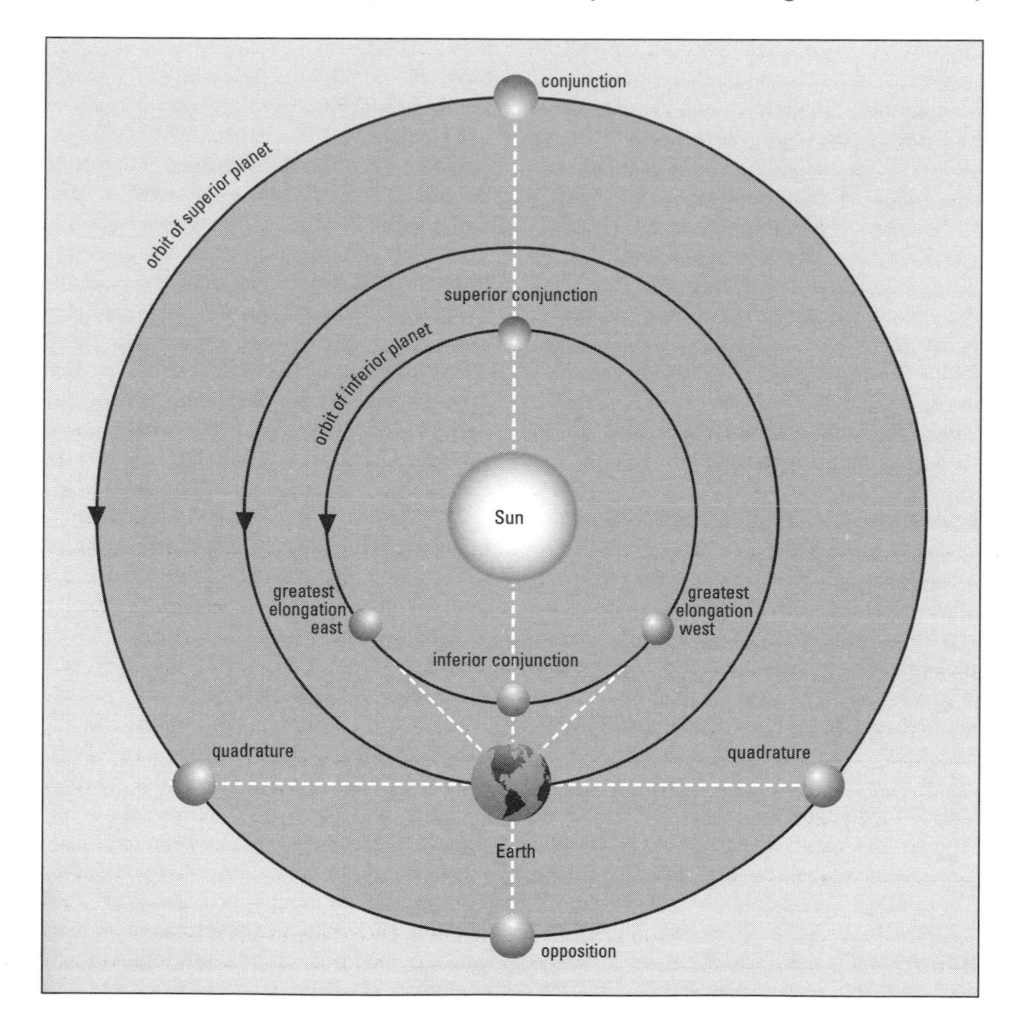

Elongation, conjunction, quadrature and opposition: planetary aspects

causing its atoms to become ionized (lose their electrons). These electrons gain energy from the ultraviolet radiation. When they eventually recombine with the atoms they re-emit the energy, some of it at visible wavelengths. The free electrons in the gas also lose energy in the form of radio waves, so emission nebulae can also be detected by radio telescopes.

Interstellar gas is mostly hydrogen, which is easily ionized by ultraviolet light. The symbol for ionized hydrogen is H II, and such emission nebulae are therefore also called **H II regions**. Hydrogen gives off its strongest light in the red, so emission nebulae appear red on photographs. However, to the eye they often look green because the eye responds more readily to a prominent pair of spectral lines of this colour emitted by oxygen.

Hot stars that emit ultraviolet light are usually young, having recently formed from the gas and dust cloud around them. Thus the association between stars and gas clouds is not by chance. A famous example is the **Orion Nebula**, with the **Trapezium** stars inside it. Emission nebulae are very tenuous. Typically, every gram of material is spread through a volume of a million cubic km.

emission spectrum The **spectrum**, consisting of a series of bright lines, produced by a highly energetic source. Energy in the form of heat or electromagnetic radiation can be absorbed by an electron orbiting the atomic nucleus, causing it to jump to a higher orbit. When it returns to its old orbit, the electron emits radiation at a characteristic wavelength. The series of lines made up of emissions from different types of atom in the source comprises the emission spectrum. The hottest stars have many strong emission lines in their spectra. SEE ALSO **forbidden lines**

Enceladus A **satellite** of Saturn, diameter 500 km (310 miles), discovered in 1789 by William Herschel. It is the most reflective body of any size in the Solar System, with an albedo approaching 1. Its surface consists of older, cratered terrain and more recently formed plains traversed by straight and curved grooves, and there are indications that **tectonic** and **volcanic** processes may still be operating. Tidal interaction with the larger satellite **Dione** probably explains why such a small body has been so geologically active. Material from Enceladus may be the source of Saturn's tenuous E Ring, in which the satellite is embedded.

Encke, Johann Franz (1791–1865) German mathematician and astronomer. In 1818 he computed the orbit of **Encke's Comet**, and predicted its return, and in 1837 he was the first to record the division in **Saturn**'s rings now named after him. He also calculated the solar parallax from measurements of the 1769 transit of Venus.

Encke's Comet A periodic **comet** discovered by Pierre Méchain in 1786 and Jean Louis Pons in 1818. **Johann Encke** computed its orbit and period, and proved that it was identical with three comets that had been previously observed. It was the second periodic comet to be discovered, and has the shortest-known period, at 3.3 years.

ephemeris (plural **ephemerides**) A table giving the predicted positions of a celestial object, such as a planet or comet, at given intervals.

epicycle SEE **Ptolemy**

Epimetheus A small inner **satellite** of Saturn. It is a **co-orbital satellite** with **Janus**.

epoch (1) A point in time used as a fixed reference for comparison of astronomical data in, for example, star catalogues. **Precession** and the **proper motions** of stars cause their positions on the celestial sphere to change gradually, and observations made over any considerable period of time have to be reduced to a common epoch for them to be comparable. The starting point in the calculation of ephemerides is at a certain epoch. SEE ALSO **standard epoch**

epoch (2) One of the elements of an **orbit**, defined as the time of perihelion passage.

Epsilon Aurigae An **eclipsing binary** star with the longest known period, 27 years. The primary component is a supergiant of type F0, magnitude 2.9 at maximum and a minimum of 3.8 during eclipse. The eclipses last 700 days, the total phase taking 400 days; the last one was in 1982–4. No light is detected from the secondary component, which is thought to be surrounded by an obscuring disk of gas and dust which causes the eclipses; the secondary itself may actually be a close binary.

equant SEE **Ptolemy**

equation of the centre An irregularity in the motion of a body in an elliptical orbit. It is the difference between the true anomaly and the mean anomaly (SEE **anomaly**).

equation of time The difference between mean solar time, as shown on a clock, and apparent solar time, as shown on a sundial, caused by the combined effect of the eccentricity of the Earth's orbit and the obliquity of the ecliptic. The difference is zero four times a year: on 15 April, 14 June, 1 September, and 25 December. The maximum difference is 16 minutes and occurs in early November.

equatorial coordinates SEE **coordinates**

equatorial telescope A telescope mounted such that it has one axis, the *polar axis*, parallel to the Earth's axis and therefore pointing to the celestial pole, and the other axis, the *declination axis*, at right angles to the polar axis and therefore parallel to the plane of the Earth's equator. Once an object is in the field of view the telescope needs only to be moved about the polar axis to follow the object. If the telescope is driven at the sidereal rate to counter the apparent rotation of the celestial sphere, the object will remain in the field of view (SEE the diagram on page 70).

There are various forms of equatorial mounting. The *German mounting*, so named because the first of the type was made by **Joseph von Fraunhofer**, is still used for small instruments. The telescope is attached at its centre of mass to one end of the declination axis, which pivots about the top end of the polar axis; balance is achieved by attaching counterweights to the other end of the declination axis. Large, professional reflecting telescopes before the age of computer-controlled **altazimuth** mountings required massive equatorial mounts in which the telescope could be supported on both sides of the declination axis and sometimes at the top and bottom of the polar axis, yet still have enough freedom of movement to access most of the sky. Such designs included the *fork mounting* and the *horseshoe mounting*.

A recent and radical departure from traditional equatorial mounts is the *equatorial table*, first conceived by French poet Adrien Poncet in the 1970s. This is a flat surface on whose underside are curved casters which sit on motor-driven rollers. It allows amateur instruments like the traditionally altazimuth **Dobsonian telescope** to be given an equatorial capability.

equinoctial colure The **great circle** passing through the north and south celestial poles and the equinoxes (which are also called the equinoctial points).

equinox Either of the two points at which the Sun crosses the celestial equator. The *vernal equinox* (or *spring equinox*) occurs when the Sun crosses the equator from south to north on or near 21 March each year. This point is also called the *First Point of Aries*. The other equinox is the *autumnal equinox*, which occurs on or near 23 September

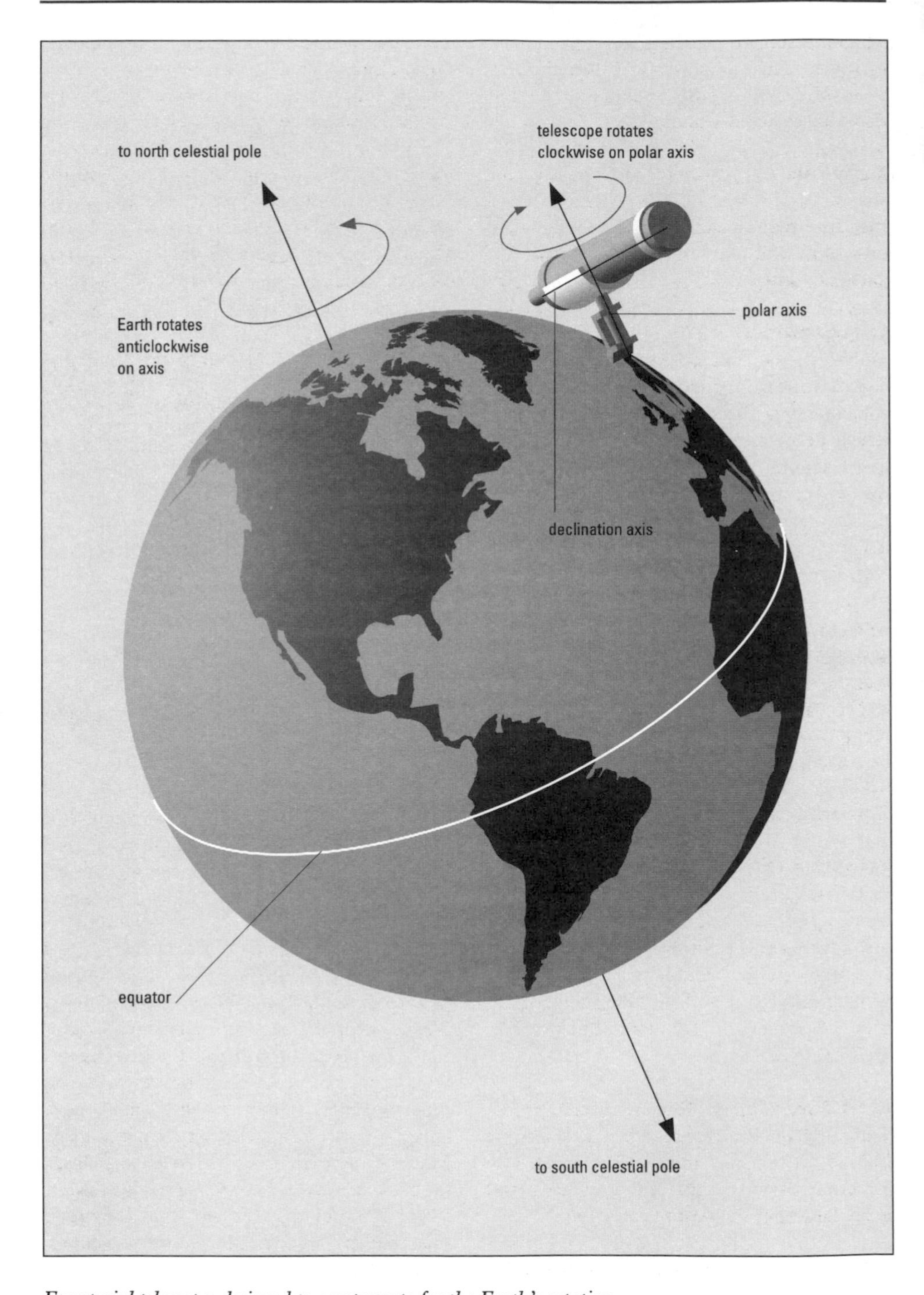

Equatorial telescope: designed to compensate for the Earth's rotation

when the Sun crosses the equator from north to south. This point is also called the *First Point of Libra*. At the equinoxes, the Sun rises due east and sets due west.

Equuleus A very small and obscure constellation, representing a foal, next to Pegasus. Its brightest star, Alpha (Kitalpha), is only of magnitude 3.9. It has no features of particular interest.

Eratosthenes (*c.* 275–195 BC) Greek geographer, mathematician and astronomer. He is best known for making the first scientific calculation of the Earth's circumference, in which he measured the altitude of the Sun at Alexandria on a day when he knew the Sun to be overhead at a place a considerable distance away. How accurately he knew this distance, and hence how accurate his result was, are uncertain, but the method was sound. Eratosthenes also compiled a star catalogue, and measured the **obliquity of the ecliptic**.

Erfle eyepiece A telescope eyepiece with low power and wide field of view, over 60°. It consists of three elements, at least one of which is a doublet. (After German optician Heinrich Erfle, 1884–1923.)

Eridanus One of the largest constellations in the sky, representing a river; it extends from near Rigel in Orion to the south polar area. **Achernar** is its brightest star. Theta (Acamar) is a fine double, with components of magnitude 3.2 and 4.3. Epsilon, magnitude 3.7, is a nearby K2 main-sequence star, 10.7 light years away.

Eros Asteroid no. 433, discovered by G. Witt in 1898. It measures 14 × 15 × 41 km (8.5 × 9.5 × 25 miles), and rotates in 5.3 hours. Its orbit crosses that of Mars, and is the largest of the **Amor asteroids**.

eruptive variables Another name for **cataclysmic variables**.

escape velocity The minimum velocity that a body, such as a rocket or a space probe, must attain to overcome the gravitational attraction of a larger body and leave on a trajectory that does not bring it back again. It depends on the mass (m) of the larger body and its radius (r), and is given by $\sqrt{(2GM/r)}$, where G is the gravitational constant. It is escape velocity rather than mass that determines the gases a body is able to retain as an atmosphere.

ESCAPE VELOCITIES OF BODIES IN THE SOLAR SYSTEM

Body	Escape velocity (km/s)
Sun	618
Mercury	4.3
Venus	10.4
Earth	11.2
Moon	2.4
Mars	5.0
Jupiter	60
Saturn	37
Uranus	21
Neptune	25
Pluto	1.1

Eta Aquarids A **meteor shower** that occurs in May. Its radiant is in Aquarius, near the star Eta Aquarii. Like the Orionids, the shower is associated with Halley's Comet.

Eta Carinae A peculiar variable star lying within the nebula NGC 3372 in the constellation Carina, about 9000 light years away. It was originally catalogued at third magnitude, but from 1833 began to vary irregularly, becoming at its brightest second only to Sirius. It then faded to just below naked-eye visibility, where it remains today. Material thrown off in the outburst forms the *Homunculus Nebula*, so called because its shape resembles a human. Eta Carinae is the most massive star known in our Galaxy, over 100 times the mass of our Sun. It has probably reached an unstable stage in its evolution, which may end in a supernova explosion a few centuries from now.

Eudoxus of Cnidus (*c.* 408–355 BC) Greek mathematician and astronomer. His theory of 27 'homocentric' (centred on the Earth) spheres was the first attempt to account for the movements of the Sun, Moon, and planets on what we would now call a properly scientific basis. This picture of 'heavenly spheres' would survive for two thousand years. According to the Roman historian Pliny, Eudoxus fixed the length of the year at 365.25 days, and according to the Roman architect Vitruvius he invented the sundial.

Euler, Leonhard (1707–83) Swiss mathematician. He made outstanding contributions to many areas of applied mathematics, especially theoretical astronomy. His analysis of the Moon's motion laid the foundations of lunar theory, and he made important advances in the study of the tides, planetary perturbations, and the orbits of comets. Euler's work on optical systems strongly influenced the technical development of telescopes and microscopes.

Europa The smallest of Jupiter's **Galilean satellites**, with a diameter of 3138 km (1950 miles), and the second-closest to the planet. Its density of 2.97 g/cm^3 indicates that it is a predominantly rocky body. Europa's smooth crust of water-ice is criss-crossed by a network of light and dark linear markings. There are very few craters, so old ones must have been removed by some form of geological activity. A kind of ice **tectonics** might be operating on Europa. Beneath the crust may lie an ocean of water. SEE ALSO **satellite**

European Southern Observatory (ESO) An observatory at La Silla, Chile, operated by a consortium of eight European countries: Belgium, Denmark, France, Germany, Italy, the Netherlands, Sweden, and Switzerland. Its telescopes include a 3.6 m (142-inch) reflector, completed in 1975, and the **New Technology Telescope**. ESO is also building the Very Large Telescope on Cerro Paranal mountain, due for completion in 1999, which will consist of four 8 m (315-inch) reflectors.

European Space Agency (ESA) An organization of European nations to promote space research and technology for peaceful purposes, founded in 1975. ESA has 14 member states: Austria, Belgium, Denmark, Finland, France, Germany, Ireland, Italy, the Netherlands, Norway, Spain, Sweden, Switzerland, and the UK. Its headquarters are in Paris. ESA has designed and built the Ariane series of launch rockets, and operates a launch site at Kourou in French Guiana, on the Atlantic coast of South America.

EUVE ABBREVIATION FOR **Extreme Ultraviolet Explorer**

evection An irregularity in the Moon's motion caused by perturbations by the Sun and planets. It amounts to a maximum of 1° 16′ in a period of 32 days.

event horizon The boundary surface of a **black hole** from within which nothing can escape. Observers outside the event horizon can therefore obtain no information about the black hole's interior. The radius of the event horizon is known as the **Schwarzschild radius**.

Evershed effect The radial outflow of gases within the penumbra of a **sunspot**. The effect was discovered in 1908 by English astronomer John Evershed (1864–1956), from the Doppler shift in the spectrum of a sunspot.

exit pupil For a telescope, the image of the primary mirror or lens as seen through the eyepiece. The point at which this image is formed is the optimum position for the observer's eye, so for comfort it should not be too close to the eyepiece itself (SEE **eye relief**).

exobiology The study of the possible existence of life beyond the Earth. This includes looking for life forms on other planets of the Solar System; the study of organic molecules in nebulae, meteorites, and comets; the study of planetary surfaces and atmospheres; questions relating to the origin of life and the range of conditions under which it can survive; and the search for signs of life in space, such as incoming messages at radio or other wavelengths.

Exosat A European Space Agency satellite for X-ray astronomy, launched in 1983. It operated for three years, studying objects such as supernova remnants, X-ray binaries, and galactic nuclei.

expanding Universe The observation that each galaxy, or cluster of galaxies, is moving apart from the others, as revealed by the **redshift** in their light. The redshift is attributed to the Doppler effect and means that these objects are receding from us at velocities that are proportional to their distance (this is the **Hubble law**). The expansion of the Universe was discovered in 1929 by Edwin Hubble, although the possibility had been suggested earlier on theoretical grounds by **Georges Lemaître**. SEE ALSO **cosmology**

extinction SEE **atmospheric extinction**, **interstellar matter**

Extreme Ultraviolet Explorer (EUVE) A NASA satellite launched in 1992 to survey the sky and study individual sources at very short ultraviolet wavelengths, a region of the spectrum not covered by previous satellites.

extrinsic variable SEE **variable star**

eyepiece (ocular) The system of lenses in an optical instrument nearest the observer's eye. Its function in a telescope used for visual observation is to magnify the image formed at the focus by the primary lens or mirror. Telescope eyepieces usually have two lenses, the *field lens*, furthest from the eye, and the *eye lens*. The magnification yielded by an eyepiece is given by the ratio of the focal length of the telescope's primary lens or mirror to the focal length of the eyepiece. There are numerous designs of eyepiece, varying in focal length, field of view, and magnification: SEE **Erfle eyepiece**, **Huygenian eyepiece**, **Kellner eyepiece**, **Nagler eyepiece**, **orthoscopic eyepiece**, **Plössl eyepiece**, **Ramsden eyepiece**. SEE ALSO **Barlow lens**

eye relief In an optical instrument such as a telescope or binoculars, the distance between the eye lens of the **eyepiece** and the **exit pupil**. When the observer's eye is placed at the exit pupil, the whole of the eyepiece's field of view is visible. The eye relief must be sufficiently large for comfortable viewing.

F

Fabry–Pérot interferometer A type of **optical interferometer** for making accurate measurements of close spectral lines (SEE **spectrum**). The two parts of the incoming beam are recombined after passing through an arrangement of parallel plates called a *Fabry–Pérot etalon*. The beams undergo multiple reflections between the plates, which are highly reflective and accurately parallel. The instrument was developed by the French physicists Charles Fabry (1867–1945) and Alfred Pérot (1863–1925).

faculae (singular **facula**) Bright areas in the Sun's upper photosphere that herald the appearance of sunspots. They appear some hours before the spots, in the same place, but can remain for months after the sunspots have gone. Faculae are at a higher temperature than their surroundings, and so appear brighter. The word is Latin for 'little torches'.

False Cross A pattern of stars made up of Iota and Epsilon Carinae, and Kappa and Delta Velorum. It is often mistaken for the Southern Cross, but it is larger, less bright, and more symmetrical.

Far Infrared Space Telescope (FIRST) A European Space Agency satellite planned for launch in 2006. It will carry a large telescope to observe the sky at far-infrared wavelengths.

F corona SEE **corona**

field of view The angular diameter of the area of sky visible through an optical instrument. For a telescope or binoculars it depends on the eyepiece that is used: a higher-power eyepiece that increases the magnification will decrease the field of view. Wide-angle eyepieces such as the **Nagler eyepiece** have fields of 80° or more.

filament A solar **prominence** seen in silhouette against the Sun's bright disk as a dark, thread-like marking in photographs taken in the light of hydrogen alpha or the K line of calcium.

filar micrometer A device used in conjunction with the eyepiece of a telescope that incorporates cross-wires for measuring the angular size of an object or the separation between two objects. Superimposed on the field of view the observer sees two parallel wires and a third wire perpendicular to them. One of the parallel wires is movable, and the observer adjusts its position until the intersections of the two parallel wires with the third wire mark off the distance being measured. The angular distance may then be read from the micrometer scale.

filter A device used in conjunction with an instrument for receiving electromagnetic radiation which allows some wavelengths to pass, but blocks others. Filters for use with optical instruments such as telescopes are at their simplest a thin sheet of material placed over the full aperture of the instrument. Simple colour filters made of gelatin bring out planetary features like Jupiter's Great Red Spot. More sophisticated filters that transmit only a narrow band of wavelengths (said to have a narrow *passband*) are used for observing the Sun, or objects such as nebulae. Multi-layer or interference filters are used to combat **light pollution**. In **astrophotography**, colour pictures can be created by combining separate exposures on suitable black-and-white emulsions taken through blue, green, and red filters. SEE ALSO **Sun**

finder A small, low-power telescope mounted on a larger one and having a much larger field of view than the main telescope, enabling the observer to locate celestial objects more easily.

fireball An exceptionally bright **meteor**. Fireballs have been defined only loosely as meteors brighter than the planets; with the modern estimate of the maximum brightness of Venus, this would mean that all meteors brighter than magnitude −4.7 should be classified as fireballs.

FIRST ABBREVIATION FOR **Far Infrared Space Telescope**

first contact In a **solar eclipse**, the moment when the leading edge of the Moon's disk first touches the Sun's disk. In a **lunar eclipse**, it is the moment when the Moon begins to enter the umbra of the Earth's shadow. The term is also used for the corresponding stage in eclipses involving other bodies.

First Point of Aries An alternative name for the vernal **equinox**, so called because it originally lay in the constellation Aries. Because of **precession**, it has now moved into neighbouring Pisces.

First Point of Libra An alternative name for the autumnal **equinox**, so called because it originally lay in the constellation Libra. Because of **precession**, it has now moved into neighbouring Virgo.

first quarter One of the Moon's **phases**, when it is at quadrature on the way to being full and is half illuminated as seen from the Earth.

Flamsteed, John (1646–1719) English astronomer. He was appointed the first Astronomer Royal by King Charles II, with the task of obtaining accurate measurements of the Moon and stars for use in navigation. From Greenwich Observatory he used equipment fitted with sighting telescopes to measure with unprecedented precision the positions of thousands of stars. Flamsteed was a perfectionist and slow to release his results. Much to his anger, The Royal Society published his uncorrected observations. The catalogue of his corrected observations was not published until 1725, after his death; the numbers assigned to stars in this catalogue are still used (SEE **Flamsteed numbers**).

Flamsteed numbers A series of numbers allocated to the stars in each constellation in order of right ascension, for purposes of identification. The numbers were allocated by later astronomers to stars in **John Flamsteed**'s star catalogue.

flare A sudden and violent release of matter and energy from the Sun's surface, usually from the region of an active group of **sunspots**. In the *flash stage* a flare builds to a maximum in a few minutes, after which it gradually fades within an hour or so. Flares emit radiation right across the electromagnetic spectrum, from gamma-rays to kilometre-wavelength radio waves. Particles are emitted, mostly electrons (some at half the speed of light) and protons, and smaller numbers of neutrons and atomic nuclei. A flare can cause material to be ejected in bulk, most spectacularly in the form of **prominences**, at speeds that can exceed the Sun's escape velocity. When energetic particles from flares reach the Earth they cause radio interference, magnetic storms, and more intense **aurorae**. (It may be that *coronal mass ejections*, in which billions of tonnes of gas are ejected into space from the corona, are responsible for these phenomena, and not flares.) Although not well understood, the origin of flares is believed to be connected with local discontinuities in the Sun's magnetic field. SEE ALSO **flare star**

flare star A variable star, usually a red dwarf, whose luminosity from time to time increases unpredictably and by as much as several magnitudes in a very short time (only a few seconds), and then decreases to its normal value in about one minute to one hour. Flare stars are also known as *UV Ceti stars*, after the best-known example,

UV Ceti, which is a pair of 13th-magnitude red dwarfs. Outbursts of flare stars are sometimes accompanied by an increase in radio emission. Large flares may give out from 10 to 100 times as much light as the rest of the star. Flare stars are often binary, notably the *BY Draconis stars*, which show additional variations of a few tenths of a magnitude attributed to large 'starspots' passing across their disk as they rotate. The low mass and relative youth of red dwarf flare stars means that steady hydrogen burning in their interiors may not yet have established itself. Flaring appears to be a consequence of instabilities in the young stars, allied with complex magnetic fields.

flash spectrum The **emission spectrum** of the solar chromosphere which is observable for a few seconds just before and after totality during an eclipse of the Sun.

flocculi (singular **flocculus**) Small features which give a mottled or granulated appearance to the Sun's chromosphere.

focal length The distance between a lens or curved mirror and its focus.

focal ratio The ratio of the **focal length** of a lens or curved mirror to its diameter. A focal ratio of, say, 8 is written as *f*/8.

focus (1) For a lens or curved mirror, the point at which is formed the image of a distant source lying on the axis of the lens or mirror.

focus (2) SEE **ellipse**

following (abbreviation *f*) Describing the trailing edge, feature, or member of an astronomical object or group of objects. For example, the following limb of the Moon is the edge facing away from its direction of motion; and the following spot (or *f*-spot) of a group of sunspots is the last of the group to be brought into view by the Sun's rotation.

Fomalhaut The star Alpha Piscis Austrini, magnitude 1.16, distance 22 light years, luminosity 13 times that of the Sun. It is a main-sequence star of type A3.

forbidden lines Emission lines (SEE **emission spectrum**) in the spectra of some celestial objects which, when they were first observed, could not be identified with lines in spectra produced under laboratory conditions on Earth. It is now known that they are produced by atoms in what is called a metastable state. On Earth this state is very short-lived, but in highly rarefied nebulae, for example, it can last long enough for 'forbidden' lines to be emitted.

fork mounting A type of mounting used for a large **equatorial telescope**.

Fornax A southern constellation originated by **Lacaille**, representing a chemical furnace. The only star above fourth magnitude is Alpha, magnitude 3.9, a double. The constellation includes a cluster of galaxies.

fourth contact (last contact) In a **solar eclipse**, the moment when the trailing edge of the Moon's disk last touches the Sun's disk. In a **lunar eclipse** it is the moment when the Moon's trailing edge leaves the umbra of the Earth's shadow. The term is also used for the corresponding stage in eclipses involving other bodies.

Fraunhofer, Joseph von (1787–1826) German physicist and optician, the founder of astronomical spectroscopy. He studied the diffraction of light through narrow slits and developed the earliest form of diffraction grating. In 1814 he observed and began to map the dark lines in the Sun's spectrum now called **Fraunhofer lines**, and was the first to appreciate their significance. He later found similar but differently distributed lines in the spectra of other stars. Fraunhofer solved many of the scientific and technical problems of astronomical telescope-making.

He manufactured achromatic lenses, was the first to build an **equatorial telescope** that realized the full potential of the design, and built the heliometer with which **Friedrich Bessel** measured the first stellar parallax.

Fraunhofer lines The dark **absorption lines** in the solar spectrum, caused by absorption at specific wavelengths in the upper, cooler, layers of the Sun's atmosphere. The most prominent ones at visible wavelengths are caused by the presence of neutral hydrogen (the **hydrogen alpha** or Hα line), singly ionized calcium (the calcium H and K lines), sodium, and magnesium. Although they were first observed in 1802 by William Hyde Wollaston, they were first carefully studied from 1814 by **Joseph von Fraunhofer**.

F star A star of spectral type F, with a surface temperature between 6100 and 7400 K on the main sequence; giants and supergiants of the same spectral type are about 300 K cooler. Main-sequence F stars range in mass from 1.2 to 1.7 solar masses, but most F-type giants and supergiants have evolved from considerably higher-mass stars. Bright examples are Canopus and Procyon.

full Moon The Moon's **phase** when it is at opposition, and its illuminated hemisphere is fully visible from the Earth.

fundamental stars Reference stars whose positions and **proper motions** have been determined with the greatest accuracy. Their coordinates are published in a *fundamental catalogue* and serve as the points to which positional measurements of other bodies can be related. The IAU has adopted 1535 stars as fundamental reference stars; relevant data are contained in the *Fifth Fundamental Catalogue*, published in Germany and known for short as the *FK5* from its German title.

G

galactic cluster An **open cluster** of relatively young stars, as found in the spiral arms of our Galaxy.

galactic coordinates SEE **coordinates**

galactic halo The spheroidal distribution of old stars and globular clusters that surrounds the Galaxy. Its radius is about 50,000 light years. There appears also to be a much more extensive halo of **dark matter** extending out to a radius of at least 200,000 light years.

Galatea One of the small inner **satellites** of Neptune discovered in 1989 during the Voyager 2 mission. It appears to be the inner **shepherd moon** to the planet's Adams Ring.

galaxy A huge assembly of stars, dust and gas, an example of which is our own Galaxy. There are three main types, as originally classified by **Edwin Hubble** in 1925. *Elliptical galaxies* are round or elliptical systems, showing a gradual decrease in brightness from the centre outwards. They are given a designation from E0 to E7 in increasing degree of ellipticity. *Spiral galaxies* are flattened disk-shaped systems in which young stars, dust, and gas are concentrated in spiral arms coiling out from a central bulge, the *nucleus*; they are designated S, with lower case letters added to show how tightly the arms are wound, from tight (a) to loose (c). *Barred spiral galaxies* are distinguished by a bright central bar from which the spiral arms emerge; they are designated SB with letters from a to c appended, as for ordinary spirals. These three main classes are represented in the **tuning fork diagram**. In addition there are *lenticular galaxies*, systems intermediate between ellipticals and spirals, having a disk and nucleus, similar to spiral galaxies, but with no apparent spiral arms; they are classified S0. *Irregular galaxies* are systems with no symmetry.

Current theories suggest that all galaxies were formed from immense clouds of gas at roughly the same time, soon after the Big Bang. In elliptical galaxies star formation took place rapidly over perhaps several hundred million years, using up all the interstellar gas. Spirals, however, are the result of a two-stage formation process. Gas was left over from the initial star formation which produced the bulge at their centres. The remaining gas rapidly settled into a disk, in which density waves formed spiral arms.

Galaxies can exist singly or in clusters that contain anywhere from just a few to thousands of members (SEE **cluster of galaxies**). Galaxies in clusters have a good chance of one near-collision in their lifetimes. Interactions between gas-rich galaxies such as spirals can produce bursts of star formation, and such interactions are a possible trigger mechanism for generating active galactic nuclei, as in **Seyfert galaxies**, **N galaxies**, and **quasars**. These objects are generally believed to be galaxies with nuclei in extreme states of activity. SEE ALSO **low-surface-brightness galaxy**

Galaxy, the The star system that contains the Solar System; the capital 'G' distinguishes it from other **galaxies**. The Galaxy is spiral in shape and about 100,000 light years in diameter. Our Sun and Solar System are located at the edge of one of the spiral arms, about 30,000 light years from the centre. The spiral arms form a disk-shaped system, with a bulging core (or nucleus) in the direction of Sagittarius. Old stars, of Population II, are found in the Galaxy's core; younger, hotter Population I stars, together with interstellar dust and gas, make up the spiral arms (SEE **stellar populations**). The stars of the spiral arms form the Milky Way, running around the sky. However, it is uncertain how many spiral arms the Galaxy has, or whether it is

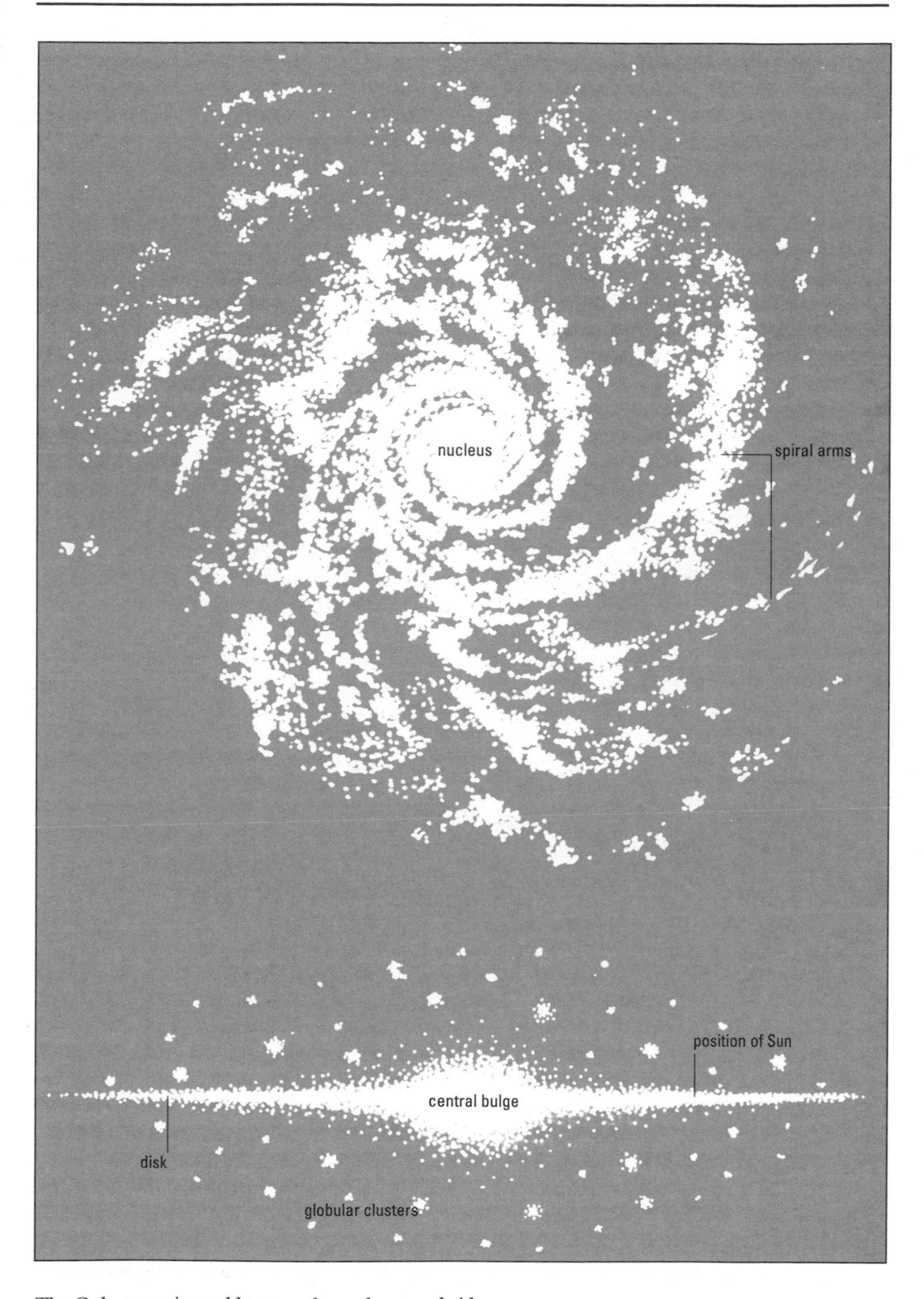

The Galaxy: as it would appear from above and side-on

THE GALAXY: DATA	
Diameter of disk	100,000 l.y.
Thickness of disk at centre	12,000 l.y.
Average density	7×10^{-24} g/cm^3
Total mass	2.2×10^{44} g = 1.1×10^{11} solar masses
North pole of galactic plane (epoch 2000)	RA 12h 51.4m, dec. +27° 7′
Point of zero longitude (epoch 2000)	RA 17h 45.6m, dec. −28° 55′
Galactic longitude of north celestial pole	123°
Distance of Sun from centre	28,000 l.y.
Distance of Sun above galactic plane	25 l.y.
Rotational velocity of Sun	250 km/s
Period of Sun's revolution around centre	220 million years
Absolute magnitude	−20.5

a normal or a barred spiral. Surrounding it all is the spheroidal **galactic halo**, which is even more extensive than the disk. In it are found very old Population II stars, including **globular clusters**.

Between us and the Galaxy's central bulge passes a prominent spiral arm, seen as the bright areas of the Milky Way in Centaurus, Crux, and Carina. This arm curls away from us in Carina where we see it virtually end-on, which accounts for the great accumulation of stars and nebulae in that constellation. Immediately outside this arm is a band of gas and dust, a region where stars have yet to form, which is visible as the dark rift in the Milky Way running from Cygnus to Crux. Beyond Carina we see a fainter arm that wraps itself round outside the Sun. It crosses the northern sky, gaining in brightness as it moves inwards towards Cygnus.

The centre of the Galaxy is marked by the radio source **Sagittarius A**, which has been adopted as the zero point of galactic longitude. The *galactic equator* is the plane of the Galaxy, running along the Milky Way. The whole Galaxy is rotating, but the rotational rate varies with the distance from the centre. Our Sun circles the centre at about 250 km/s, taking 220 million years per orbit. The Galaxy's age is thought to be about 10,000 million years, during which time the Sun would have completed approximately 50 revolutions.

The Galaxy is a member of a cluster known as the **Local Group**, which includes the Andromeda Galaxy and the Magellanic Clouds.

Galilean satellites The four chief satellites of Jupiter – **Io, Europa, Ganymede,** and **Callisto** – named collectively after **Galileo**, who observed them in 1610. All are in low-inclination, near-circular, **synchronous orbits**, and their orbital periods are in simple ratios to one another as a result of **resonance**. Apart from the Earth's Moon, they are the brightest satellites in the Solar System, and all would be naked-eye objects were it not for the glare from Jupiter itself. They pass through cycles of eclipses by Jupiter and its shadow, and transits across its disk. SEE ALSO **Jupiter, satellite**

Galilean telescope A **refracting telescope** having a planoconvex objective and a planoconcave eyepiece. It forms an erect image and gives a small field of view. This type of instrument, first employed by **Galileo** to make astronomical observations, is no longer used in astronomy, but the same optical system is still used in opera glasses.

Galileo Galilei (1564–1642) Italian astronomer, physicist, and mathematician. He was one of the first to use the telescope for astronomical observations, and he improved its design. He discovered mountains and craters on the Moon, and found the four satellites of Jupiter now known as the **Galilean satellites**. He observed the phases of Venus, and studied sunspots, from whose motion he deduced that the Sun rotates. He found that some celestial objects, such as the Milky Way and the Pleiades, are resolvable into many more stars than can be distinguished by the naked eye.

Galileo concluded that **Aristotle**'s picture of the world, still widely believed, was wrong, and championed **Copernicus**'s heliocentric theory. This brought him into conflict with the Catholic Church, and led to his trial and house arrest for the last eight years of his life. His many advances in physics included the law of the pendulum and how to create a vacuum. His chief books were *Siderius nuncius* (on discoveries with the telescope), *Dialogue on the Two Chief World Systems* (heliocentric Universe), and *Discourses and Mathematical Demonstrations on Two New Sciences* (mathematical physics).

Galileo A space probe to Jupiter, launched in October 1989. Galileo flew past Venus once and the Earth twice, getting *gravity assists* to pick up the necessary speed. The probe passed the asteroids **Gaspra** and **Ida** in October 1991 and August 1993 respectively, photographing both. Galileo is due to go into orbit around Jupiter in December 1995, after dropping off a sub-probe to enter the planet's atmosphere.

Galle, Johann Gottfried (1812–1910) German astronomer. In 1846 he and Heinrich D'Arrest were the first to locate the planet Neptune, close to positions calculated separately by **John Couch Adams** and **Urbain Le Verrier**. Galle discovered Saturn's faint C Ring, and determined the solar parallax (used to fix the scale of the Solar System) by observing the asteroid Flora.

gamma-ray astronomy The study of very high-energy **electromagnetic radiation** from space with wavelengths of about 0.01 nm and shorter. Gamma rays are absorbed by the Earth's atmosphere, so all studies have to be made by high-altitude balloons and spacecraft. Surveys of the sky at gamma-ray wavelengths were made by the satellites SAS II, launched in 1972, and COS-B, launched in 1975. They found a general concentration of emission in the galactic plane, with a peak towards the galactic centre, plus several strong individual sources such as the Crab Nebula, the Vela Pulsar, and the quasar 3C 273.

Most of the gamma radiation from the Galaxy is produced by cosmic rays interacting with gas in the interstellar medium. Certain gamma-ray lines at specific energies have been discovered, one attributed to the mutual annihilation of electrons and positrons in the galactic centre and another to the isotope aluminium-26 produced by supernovae.

Gamma-ray bursts lasting only a few seconds were discovered in 1967 by satellite-borne detectors meant to monitor nuclear explosions on Earth. In 1991 NASA launched a major satellite for gamma-ray studies called the **Compton Gamma Ray Observatory**. It has recorded over a thousand gamma-ray bursts, which seem to come from far outside the Galaxy, but their origin is still unknown.

Gamow, George (1904–68) Russian–American nuclear physicist and cosmologist. He applied nuclear physics to astronomical problems such as stellar evolution and nucleosynthesis (the origin of chemical elements). Gamow's most famous work, co-authored with Ralph Alpher and **Hans Bethe**, pictured a hot and dense early Universe in which the nuclei of elements were built up from protons, and was a cornerstone of **Big Bang** theory. His greatest prediction, borne out by the detection of the **cosmic microwave background** radiation, was that the Universe should by now have cooled to a few degrees above absolute zero.

Ganymede The largest of Jupiter's **Galilean satellites**, with a diameter of 5262 km (3270 miles), and the largest satellite in the Solar System – bigger, though less massive, than the planet Mercury. With a density of 1.94 g/cm^3, Ganymede contains a high proportion of water-ice. Its surface is a mixture of dark, heavily cratered terrain and a brighter, more lightly cratered terrain covered with meandering, intersecting grooves that suggest recent geological activity. The most prominent feature is the dark Galileo Regio, some 4000 km (2500 miles) across. SEE ALSO **satellite**

gas giant SEE **giant planet**

Gaspra Asteroid no. 951, discovered by Grigorii Neujmin in 1916. In 1991 it became the first asteroid to be imaged from close quarters when the probe Galileo passed it en route to Jupiter. It is an irregular body, with a maximum dimension of about 20 km (12 miles), and is pitted with small craters.

Gauss, Carl Friedrich (1777–1855) German mathematician. His many achievements include valuable contributions to astronomy, geodesy, and physics. In 1801 he worked out how to calculate the orbit of the newly found asteroid Ceres from observations spanning just 3° of its orbit. This led him to the study of celestial mechanics, and on to a comprehensive study of the determination of cometary and planetary orbits, taking into account the effects of perturbations, and then to the theory of errors of observation, for which he developed the mathematical method of least squares. From 1807 he was director of the observatory at Göttingen.

gegenschein A faint luminous patch on the ecliptic, visible usually only in the tropics on dark clear nights, directly opposite the Sun. It is caused by the scattering of sunlight back towards the Earth by particles in the zodiacal dust cloud, and is part of the **zodiacal light**. The word is German for 'counterglow', by which name it was formerly also known.

Gemini A large and prominent constellation of the zodiac. Its brightest stars are **Castor** and **Pollux**, named after the twins of Greek mythology. Eta is a semiregular variable, magnitude 3.2 to 3.9, with a period averaging 233 days. Zeta is a Cepheid variable, range 3.6 to 4.2, period 10.15 days. The open cluster M35 is easily visible with binoculars.

Geminids A **meteor shower** which occurs in December. Its radiant is in the constellation Gemini, near the star Castor. The shower, one of the most prolific, is associated not with a comet but with **Phaethon**, an asteroid.

Gemini Telescopes A pair of 8.1 m (319-inch) reflectors jointly funded by the USA, UK, Canada, Chile, Brazil, and Argentina. One will be sited at Mauna Kea, Hawaii, and is due for completion in 1999. The other will be at Cerro Pachón, Chile, and is due to open in 2001. Together they will provide complete sky coverage.

geocentric As viewed from or related to the Earth's centre, as in *geocentric parallax* or *geocentric system*. COMPARE **heliocentric**

geocentric parallax SEE **parallax**

geodesic SEE **curvature of space**

German mounting A type of mount used for an **equatorial telescope**.

Giacobini–Zinner, Comet A short period **comet**, orbiting the Sun in 6.59 years. It was discovered in 1900 by M. Giacobini and recovered in 1913 by E. Zinner. In 1985 it became the first comet to be studied at close hand when the International Cometary Explorer space probe passed through the comet's plasma tail.

giant molecular cloud (GMC) SEE **infrared astronomy**

giant planet The planets Jupiter, Saturn,

Uranus, and Neptune, also known as the *gas giants*, so called because their masses are great compared with the Earth's.

giant star A star that is placed well above the **main sequence** in the **Hertzsprung–Russell diagram**. Giant stars are larger than main-sequence (**dwarf**) stars of the same temperature, but giants of intermediate temperatures are smaller than the hottest dwarfs. Thus the term 'giant' really refers to the relative luminosity of a star rather than its dimensions. Giant stars evolve from dwarf stars that have run out of hydrogen in their central regions, and so represent later stages in the lives of stars. Because of their high luminosity, giants are quite common among naked-eye stars, examples being Arcturus, Aldebaran, and Capella, but they are relatively rare overall. SEE ALSO **red giant**

gibbous A **phase** of the Moon or a planet when more than half but less than all of its illuminated hemisphere is visible from the Earth.

Giotto A space probe built by the European Space Agency (ESA) to investigate **Halley's Comet**. It was launched on 2 July 1985 from the ESA launch site at Kourou, French Guiana, and flew to within 600 km (375 miles) of the comet's nucleus on 14 March 1986. Giotto photographed the nucleus and analysed the gas and dust given off. After the encounter Giotto was sent on a new trajectory to fly past Comet **Grigg–Skjellerup**, on 10 July 1992. Although its camera had been put out of action by dust impacts during the Halley encounter, Giotto's other instruments were able to study the composition of this second comet.

glitch A jump in the period of a pulsar. Pulsars slow down with age, but a few pulsars, particularly the Vela and Crab pulsars, occasionally show an abrupt speeding up. The largest of these glitches so far observed have been in the **Vela Pulsar**, and occur every few years. Typically, they make up for the slowdown of the previous few weeks. Glitches may be caused by small changes in the internal structure of the neutron star. After the glitch the pulsar slows more rapidly than normal, settling back to the same rate as before the glitch occurred. This suggests that there are several fluid layers within the pulsar, which take a few weeks to return to the same spin rate after a glitch.

globular cluster A near-spherical cluster of very old (**Population II**) stars in the halo of our Galaxy. Globular clusters contain anything from 100,000 to several million stars, concentrated so tightly near the centre that they cannot be fully separated by ground-based telescopes. Globulars are at least 10,000 million years old, as deduced from the highly evolved state of their stars. This extreme age, combined with their distribution in the galactic halo, indicates that globulars formed while our Galaxy was condensing from a huge cloud of gas. Globulars can also be seen around other galaxies. They are more plentiful around elliptical galaxies than spirals – our Galaxy has about 140 known globular clusters, whereas the largest elliptical galaxies may have thousands. Although they occupy the spheroidal galactic halo, most globulars lie no further from the centre of the Galaxy than does the Sun. As a result they congregate in the general area of sky containing the centre of our Galaxy, in particular in the constellations Sagittarius, Ophiuchus, and Scorpius. Globular clusters orbit the galactic centre, making periodic passages through the plane of our Galaxy. SEE ALSO **star cluster**

globule SEE **Bok globule**

GMT ABBREVIATION FOR **Greenwich Mean Time**

Gold, Thomas (1920–) Austrian-American astronomer. With Hermann Bondi and Fred

Hoyle he developed the **steady-state theory**, in which the continuous creation of matter drives the expansion of the Universe. With Franco Pacini, he explained that the signals from pulsars were beams of synchrotron radiation from rapidly rotating neutron stars.

Goodricke, John (1764–86) English astronomer, born in Holland. With the encouragement and collaboration of his neighbour Edward Pigott, the deaf-mute Goodricke began at the age of seventeen to study the brightest variable stars. In 1782 he announced the cause of Algol's variability (that it was what is now called an eclipsing binary), and the length of its period. Two years later he discovered the variability and periods of Beta Lyrae and Delta Cephei, the prototype Cepheid. At twenty-one he died from pneumonia contracted while observing Delta Cephei.

Gould's Belt A region of bright stars and gas inclined at about 20° to the galactic plane. It contains the greatest concentration of naked-eye stars of spectral types O and B, and probably represents part of the local spiral arm of which the Sun is a member. It is named after the American astronomer Benjamin Apthorp Gould (1824–1896), who studied the region.

grand unified theory (GUT) A theory that attempts to 'unify' three of the four forces of nature, and thus to demonstrate that they are different manifestations of a single force. In descending order of strength, the four forces of nature are: the strong nuclear force (which binds atomic nuclei); the electromagnetic force (which holds atoms together); the weak nuclear force (which controls the radioactive decay of atomic nuclei); and gravitation. The weak and strong nuclear forces operate over tiny distances within atomic nuclei, while the other two are infinite in range.

In the 1970s and 1980s, the unification of the electromagnetic and weak nuclear forces was first predicted and then demonstrated experimentally. GUTs are attempts to unify this *electroweak force* with the strong nuclear force. Many physicists believe that a more elaborate theory, a so-called *theory of everything*, will eventually show how all four forces are unified.

granulation The mottled appearance of the Sun's **photosphere**, caused by gases rising from the Sun's interior. When seeing conditions are very good the granulations can be resolved into small, nearly circular patches, called *granules*, surrounded by darker areas. Individual granules last only a few minutes.

graticule A system of reference marks or a measuring scale consisting of parallel vertical wires, or cross-wires, or a reticule (grid squares) placed in the focal plane of a telescope and covering the entire field of view. SEE ALSO **filar micrometer**

grating SEE **diffraction grating**

gravitation (gravity) The universal force of attraction between all particles of matter. *Newton's law of gravitation* states that the force of attraction F between two masses m_1 and m_2 is inversely proportional to the square of their separation r:

$$F = Gm_1m_2/r^2,$$

where G is the *gravitational constant*. This law is the basis of celestial mechanics.

Although at close range gravity is far, far weaker than the nuclear or electromagnetic force, it has a greater range and is the most important force on the large, cosmological scale. The whole concept of gravitation was reinterpreted by **Albert Einstein** in his General Theory of Relativity in terms of the curvature of **spacetime**.

gravitational lens The bending of light and other radiation by a massive body such as a star, black hole, or galaxy. The effect, which is predicted by Einstein's Theory of

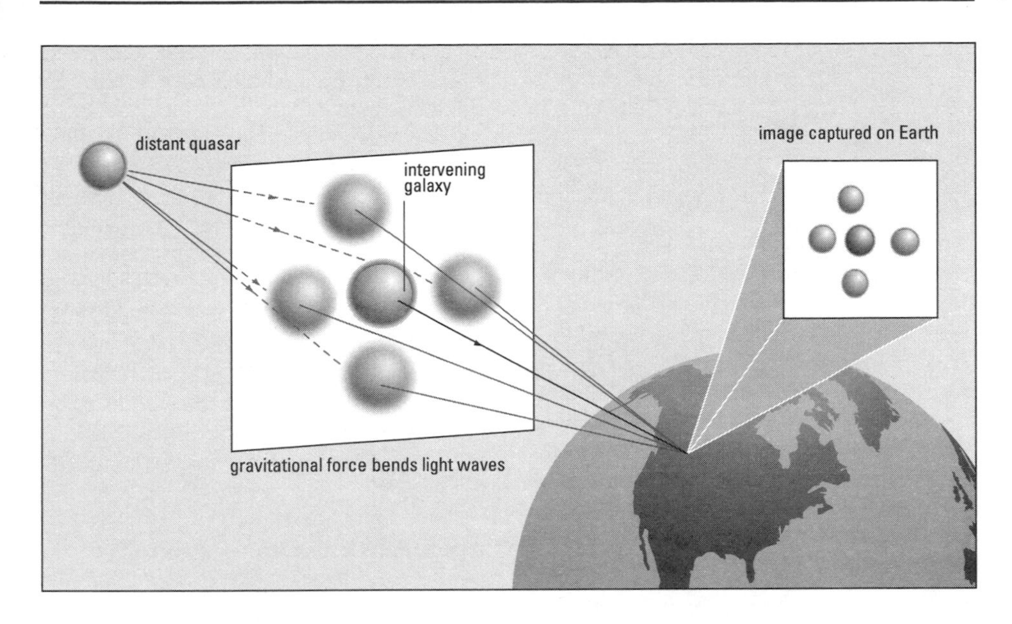

Gravitational lens: a nearby galaxy's gravity multiplies a distant quasar's image

Relativity, was first detected during the total solar eclipse of 1919 by **Arthur Eddington**, who found that light from distant stars was bent as it passed the Sun. In 1979, the first example of gravitational lensing of light from a quasar was discovered; in such cases, the quasar's light is split into two or more separate images by an intervening massive galaxy or cluster of galaxies. Gravitational lensing of quasars is now being used to measure their distances.

gravitational redshift SEE **redshift**

gravitational waves Waves emitted as the result of a violent disturbance in a gravitational field, such as a massive body being accelerated or distorted with a velocity that is an appreciable fraction of the velocity of light. Gravity waves were predicted by Einstein's General Theory of Relativity, but have proved elusive. Just as an oscillating charged particle emits electromagnetic waves, so a violently disturbed mass should emit gravity waves. Likely sources of gravity waves include close binary systems where at least one component is a neutron star or black hole. Because gravity is by far the weakest of the forces of nature, gravity waves must be extremely feeble. Although several detectors have been built, detection seems beyond the sensitivity of present-day technology.

Great Attractor A distant concentration of galaxies with a mass about 100,000 times that of the Milky Way, in the direction of the constellations Hydra and Centaurus. Its gravitational pull is attracting objects within several hundred million light years of it, including our own Local Group.

great circle A circle on a sphere whose plane passes through the centre of the sphere. On the **celestial sphere** the celestial equator is a great circle, as are meridians (circles passing through both celestial poles).

greatest elongation SEE **elongation**

Great Red Spot A large oval area in the southern hemisphere of Jupiter. Since it was first observed by Robert Hooke in 1664, it has varied in size and colour, from about 11,000 to 14,000 km wide by 24,000 to 40,000 km long (7,000 to 9,000 miles by 15,000 to 25,000 miles), and from a deep red to a light pink, occasionally fading away altogether. It is an atmospheric phenomenon, a huge anticyclone, projecting above the surrounding cloud-tops. The red colour may be produced by phosphorus.

Green Bank Location in West Virginia, USA, of the **National Radio Astronomy Observatory**.

green flash An atmospheric phenomenon, observable under ideal conditions during the last seconds before sunset or the first seconds after sunrise, in which the uppermost segment of the Sun separates and flashes a green colour. At these moments most of the sunlight has been absorbed in its long passage through the atmosphere, and only two narrow bands of wavelengths in the red and green are left. The atmospheric refraction of light is greater for short wavelengths (green rather than red), so the green flash appears.

greenhouse effect The warming of a planet's surface by heat trapped by its atmosphere. Incoming sunlight is absorbed at the surface and re-radiated at longer wavelengths, in the infrared – that is, as heat. If the atmosphere contains *greenhouse gases*, such as carbon dioxide, which are not transparent to the infrared, the heat is reflected back to the ground. The effect is the same as is produced by the glass enclosing a greenhouse.

The result is to raise the planet's average surface temperature by an amount that depends on the atmospheric composition and density. On **Venus** the dense carbon dioxide atmosphere has raised the surface temperature to around 750 K. On Earth a smaller greenhouse effect has led to a surface temperature that suits many life forms, but some people believe that industrial emissions of carbon dioxide will increase the greenhouse effect, producing *global warming*, and should be halted.

Greenwich Mean Time (GMT) Mean solar time for the meridian of Greenwich. In 1928 the International Astronomical Union recommended that GMT should be known as **Universal Time** for scientific purposes. Today, that usage has extended to legal and other purposes, although the term GMT is still used in many contexts. Greenwich was chosen as the world's prime meridian for timekeeping and navigation by the International Meridian Conference in 1884.

Gregorian calendar SEE **calendar**

Gregorian telescope A reflecting telescope described by Scottish mathematician James Gregory (1638–75) in 1663, five years before Isaac Newton constructed his first reflector. Light received by the parabolic primary mirror is reflected to a concave ellipsoidal secondary mirror, and then reflected back through a hole in the centre of the primary to an eyepiece. It differs from the Cassegrain telescope in having a concave instead of a convex mirror for the secondary. For astronomical purposes this form of telescope has fallen into disuse as it is difficult to mount and handle.

Grigg–Skjellerup, Comet The periodic **comet** with the second-shortest period known, 5.09 years. It was discovered in 1902 by New Zealander John Grigg, and recovered in 1922 by Australian John Skjellerup. In 1992 it became the third comet to be encountered by a spacecraft when the Giotto probe passed within 200 km (125 miles) of it.

Grubb, Howard (1844–1931) Irish telescope-maker. In 1868 he took over the telescope-manufacturing business established by his father, **Thomas Grubb** (1800–1878), and went on to make seven of the **astro-**

graph refractors used for the *Carte du Ciel* photographic sky survey, besides many other medium-sized and large telescopes for leading observatories. On his retirement in 1925 Grubb's business was acquired by Charles Parsons, son of the Third Earl of **Rosse**, who set up a new firm under the name of Sir Howard Grubb, Parsons & Co.

Grus A southern constellation, representing a crane. Its brightest star is Alpha (Alnair), magnitude 1.7. Mu and Delta are both optical pairs, wide enough to be separated with the naked eye.

G star A star of spectral type G, with a surface temperature of between 5000 and 6050 K if it is on the **main sequence**, and between 4600 and 5600 K if it is among the giants. The coolest G-type supergiants are about 4200 K. G-type dwarfs range in mass from 0.8 to 1.1 solar masses. The giants and supergiants have generally evolved from more massive stars further up the main sequence. Almost all G stars are dwarfs with compositions similar to the Sun. A few are subdwarfs with lower metal abundances. A small fraction of the G-type giants have peculiar chemical compositions that give them the name *barium stars*. Examples of G stars are the Sun, Alpha Centauri, and Capella.

Gum Nebula An immense emission nebula spanning 30 to 40° of sky in the constellations Puppis and Vela, containing the Vela Pulsar. It is named after the Australian astronomer Colin Gum (1924–1960), who made a photographic survey of the southern Milky Way in a search for nebulae.

GUT ABBREVIATION FOR **grand unified theory**

HI region A cloud of neutral (i.e. not ionized) atomic hydrogen gas in interstellar space. HI (pronounced 'H one') regions do not emit light and so are invisible, but they can be detected by their radio line emission at a wavelength of 21 cm (SEE **twenty-one centimetre line**).

In hydrogen atoms, both the proton and its orbiting electron have spins, rather like the axial spin of the Sun and Earth. The spin on the proton may either add energy to the electron or subtract energy from it, depending on their relative orientations. Collisions between atoms can alter this condition and, when the spins on electron and proton are aligned, the atom has slightly more energy than usual. When the atom returns to normal, it emits a photon of 21 cm wavelength which can be detected by radio telescopes.

Although this is a rare event, there is so much hydrogen present that the 21 cm radiation can be detected from quite small clouds. Since the propagation of 21 cm radiation is hardly affected by intervening matter, much of the neutral hydrogen in the Galaxy is detectable. Observations show that the hydrogen in the Milky Way is distributed in a flat layer, about 1000 light years thick, throughout the disk of the Galaxy. We can detect the Doppler shift of the radiation both from the rotation of the Galaxy and from smaller random motions within individual clouds.

Radio observations of other galaxies show that spiral or irregular galaxies contain an appreciable fraction of their total mass in the form of HI. Ellipticals contain little or no detectable HI. Vast HI clouds, apparently devoid of luminous matter, have also been detected in the space between galaxies (SEE **Magellanic Clouds**).

HII region A cloud of ionized hydrogen in interstellar space. HII (pronounced 'H two') regions are produced by the action on neutral hydrogen (i.e. un-ionized hydrogen, or HI) of ultraviolet radiation from hot stars, or by another process such as a shock wave. A prominent example is the **Orion Nebula**.

The simplest HII regions are those that arise around single hot stars with surface temperatures in excess of 30,000 K. The zone in which the gas is ionized is termed a *Strömgren sphere*. The star raises the temperature of the hydrogen to around 10,000 K, but it is not its temperature that causes an HII region to glow. Rather, it is the energy emitted as electrons that were released when the hydrogen was ionized recombine with protons to form hydrogen atoms. The energy is restricted to one of several emission lines (SEE **emission spectrum**). In the visible part of the spectrum, hydrogen shows strong recombination lines in the blue-violet and blue, and a very prominent red line. Colour photographs show this mixture of emission lines as a pinkish-red. Ionized hydrogen also produces radio emission, either continuous (*free–free radiation*, emitted as liberated electrons curve around protons) or at specific frequencies (*recombination line radiation*, emitted as some electrons recombine with protons to form new hydrogen atoms).

Other HII regions can be created by the impact of a high-speed gas stream on the surrounding medium. This process of collisional ionization may account for the presence of HII regions around cool, low-luminosity objects which emit no ultraviolet radiation to cause ionization, but which possess appreciable stellar winds. SEE ALSO **emission nebula**

Hale, George Ellery (1868–1938) American astronomer. He planned and secured finance for the 1 m (40-inch) refracting telescope at **Yerkes Observatory**, completed in 1897, the 2.5 m (100-inch) Hooker reflector at **Mount Wilson Observatory**, completed in 1917, and the 5 m (200-inch) reflector at the **Palomar Observatory**, which entered regular service in 1949. Each was in turn the largest telescope of its day. The 5 m (200-inch) was named the Hale telescope in his honour. Hale was an accomplished solar observer, discovering among other things the magnetic fields of sunspots via the Zeeman effect in their spectra. He also invented the spectroheliograph.

Hale Observatories The name by which **Mount Wilson Observatory** and **Palomar Observatory** were jointly known from 1970 to 1980.

Hall, Asaph (1829–1907) American astronomer. His planetary studies covered the orbital elements of satellites, the rotation of Saturn, and the mass of Mars. On 11 and 16 August 1877 he discovered the two tiny satellites of Mars, which he named Deimos and Phobos. The largest craters on Phobos have been named Hall and Stickney (the maiden name of Hall's wife).

Halley, Edmond (1656–1742) English scientist. His most famous achievement in astronomy was to realize that comets could be periodic, following his observation of **Halley's Comet** in 1682. He catalogued the stars of the southern hemisphere, discovered the proper motions of stars, suggested that the nebulae were clouds of gas, and speculated on the infinity of the Universe. He made the first complete observation of a transit of Mercury and showed how the results could be used to measure the Sun's distance. As the second Astronomer Royal, he embarked on a long series of lunar studies. It was Halley who persuaded **Isaac Newton** to write the *Principia*, and he also financed its publication. Halley's other achievements are numerous. He founded modern geophysics, charting variations in the Earth's magnetic field, and establishing the magnetic origin of the aurora borealis. In meteorology he showed that atmospheric pressure decreases with altitude, and studied monsoons and trade winds. In mathematics, he showed how to use mortality statistics to cost life assurance policies.

Halley's Comet The brightest and best-known periodic **comet**. It takes 76 years to complete an orbit that takes it from within Venus's orbit to outside Neptune's. The orbit's inclination is 162°, making it retrograde. It was observed by **Edmond Halley** in 1682; later he deduced that it was the same comet as had been seen in 1531 and 1607, and predicted it would return in 1758. There are historical records of every return since 240 BC. Several space probes were sent to Halley on its 1986 return. The closest approach was by the probe **Giotto**, whose cameras showed the nucleus to be an irregular object measuring 15 × 8 km (9 × 5 miles). The main constituent was found to be water-ice, but the surface was covered with a dark deposit, giving an albedo of only about 0.03. Some craters were imaged.

halo (1) A bright whitish ring, complete or partial, around a celestial body, usually the Sun or the Moon. It is caused by refraction of light by ice crystals suspended in the atmosphere.

halo (2) The **galactic halo**, a spheroidal distribution of stars and globular clusters around our Galaxy.

halo population stars SEE **Population II stars**

harvest Moon The full Moon nearest the autumnal **equinox** (on 23 September). The Moon's path is then almost parallel to the horizon, and so it rises only about 15 minutes later each evening instead of the usual half-hour or more, providing a succession of moonlit evenings formerly of great help to harvesters. In the southern hemisphere the harvest Moon is the full Moon nearest the vernal equinox (on 21 March). SEE ALSO **hunter's Moon**

Hawking, Stephen William (1942–) English theoretical physicist. He has used the General Theory of Relativity and quantum mechanics to study the **Big Bang** and **black holes**. He found that small black holes should lose energy – by emitting *Hawking radiation* – and eventually 'evaporate'.

Hayashi track The evolutionary path on the **Hertzsprung–Russell diagram** of a star beginning to form (SEE **protostar**) after it has collapsed from a gas cloud and before it reaches the main sequence. The Hayashi track lies above and to the right of the main sequence, and is nearly vertical. The star descends its Hayashi track, decreasing in luminosity but with little change in effective temperature. (After Japanese astrophysicist Chushiro Hayashi, 1920– .)

heavy elements SEE **metals**

Hebe Asteroid no. 6, diameter 192 km (119 miles), discovered in 1847.

Hektor Asteroid no. 624, discovered in 1907 by August Kopff. It measures roughly 300 × 150 km (200 × 100 miles) and is the largest and brightest of the **Trojan asteroids** sharing Jupiter's orbit. Its elongated shape may indicate that it is a binary asteroid.

Helene A small **satellite** of Saturn, discovered in 1980 by P. Lacques and J. Lecacheux. It is an irregular body, and is a **co-orbital satellite** with Dione, orbiting near the L_4 **Lagrangian point** of Dione's orbit around Saturn.

heliacal rising In modern usage, the rising of a celestial body simultaneously with the rising of the Sun. Formerly it meant the moment when a bright star, such as Sirius, could just be seen rising before the Sun. Observation of this event was once of great importance for agricultural purposes, the heliacal rising of Sirius heralding to the ancient Egyptians the annual flooding of the Nile.

heliocentric As viewed from or in relation to the Sun's centre, as in *heliocentric coordinates* and the *heliocentric system*. COMPARE **geocentric**

heliocentric parallax (annual parallax) The **parallax** of a body determined from observations of its position made six months apart, using the diameter of the Earth's orbit as the baseline.

heliometer A refractor with the objective lens split across its diameter to give two images, formerly widely used for measuring very small angular distances. Two close stars or the opposite limbs of a planet can be brought to coincidence and the separation measured by a micrometer. The first was made in 1754 by **John Dollond**. **Friedrich Bessel** used one in 1838 to make the first parallax measurement of a star (SEE **Sixty-one Cygni**). The heliometer has been superseded by photographic methods.

heliosphere The region of space in which the Sun's magnetic field and the solar wind dominate the interstellar medium. It is similar to a planet's **magnetosphere**, with an outer *bow shock*, and a boundary called the *heliopause* where the energy of the solar wind particles has fallen to the level of galactic cosmic ray particles. Measurements continuing to be returned by the Voyager probes suggest that the heliosphere extends for 100 to 150 AU.

heliostat A mirror on an equatorial mount driven so that it follows the Sun's apparent motion across the sky and directs its light into a fixed instrument such as a solar telescope. A more sophisticated version is the **coelostat**.

helium flash The sudden commencement of helium fusion in the core of a red giant after the star's supply of hydrogen fuel has been exhausted. The helium flash is restricted to stars of less than about 2 solar masses; in more massive stars, helium fusion commences more gradually.

Helix Nebula A planetary nebula in Aquarius, NGC 7293. It has the largest angular diameter of any planetary nebula, nearly half a degree, and at 300 light years away it is the nearest known planetary to the Sun.

Henry Draper Catalogue A catalogue of stellar spectra, compiled by **Annie Cannon** at Harvard College Observatory. It was named after Henry Draper(1837–1882), an American pioneer of astrophotography, whose widow supported the work financially. The catalogue, completed in 1924, introduced the Harvard system of classifying stellar spectral types: O, B, A, F, G, K, M, in order of decreasing surface temperature. Stars are still widely known by their HD numbers from this catalogue. SEE ALSO **spectral classification**

Herbig–Haro object (HH object) One of a class of small, faint nebulae independently discovered by George Herbig (1920–) and Guillamero Haro (1900–90) in the 1940s. They are irregular in outline, contain bright knots, and are found in regions rich in interstellar material. HH objects are thought to be very young stars hidden behind clouds of gas and dust. Powerful stellar winds from the young stars excite the gas, producing emission lines, while the light from the star is reflected by dust in other parts of the cloud.

Hercules A constellation named after the great hero of Greek mythology, but not very conspicuous. Its brightest star is Beta (Kornephoros), magnitude 2.8. Alpha (Rasalgethi) is a semiregular red giant variable, range 2.7 to 4.0, and with a period of roughly 100 days. It lies 500 light years away and has a diameter about 500 times that of the Sun. It is also a double star, with a magnitude 5.4 companion, period about 3600 years. Zeta is a fine close binary, with a period of 34 years, magnitudes 2.8 and 5.5. The most interesting objects for small instruments are the two globular clusters M13 and M92.

Hermes An unnumbered asteroid, about 1 km (0.6 miles) in diameter, discovered by Karl Reinmuth in 1937. On 28 October of that year it came within 800,000 km (500,000 miles) of the Earth, up to then the closest approach of any asteroid. Although its period is less than 3 years, it has never been observed since.

Herschel, Caroline Lucretia (1750–1848) German-born English astronomer, sister of **William Herschel**. In 1772 she joined her brother at Bath in England, becoming his housekeeper and astronomical assistant, editing and copying his papers, recording his observations, and preparing his catalogues. In the 1780s she became an observer in her own right, discovering many nebulae (as they were then called, including the galaxy NGC 253 in Sculptor) and eight comets. When her brother died in 1822, Caroline returned to her native Hanover, where she received many honours. She was the first woman to achieve real distinction in astronomy.

Herschel, John Frederick William (1792–1871) English scientist and astronomer, son of **William Herschel**. At first assistant to his father with observations and telescope-making, he went on to extend his father's work on double stars and nebulae, discovering over 500 more nebulae and clusters. In 1834 he took one of William's telescopes to the Cape of Good Hope and undertook a systematic survey of the southern sky, discovering over 1200 doubles and 1700 nebulae and clusters. He combined these and his father's observations into a *General Catalogue of Nebulae and Clusters*, which formed the basis for the later **NGC**. He made the first good direct measurement of solar radiation, and inferred the connection between solar and auroral activity. His *Outlines of Astronomy* (1849) was a standard textbook for many decades. He was also a pioneer of photography and its application in astronomy.

Herschel, (Frederick) William (originally Friedrich Wilhelm Herschel) (1738–1822) German-born English astronomer. In 1757 he came to England as a musician and became a naturalized Englishman. He turned to astronomy in the 1770s, making telescopes and mirrors for his own use and later for sale. Herschel became famous in 1781 by discovering the planet Uranus, which he wanted to name after King George III; the next year he was appointed the king's private astronomer. He discovered two satellites of Uranus (1787), and two of Saturn (1789). He observed and catalogued many double stars, hoping to measure their distances by detecting the parallactic movement of the nearer against the further if they were optical doubles. He observed over 2000 nebulae and clusters, and published catalogues of them which later would be incorporated in the **NGC**. Herschel realized that the Milky Way is the plane of a disk-shaped stellar universe, whose form he calculated by counting the numbers of stars visible in different directions, and also noted the motion of the Sun towards a point in the constellation Hercules. In 1800 he discovered and investigated the properties of infrared radiation.

Herschelian telescope A reflecting telescope in which the primary parabolic mirror is tilted so as to bring rays of light to a focus at the upper end of the tube, where an eyepiece is placed. It was devised by William Herschel to reduce the high loss of light from the secondary mirror of speculum metal used in the Cassegrain and Newtonian reflectors. This type of instrument is not now in use.

Hertzsprung, Ejnar (1873–1967) Danish astronomer. He showed that the colours of the stars were related to their brightnesses, plotting one against the other on what is now called the **Hertzsprung–Russell diagram** and discovering the existence of giant and dwarf stars. All studies of stellar evolution stem from this finding, but Hertzsprung's work remained largely unknown until **Henry Norris Russell** independently developed the diagram in a slightly different form. Hertzsprung also did work on double and variable stars, and applied **Henrietta Leavitt**'s methods of stellar distance measurement.

Hertzsprung gap The region on the **Hertzsprung–Russell diagram** between the main sequence and the giant branch where there are noticeably few stars. The gap exists because stars evolve rapidly through this region.

Hertzsprung–Russell diagram (HR diagram) A plot of the **absolute magnitude** of stars against their **spectral type**; this is equivalent to plotting their **luminosity** against their surface temperature or **colour index**. Brightness increases from bottom to top, and temperature increases from right to left. The diagram was devised by **Henry Norris Russell** in 1913, independently of **Ejnar Hertzsprung**, who had had the same idea some years before.

The HR diagram reveals a pattern in which most stars lie on a diagonal band, the **main sequence**, which includes our Sun. There is another region of **giant stars**, mainly K and M spectral types, which lies above the main sequence. There are further groupings above the main sequence called **supergiants** and **subgiants**. Below the main sequence are **subdwarfs** and **white dwarfs**.

The HR diagram is an important aid for interpreting astrophysical data. For example, the total range of stellar brightness is 27 magnitudes, corresponding to a factor of 10^{11} in luminosity, and stars' surface temperatures range from 2200 to 50,000 K. Luminosity and surface temperature are

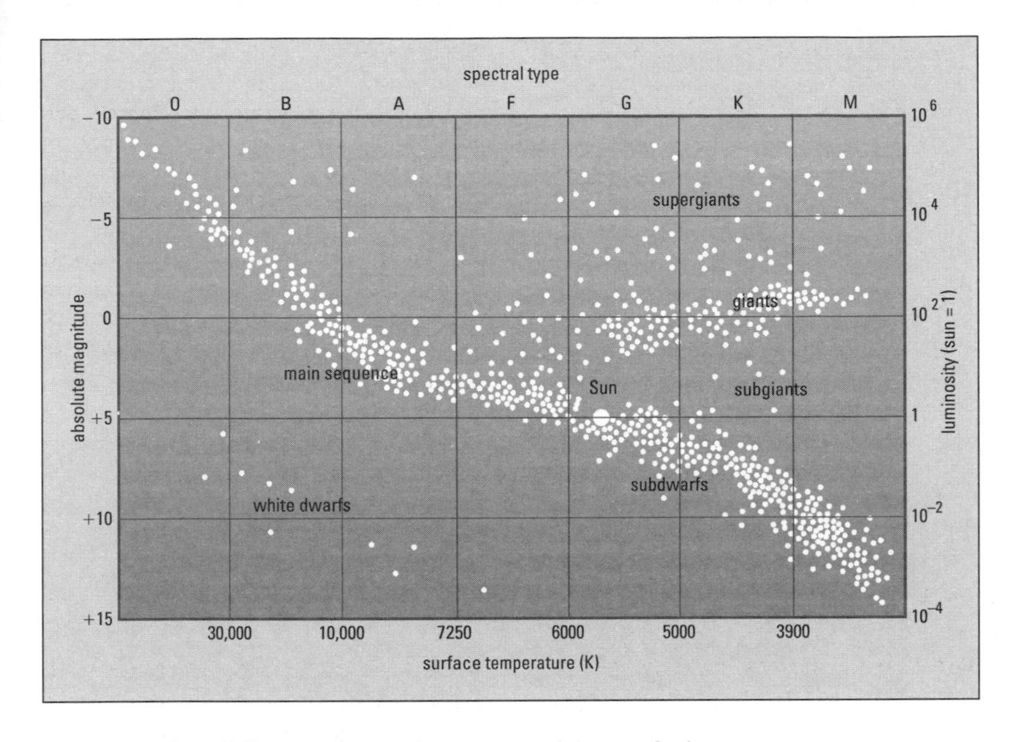

Hertzsprung–Russell diagram: the most important graph in astrophysics

both related to the size of a star. Most main-sequence stars are about the size of the Sun. White dwarfs are about the size of the Earth, while red giants and supergiants can be as big as the Earth's orbit. About 90% of the stars in our neighbourhood are main-sequence stars, about 10% are white dwarfs, and only about 1% are red giants or supergiants.

The existence of the three main groupings on the HR diagram means that a typical star passes through three very different stages during its life, and the progress of an evolving star can be followed on an HR diagram, from **protostar** to main sequence, then to giant star, and ending as a white dwarf (SEE **stellar evolution**). The diagram can also provide information about the ages of stars, and such dating is especially useful for establishing the ages of star clusters. Since the main sequence is a progression in stellar mass, the stars at the upper (most massive) end of the main sequence are the first to leave. As the cluster ages, its main sequence burns down like a candle as stars of progressively decreasing mass and luminosity become red giants. The stage which this process has reached reveals the cluster's age. SEE ALSO **Hayashi track**, **Hertzsprung gap**, **instability strip**

Hevelius, Johann (1611–87) German astronomer. He was the last great astronomer to make observations with naked-eye sighting instruments for positional measurements, from which he produced a catalogue of 1500 stars. He also made a map of the Moon, and discovered the Moon's **libration** in longitude.

Hewish, Antony (1924–) English radio astronomer. He and Martin Ryle developed the technique of **aperture synthesis** in 1960. In 1967 his student **Jocelyn Bell**

Burnell obtained the first signal from a pulsar, now known as CP 1919. For subsequent work on pulsars, Hewish shared (with Ryle) the 1974 Nobel Prize for Physics.

Hidalgo Asteroid no. 944, discovered by Walter Baade in 1920. Its diameter is about 50 km (30 miles), and its eccentric comet-like orbit takes it from just outside Mars's orbit out as far as Saturn. Although it has an asteroidal designation, it could well be a large extinct cometary nucleus.

high-velocity star A star in the galactic halo which does not share the rotation of the majority of stars in the Galaxy, including the Sun, around the galactic centre. Consequently, it seems to have a large proper motion and a high velocity relative to the Sun. The orbits of high-velocity stars are usually eccentric, with a large inclination to the galactic plane. Most of them are **Population II stars**.

Himalia The largest of the four small **satellites** making up Jupiter's intermediate group. It was discovered by Charles Perrine in 1904.

Hipparchus of Nicea (*c.* 190–125 BC) Greek astronomer, geographer and mathematician. He made many accurate astronomical observations and compiled a star catalogue (now lost but incorporated in Ptolemy's ***Almagest***) giving coordinates and magnitudes. He discovered the **precession** of the equinoxes. He found irregularities in the motion of the Sun and Moon, and from observations of eclipses made commendable estimates of their distances and sizes.

Hipparcos A satellite launched by the European Space Agency, designed for **astrometry**. The name is a contraction of High Precision Parallax Collecting Satellite, and is also an allusion to **Hipparchus of Nicea**, who recorded accurate star positions more than two millennia previously. Hipparcos was launched in August 1989 and operated until August 1993, measuring parallaxes and proper motions of 120,000 stars to a precision of 1 to 2 thousandths of an arc second, up to 100 times better than possible from the ground. In addition, it performed two-colour photometry of a million stars and measured their parallaxes and motions (though not to such a high accuracy).

Hirayama family A group of **asteroids** having closely similar orbital elements (SEE **orbit**), in particular the **semimajor axis**. About a hundred such families are known, named usually after the largest asteroid in them. The *Flora*, *Eos*, and *Themis* families are three of the main ones. The fact that in many families the members have similar spectra suggests that they originate from one larger body which has been broken up by impact. These groupings were discovered by Japanese astronomer Kiyotsugu Hirayama (1874–1943) in 1928.

Homunculus Nebula SEE **Eta Carinae**

horizon The great circle on the **celestial sphere** 90° from the observer's overhead point, the zenith.

horizontal coordinates SEE **coordinates**

Horologium An obscure constellation of the southern sky introduced by **Lacaille**, representing a pendulum clock. Alpha, which is of magnitude 3.9, is the only star above magnitude 4.

Horsehead Nebula A dark nebula, with a very distinctive shape like the head and mane of a horse. It is silhouetted against the bright nebula IC 434 south of Zeta Orionis. IC 434 is colliding with an extensive dusty cloud on the east side of Orion, and the Horsehead is the most distinctive intrusion of that dark cloud. The Horsehead measures some 3 light years across.

horseshoe mounting A type of mounting used for a large **equatorial telescope**.

hot dark matter SEE **dark matter**

hour angle The angle that the **hour circle** of an object makes with the **celestial meridian**. It is measured westwards from the meridian, along the celestial equator to the ponit where the equator and hour angle first intersect.

hour circle A great circle on the **celestial sphere** passing through both celestial poles. **Declination** is measured along hour circles. On each half of an hour circle running from pole to pole, all points have the same **right ascension**.

Hoyle, Fred (1915–) English astrophysicist and cosmologist. With Hermann Bondi and Thomas Gold he developed the **steady-state theory**, in which the continuous creation of matter drives the expansion of the Universe. Although it was subsequently displaced by the Big Bang theory (which owes its name to a disparaging remark by Hoyle), it sparked research by Hoyle and others into **nucleosynthesis** in stars which has been of lasting value. Hoyle has often attracted controversy with unorthodox ideas, such as his suggestion that life was brought to Earth by comets.

HR diagram ABBREV. FOR **Hertzsprung–Russell diagram**

Hubble, Edwin Powell (1889–1953) American astronomer. At Mount Wilson Observatory he used the 2.5 m (100-inch) telescope to examine nebulae. He found that although some nebulae were clouds of gas lying within our Galaxy, many others – the so-called spiral nebulae – were resolvable as independent star systems. These results were published in 1925, along with the **Hubble classification** of galaxies, still in use. Also in 1925, he measured the distance to the Andromeda Galaxy by identifying **Cepheid variable** stars in it. He attributed the **redshift** of the spectral lines of galaxies to the recession of galaxies, and hence to the expansion of the Universe, upon which modern cosmology is based. He established the velocity–distance relationship of recession (SEE **Hubble constant, Hubble law**).

Hubble classification A classification of **galaxies** according to their shape or structure, introduced by **Edwin Hubble** in 1925. The three major categories are elliptical (E), spiral (S), and barred spiral (SB); each category has several subdivisions depending on the observable shape or structure of the galaxy. Irregular galaxies (Ir) were not included in Hubble's original classification. SEE ALSO **tuning fork diagram**

Hubble constant (symbol H_0) The rate at which the velocity of recession of galaxies increases with distance from us, according to the **Hubble law**. The zero subscript specifies that it is the expansion rate at the present time that is meant, for the rate changes with time (SEE **deceleration parameter**). The value is currently estimated to lie in the region of 50 to 100 km/s/megaparsec. The inverse of the Hubble constant is called the *Hubble time*, which gives a maximum age for the Universe on the assumption that there has been no slowing of the expansion. The Hubble times for the above values of H_0 are respectively 20×10^9 and 10×10^9 years, assuming that all the galaxies started to recede from a common point in the past, at the time of the **Big Bang**.

Hubble diagram A graph in which the apparent magnitude of galaxies is plotted against the redshift of their spectral lines. It is a straight line, demonstrating the linear relation between redshift and distance, as embodied in the **Hubble law**.

Hubble law The law proposed by **Edwin Hubble** in 1929 claiming a linear relation between the distance (*r*) of galaxies from us and their velocity of recession (v), deduced from the redshift in their spectra. The figure linking velocity with distance is the **Hubble constant**. Hence $v = H_0 r$.

Hubble Space Telescope The largest astronomical telescope ever put into space, of Cassegrain design with a main mirror 2.4 m (94 inches) in diameter. It was launched by the space shuttle *Discovery* in April 1990 into an orbit about 600 km (375 miles) high. However, it was found to suffer from spherical **aberration** (2) as a result of a manufacturing error in its main mirror. Astronauts fitted corrective optics to the telescope during a servicing mission in December 1993. Its main instruments are a Wide-Field and Planetary Camera, a Faint Object Camera, a Faint Object Spectrograph, and a High Resolution Spectrograph.

Huggins, William (1824–1910) English amateur astronomer and spectroscopist. He used an inheritance to build an observatory and the first stellar spectroscope, and began a series of investigations, aided by neighbour and chemist **William Miller** (1817–70). In 1863 he demonstrated that the stars contain the same chemical **elements** that are present on Earth, and the next year confirmed the gaseous nature of a diffuse nebula. He investigated the motion of Sirius, and made the first discovery of a **moving cluster** (the Ursa Major cluster). In 1876 he took one of the first photographs of a star's (Vega's) spectrum. Huggins used spectroscopy to investigate many other objects, including comets, meteors, and the 1892 nova in Auriga.

Hulst, Hendrik Christoffel van de (1918–) Dutch astronomer. He carried out research on interstellar matter and the solar corona. In 1944 he predicted that hydrogen should emit radio waves with a wavelength of 21 cm (SEE **twenty-one centimetre line**), and in 1951 he and **Jan Oort** used Doppler shifts at this wavelength to map the Galaxy.

hunter's Moon The first full Moon after the **harvest Moon**, providing a succession of moonlit evenings in early October. In the southern hemisphere the hunter's Moon is in April or May.

Huygenian eyepiece A basic telescope lens commonly found on small refractors. It consists of two simple elements and is relatively free of **chromatic aberration**. (After **Christiaan Huygens**.)

Huygens, Christiaan (1629–95) Dutch mathematician, physicist, and astronomer. In 1655 he discovered Saturn's largest satellite, Titan, and the next year explained that the planet's telescopic appearance was due to a broad ring surrounding it. He was concerned with the development of the telescope and introduced the convergent eyepiece. Huygens' many contributions to physics include the idea that light is a wave motion, and the theory of the pendulum; he built the first pendulum clock.

Hyades A large open cluster several degrees across, 150 light years distant in the constellation Taurus. The brightest members are visible to the naked eye; in all it contains about 200 stars, most within a radius of 20 light years. Its estimated age is 650 million years. It is a **moving cluster** receding from us at 43 km/s. The star Aldebaran appears to be a member, but is in fact much closer to us and not part of the cluster.

Hydra The largest constellation in the sky, representing the water snake killed by Hercules; it extends from Canis Minor to the south of Virgo. Its brightest star is Alpha (Alphard), magnitude 2.0. The Mira variable R Hydrae can reach magnitude 3.5 at its brightest, and fades to 10.9; its period is 389 days. There is one bright open cluster, M48.

hydrogen-alpha line (Hα line) The absorption or emission line of hydrogen in the red portion of the spectrum at a wavelength of 656.3 nm. The first line in the **Balmer series**, it is the third most conspicuous absorption line in the Sun's spectrum, and is responsible for the pink colour of the solar chromosphere when viewed during a total eclipse. Light of this wavelength is often used for monochromatic study of the Sun with the spectrograph, and Hα filters are used by amateur observers for solar photography.

hydrogen spectrum The **spectrum** of atomic hydrogen. The hydrogen atom has the simplest structure of any atom (one proton as nucleus with one single orbiting electron), and the hydrogen spectrum is the simplest to interpret. As over 90% of the atoms in the Universe are hydrogen atoms, hydrogen lines feature prominently in the spectra of many celestial objects. The spectrum contains five principal series of lines associated with different levels of energy in the hydrogen atom, among them the **Balmer series** and the **Lyman series**. Each line corresponds to a quantum-mechanical process called a transition in which the electron 'jumps' between orbits (SEE **emission spectrum**).

Hydrus A far southern constellation, representing a small water snake. There are few interesting telescopic objects. Beta Hydri, magnitude 2.8, is the brightest star in the constellation and the nearest fairly bright star to the south celestial pole.

Hygeia **Asteroid** no. 10, discovered by Annibale de Gasparis in 1849. It orbits in the main belt, and is the fourth-largest asteroid, with a diameter of 429 km (267 miles).

hyperbola An open curve. In mathematics the hyperbola is a type of conic section, so called because it is one of the intersections of a plane with a cone, and is formed when a plane cuts a cone at an angle to the cone's axis that is less than its slope. In astronomy a hyperbolic **orbit** is the path followed by a celestial body which passes another body but is not captured into an orbit around it. For a hyperbolic orbit the **eccentricity** is greater than 1.

hyperboloid The figure obtained by rotating a **hyperbola** about its axis. It is the shape used for the convex secondary mirror in a **Cassegrain telescope**.

Hyperion An outer **satellite** of Saturn, discovered independently by William Cranch Bond and his son George Philips Bond, and independently by **William Lassell** in 1848. It is irregular in shape, measuring 410 × 260 × 220 km (255 × 160 × 135 miles), and displays chaotic rotation – tumbling about no fixed axis as it follows its eccentric orbit. Although it is a low-density, icy body, its surface is dark. There are several large craters and a scarp, named Bond-Lassell, which extends for 300 km (200 miles). It seems most likely that Hyperion is a fragment of a larger body.

I

Iapetus A large outer **satellite** of Saturn, discovered by **Giovanni Cassini** in 1671. Its diameter is 1460 km (910 miles), and its density of 1.21 g/cm^3 indicates that it is predominantly icy. However, although its trailing hemisphere is bright, its leading hemisphere (the one facing the direction of orbital travel) is extremely dark, being completely coated by a deep-red material. (Like most of Saturn's satellites, Iapetus has **synchronous rotation**.) It seems more likely that the dark material was somehow deposited from outside rather than extruded from within, and it may be that it is dust ejected from **Phoebe**, the next satellite out.

IAU ABBREVIATION FOR **International Astronomical Union**

IC ABBREVIATION FOR *Index Catalogue*, consisting of two supplements to the *New General Catalogue* (SEE **NGC**) produced in 1895 and 1908 by **J.L.E. Dreyer**. Objects in these Index Catalogues are still referred to by their IC numbers.

Icarus **Asteroid** no. 1566, discovered by **Walter Baade** in 1949. It is an **Apollo asteroid**, and has a perihelion distance of 30 million km (19 million miles), within the orbit of Mercury. No other asteroid except **Phaethon** ventures this close to the Sun.

Ida Asteroid no. 243, discovered by Johann Palisa in 1884. In 1993 it became the second asteroid to be examined from close quarters, by the probe Galileo en route to Jupiter. It is an elongated, cratered body, measuring 56 × 24 × 21 km (35 × 15 × 13 miles), and has a 1.6 × 1.3 × 1 km (1 × 0.8 × 0.6 mile) satellite, named Dactyl.

immersion The entry of a celestial body into the shadow of another at the beginning of an eclipse or occultation.

impact feature Any surface feature of a planet, satellite, asteroid, or other body with a solid surface that results from the past impact of another body. The most obvious are impact **craters**. *Basins* are large shallow depressions marking the impact of a large body, which have often been filled in by upwelling fluid from below, like the lunar maria (SEE **mare**). *Multiringed basins*, like **Valhalla** on Callisto, are surrounded by concentric ringed structures representing 'ripples' from the impact. Other impact features include **rays** radiating from craters, and *palimpsests* – the 'ghosts' of craters flattened out or filled in since the impact that caused them. The term has also been applied to the visible atmospheric disturbances caused by the collision of Comet **Shoemaker–Levy 9** with Jupiter.

inclination (1) (symbol i) The angle between the orbital plane of a planet and the plane of the ecliptic.

inclination (2) (symbol i) The angle between the orbital plane of a satellite and the equatorial plane of the planet it orbits.

inclination (3) (symbol i) The angle between the orbital plane of a double star and the plane at right angles to the line of sight. The angle is zero if the orbit is seen in plan, and 90° if seen in profile.

inclination (4) (symbol i) The angle between the equatorial plane of a planet or satellite and its orbital plane; this is equivalent to the angle between the axis of rotation and the perpendicular to the orbital plane.

Indus A small southern constellation representing an American Indian. The only star above magnitude 3.5 is Alpha (3.1). Epsilon is the least luminous star visible to the naked eye: magnitude 4.7, distance 11.2 light years, type K3, absolute magnitude 7.0.

inequality The departure from uniform orbital motion of a body, due to **perturbations** by other bodies and the eccentricity of its orbit.

inferior conjunction SEE **conjunction**

inferior planet Either of the planets Mercury and Venus, which orbit inside the Earth's orbit.

inflation A proposed early stage of the Universe, immediately after the Big Bang, in which for a brief fraction of a second the Universe expanded faster than the speed of light. Such a period of inflation explains the uniformity of the **cosmic microwave background**. SEE ALSO **cosmogony**

Infrared Astronomical Satellite (IRAS) A satellite, a collaboration between the USA, Netherlands, and UK, that surveyed the sky at infrared wavelengths. It was launched in January 1983 and operated for nine months, cataloguing nearly 250,000 sources. Its discoveries included several comets and asteroids, possible planetary systems forming around nearby stars, and **starburst galaxies**.

infrared astronomy The study of **electromagnetic radiation** falling between the red end of the visible spectrum and radio wavelengths, between about 700 nm, or 0.7 mm, and 1 mm. The Earth's atmosphere is opaque to much of this wavelength range, although there are a number of atmospheric *windows* through which infrared observations can be made from high mountains. To avoid the Earth's atmosphere, observations are made from high-flying aircraft, notably the **Kuiper Airborne Observatory**, from unmanned balloons, and from satellites such as the **Infrared Astronomical Satellite** (IRAS).

Astronomical sources of infrared radiation include the planets and other Solar System bodies, the stars and dusty regions of the Milky Way, and other galaxies (including **starburst galaxies**). Most of the interstellar dust and gas in our Galaxy is contained in *giant molecular clouds* (GMCs), such as the one in the Orion Nebula. Maps of GMCs at infrared wavelengths show the warm or hot regions where star formation is in progress or is about to begin. On a much smaller scale, highly compact and opaque **Bok globules** with just a few tens of solar masses are also potential star-forming regions. Infrared observations show the temperatures of the globules to be very low, 10–20 K, indicating that they are at an early stage of gravitational collapse. Faint, wispy structures in interstellar space termed *infrared cirrus*, discovered by IRAS, are probably caused by infrared emissions from dust grains. These grains are composed of carbon in the form of graphite at temperatures of 30–40 K, ejected by the stellar winds of cool stars or supernova explosions.

Infrared wavelengths can penetrate the otherwise opaque clouds of dust in the plane of our Galaxy, so we can see sources towards the galactic centre which are completely obscured at optical wavelengths. The observations show a high density of late-type stars and a large ring of molecular gas and dust, as well as many regions of active star formation. Beyond the Milky Way, infrared observations are a key to understanding the energy output of active galaxies and quasars and the physical processes operating within them.

In the Solar System, infrared observations allow us to work out the basic chemistry of the planets, their satellites, and the asteroids. Molecules in planetary atmospheres exhibit detailed infrared absorption spectra, which permit the composition and temperature of the atmospheres to be established. Spectral reflectance observations of asteroids, planetary satellites, and small planets without substantial atmospheres have been used to discover the mineralogical compositions of their surfaces. Infrared techniques have established the presence of water and carbon dioxide ices on Mars, water-ice on many of the satellites of Jupiter, Saturn, and Uranus, and methane ice on Triton and Pluto. Measurement of the black-body emission of small Solar System bodies has given us information on their sizes and surface albedos where no other technique would succeed.

Infrared Space Observatory (ISO) A European Space Agency satellite with a 0.6 m (24-inch) telescope for infrared astronomy, planned for launch in autumn 1995.

inner planet Any of the planets Mercury, Venus, Earth, and Mars, which orbit inside the main asteroid belt.

instability strip The narrow region of the **Hertzsprung–Russell diagram** where pulsating stars are located. It extends from the brightest Cepheids down to pulsating white dwarf stars and also includes RR Lyrae stars, Delta Scuti stars, and dwarf Cepheids.

Integral A joint European, US and Russian satellite for gamma-ray astronomy; the name is short for International Gamma-Ray Astrophysics Laboratory. Its launch is planned for 2001.

interferometry The study of a celestial object by analysing its interference pattern. When electromagnetic radiation emitted by a source is split into two and recombined, a pattern of interference fringes is produced that can be used to reveal more detailed information about the source than is possible from a single beam. Various types of **optical interferometer** are used to measure the diameters of stars and resolve fine detail in their spectra. Interferometry is particularly important in radio astronomy, where the long wavelengths of the radiation mean that a single detector gives only a coarse resolution (SEE **radio interferometer**).

intergalactic matter Matter in the space between galaxies. There are no significant quantities of intergalactic dust, but there are clear signs of intergalactic gas. X-ray emission from clusters of galaxies comes from a hot, tenuous gas with a temperature of 10 to 100 million K between the galaxies in the cluster. Radio-emitting clouds of electrons ejected by radio galaxies are drawn out into a tadpole shape as if they were ploughing through a resisting medium. How much intergalactic gas is primordial hydrogen and helium from the Big Bang falling into the clusters, and how much is matter expelled from the galaxies, is uncertain. Tidal interactions (SEE **tides**) between galaxies are common, and radio observations reveal streams of neutral hydrogen around interacting galaxies such as M81 and M82. A similar gas stream, the Magellanic Stream, links the **Magellanic Clouds** with our Galaxy.

International Astronomical Union (IAU) The controlling body of world astronomy, founded in 1919. It is the authority for naming celestial bodies and their features.

International Atomic Time (TAI) The international reference scale of atomic time, formed by the Bureau International de l'Heure (BIH) in Paris from atomic clock data supplied by establishments worldwide. (The abbreviation TAI is derived from the name in French.) The fundamental unit of time is the atomic second, which is defined in terms of the resonance of the caesium atom. SEE ALSO **Universal Time**

International Ultraviolet Explorer (IUE) A joint NASA, European Space Agency, and UK satellite, carrying a 0.45 m (18-inch) telescope for ultraviolet astronomy. It was launched in 1978 and takes ultraviolet spectra of stars and galaxies.

interplanetary matter The material in the space between the planets. It is made up of atomic particles (mainly protons and electrons) ejected from the Sun via the solar wind, and dust particles (mainly from

comets, but some possibly of cosmic origin) in the plane of the ecliptic. The particles in the inner Solar System constitute the *zodiacal dust cloud*, scattered sunlight from which is responsible for the **zodiacal light**. An extremely tenuous ring of dust, over 50 million km (30 million miles) wide, orbits the Sun just outside the Earth's orbit. There is also the dust in **meteor streams**.

interstellar matter The matter contained in the regions between objects in the Galaxy. It includes clouds of *interstellar gas*, mainly hydrogen with some helium, and a small percentage (about 1% by mass) of interstellar dust grains. The hydrogen is mainly in a relatively cool neutral form at a temperature of 10–100 K (SEE **H I region**). Some very much hotter and more diffuse regions of ionized gas (**H II regions**) also exist, together with dense clouds of molecular hydrogen and other molecules. The dust grains appear to be elongated in shape, about 100 nm long, and aligned by the galactic magnetic field. Most grains are thought to be composed of silicates or forms of carbon, and some have icy coatings.

The clouds of gas and dust are largely confined to the plane of the Galaxy and tend to be concentrated in the spiral arms. Our Galaxy contains about 10^{10} solar masses of material between the stars, making up about 10% of its total mass. The density of these clouds is less than that of a laboratory vacuum on Earth, but they nevertheless have an obscuring effect on the light of stars behind them. The clouds scatter and absorb starlight passing through them, a phenomenon known as *interstellar extinction*. On average, starlight is dimmed by one magnitude for every kiloparsec it travels in the plane of the Galaxy. Red light is dimmed less than blue light, which causes a reddening of starlight.

The spectra of remote stars show **absorption lines** superimposed on the stellar spectra, produced by interstellar dust and gas. Sodium, calcium, potassium, iron, titanium, and other elements have been detected from these absorption lines. These elements are ejected from stars. Heavier elements, including lead, gallium, and krypton, were detected in 1994 by the refurbished Hubble Space Telescope; these are manufactured in the extreme conditions of a **supernova** explosion. SEE ALSO **nebula**

interstellar molecules Molecules that exist in interstellar clouds. Most have been identified by their emission lines at millimetre and centimetre wavelengths, although the first, including the cyanogen radical (CN) and the methylidyne radical and ion (CH, CH^+), were detected in the 1940s at optical frequencies. The hydroxyl radical (OH) was the first (in 1963) to be identified by radio astronomy. The list of identified radicals has been greatly extended since then, and includes hydrogen cyanide (HCN), ammonia (NH_3), methane (CH_4), methyl alcohol (CH_3OH), and around a hundred others.

intrinsic variable SEE **variable star**

inverse-square law A reduction in the intensity of a physical quantity in proportion to the square of the distance from its source. For example, a planet three times the Earth's distance from the Sun would experience one-ninth the gravitational attraction of the Sun and receive one-ninth as much sunlight as on Earth.

Io The innermost of Jupiter's four **Galilean satellites**, and the second-smallest, with a diameter of 3630 km (2980 miles). It has the highest density, at 3.57 g/cm^3, indicating a lack of water-ice in its make-up. Io's tidal interaction with Jupiter creates sufficient flexing and heating to make it by far the most volcanically active world in the Solar System. The surface, multicoloured by the presence of sulphur, is being continually renewed. The active regions contain volcanic vents and calderas, lava flows, and volcanic plumes

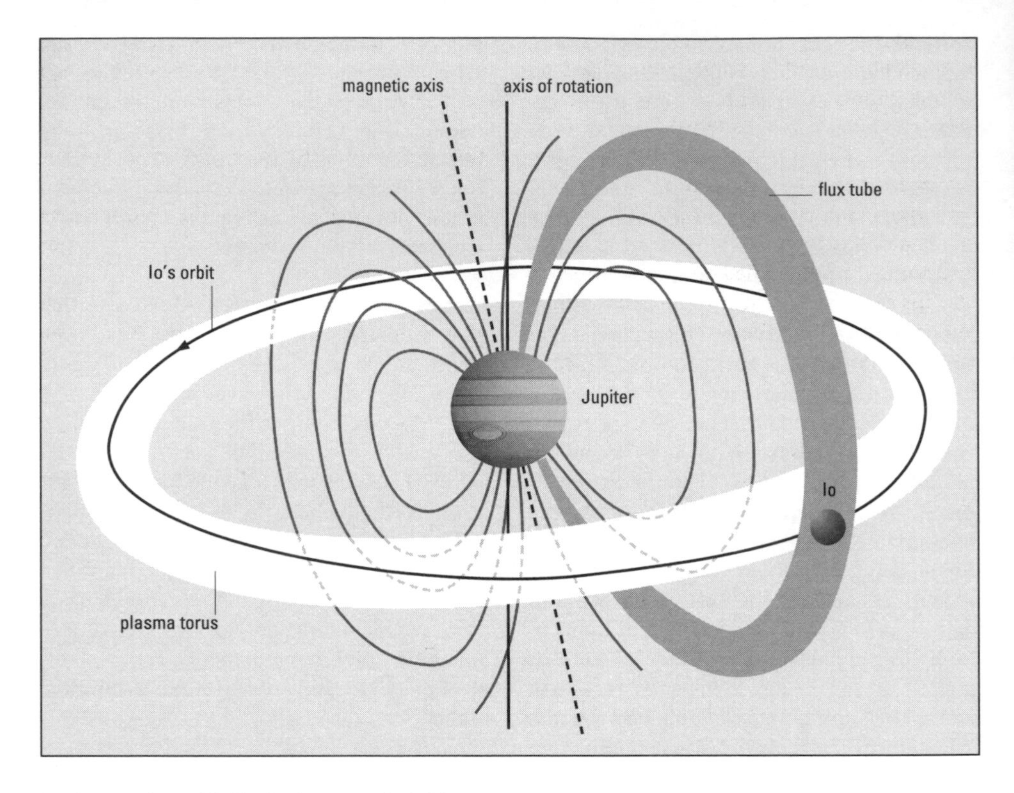

Io: interaction with Jupiter's magnetic field

that send material to heights of up to 300 km (200 miles). Plumes prominent when the Voyager probes photographed Io included those named Pele, Loki and Prometheus. In the active regions there may be lakes of molten sulphur or even molten rock; elsewhere there are plains formed by volcanic deposits, and mountains. A thin atmosphere of sulphur dioxide surrounds Io, and a doughnut-shaped *plasma torus* of ionized (electrically charged) sulphur and oxygen surrounds its whole orbit and interacts with Jupiter's **magnetosphere** to create a *flux tube* carrying an electric current of about 5 million amperes (SEE the diagram above). SEE ALSO **satellite**

ionosphere A region around a planet or other Solar System body in which there are free electrons and ions produced by ultraviolet radiation and X-rays from the Sun. The degree of ionization is greatly affected by solar activity.

The Earth's ionosphere extends from about 60–1000 km above the surface, and reflects radio waves of long wavelengths, although shorter wavelengths can pass through undisturbed. The reflecting power of these layers makes long-range broadcasting and telecommunications possible up to frequencies of about 30 MHz. The radio waves and microwaves from around 30 MHz to about 100 GHz that pass through the ionosphere are used in radio astronomy and satellite communications; they make up the *radio window*. The free electrons formed as a result of ionization do at times have serious adverse effects on the reception of radio signals.

IRAS ABBREVIATION FOR **Infrared Astronomical Satellite**

Iris Asteroid no. 7, discovered by John Russell Hind in 1847. Although it ranks only about 25th in size, with a diameter of 203 km (126 miles), it is the fourth-brightest.

iron meteorite (siderite) A meteorite consisting mainly of iron (90–95%) and some nickel, with traces of other metals. Iron meteorites are classified into a number of groups according to the proportions of these other metals that they contain, and each group is thought to correspond to a different parent asteroid.

irradiation A physiological phenomenon in which a bright object viewed against a dark background appears larger than it really is. The effect can introduce errors into measurements made by an observer using a telescope.

irregular variable A pulsating variable star whose variation in brightness does not follow any regular or predictable pattern. Irregular variables are red giants or supergiants.

ISO ABBREVIATION FOR **Infrared Space Observatory**

isotropy SEE **cosmology**

IUE ABBREVIATION FOR **International Ultraviolet Explorer**

J

Jansky, Karl Guthe (1905–50) American radio engineer. In 1931, while looking for sources of 'noise' that interfered with radio communications, he discovered one located in the Milky Way, in Sagittarius. Jansky himself did not follow this up, but it led directly to the development of radio astronomy. The unit of flux density used in radio astronomy is named the *jansky* in his honour.

Janssen, Pierre Jules César (1824–1907) French astronomer. He specialized in solar spectroscopy and developed special instruments, including a spectrohelioscope for observing prominences without waiting for an eclipse. He discovered the helium line in the solar spectrum, and produced a photographic atlas of the Sun.

Janus A small inner **satellite** of Saturn, a **co-orbital satellite** with Epimetheus. The two were discovered in 1978 when John Fountain and Stephen Larson re-examined photographs they had taken in 1966. They are both irregular in shape, and are probably fragments of the same larger body. Either satellite may be the object reported in 1966 by Audouin Dollfus and given the name Janus, but not subsequently confirmed.

Jeans, James Hopwood (1877–1946) English mathematician and physicist. In astronomy he worked on stellar dynamics, the formation and evolution of stars, double-star systems, and spiral galaxies. In 1917 he proposed that the Solar System was formed from matter pulled out of the Sun by the gravitational attraction of a passing star, a theory later developed by Harold Jeffreys (1891–1989), but since abandoned. Jeans wrote many popular books on astronomy.

jet A stream of very-high energy particles emitted by objects such as active galaxies, quasars and protostars. They are detectable by their radio emissions and may also be visible at optical wavelengths. In galaxies there are often two jets, emerging in opposite directions from the nucleus perpendicular to the galactic plane.

Jewel Box The open cluster NGC 4755 in the constellation Crux, around the star Kappa Crucis. The name derives from the varied colours of the stars in the cluster. It lies 7600 light years away.

Jodrell Bank The site at Macclesfield, Cheshire, of the University of Manchester's Nuffield Radio Astronomy Laboratories, founded after World War II by **Bernard Lovell**. Its main instrument is the 76 m (250 ft) fully steerable dish opened in 1957 and subsequently upgraded; in 1987 it was named the Lovell Telescope. A second dish, elliptical in shape, 38 × 25 m (125 × 83 ft), was opened in 1964. These telescopes can be used individually or as part of the **MERLIN** array.

Jones, Harold Spencer (1890–1960) English astronomer. He was concerned with refining values for the distances and periods of Solar System bodies, and in 1941 obtained what was then the most accurate value of the Sun's distance. As the tenth Astronomer Royal, he moved the Royal Greenwich Observatory to Herstmonceux in Sussex.

Jovian The adjective for Jupiter. *Jovian planet* is another term for **giant planet**.

Joy, Alfred Harrison (1882–1973) American astronomer. In the 1920s and 1930s, he measured the distances to over 5,000 stars. He also measured many radial velocities, and studied T associations and T Tauri stars.

Julian calendar SEE **calendar**

Julian Day A system of numbering days continuously, without division into years and months. It is used in astronomy when calculating events that recur over long periods, such as the maxima of variable stars. The Julian Day Number is the number of days that have elapsed since noon GMT on 1 January 4713 BC, which is defined as day zero. The idea was conceived in 1582 by the French chronologist Joseph Justus Scaliger (1540–1609). The name 'Julian' was given by Scaliger in memory of his father, Julius Caesar Scaliger, and is not to be confused with the Julian calendar devised by Julius Caesar.

The Julian Date is given by the Julian Day Number for the preceding noon at Greenwich followed by the fraction of the mean solar day that has elapsed. Thus, the Julian Date of an observation made at 6 p.m. on 1 January 1995 is 2,449,719.25, the Julian Day Number being 2,449,719.

Note that in historical usage the year 1 BC is followed by the year AD 1, whereas in astronomical usage AD 1 is preceded by year 0, which in turn is preceded by year −1 (corresponding to the historical year 2 BC). Thus to convert years BC to astronomical years, subtract 1 and prefix a minus sign.

Juliet One of the small inner **satellites** of Uranus discovered in 1986 during the Voyager 2 mission.

Juno **Asteroid** no. 3, discovered by Karl Harding in 1804. It is the tenth-largest, with a diameter of 244 km (152 miles).

Jupiter The fifth major planet from the Sun, and the largest of the giant planets. Not counting the Sun, Jupiter contains around 70% of the mass of the Solar System. It is one of the brightest objects in the sky, with a magnitude at opposition of between −2.3 and −2.9. Its maximum apparent diameter is 50 arc seconds, and even at its most distant it is still 30 arc seconds across, so for most of the time it is the largest planetary disk visible. Through a telescope, Jupiter's yellowish elliptical disk is seen to be crossed by a number of continually changing, brownish-red bands. Our knowledge of the planet owes much to visits by space probes: Pioneers 10 and 11, Voyagers 1 and 2, and Ulysses (SEE ALSO **Galileo**). It became the focus of public attention in 1994 when Comet **Shoemaker–Levy 9** plunged into its atmosphere. The main data for Jupiter are given in the first table.

The darker bands are called *belts*, and the lighter ones are known as *zones*. Their nomenclature is given in the second table. Changes occur in the pattern and colours of the belts and zones on a timescale of years, but in the long term the same fundamental structure persists. There is also much activity on timescales from days to weeks. Dark and light spots and streaks come and go. Most spots are short-lived. The white *ovals*, such as those that appeared on the STB in the 1940s, are rather more persistent. The most distinctive feature is the **Great Red**

JUPITER: DATA	
Globe	
Diameter (equatorial)	142,800 km
Diameter (polar)	133,500 km
Density	1.33 g/cm^3
Mass	1.90×10^{27} kg
Sidereal period of axial rotation (equatorial)	9h 50m 30s
Escape velocity	59.6 km/s
Albedo	0.73
Inclination of equator to orbit	3° 07′
Temperature at cloud-tops	125 K
Surface gravity (Earth = 1)	2.69
Orbit	
Semimajor axis	5.203 AU = 778.3×10^6 km
Eccentricity	0.048
Inclination to ecliptic	1° 18′
Sidereal period of revolution	11.86y
Mean orbital velocity	13.06 km/s
Satellites	16

JUPITER: NOMENCLATURE OF BELTS AND ZONES	
Name	Abbr
North polar regions	NPR
North north temperate zone	NNTZ
North north temperate belt	NNTB
North temperate zone	NTZ
North temperate belt	NTB
North tropical zone	NTropZ
North equatorial belt	NEB
Equatorial zone	EZ
South equatorial belt	SEB
South tropical zone	STropZ
South temperate belt	STB
South temperate zone	STZ
South south temperate belt	SSTB
South south temperate zone	SSTZ
South polar regions	SPR

Spot (GRS) in the STZ, first observed by Robert Hooke in 1664, since when it has disappeared and reappeared several times. Another prominent feature at the same latitude was the South Tropical Disturbance, first observed in 1901 as a grey bar across the STZ which grew to dominate the planet until its disappearance in 1940. A similar dark disturbance, the SEB Revival, erupts from time to time, as it did in 1993. Different types of streak are given names such as *projections*, *plumes* and *festoons*. They are particularly common on the south edge of the NEB, as are small bright and dark spots.

All these cloud features indicate a dynamic and turbulent atmosphere. The clouds are drawn into horizontal bands by the rapid rotation and wind speeds of up to 400 km/h (250 mile/h). An equatorial jet-stream keeps the region between the equatorial belts rotating in about 9 hours 51 minutes (referred to as *System I* rotation), five minutes faster than the rest of the atmosphere (*System II*). Turbulence is often evident at the boundaries between Systems I and II. Individual features may drift backwards and forwards in longitude. The bright zones are high-altitude clouds supported by upwelling gases, while the darker belts are regions of descending currents. Eddies give rise to the spots, which are cyclones or (like the GRS) anticyclones.

Jupiter's predominant atmospheric constituents are hydrogen and helium, hydrogen accounting for nearly 90%, and helium for most of the rest. But the upper atmosphere contains a considerable variety of chemical compounds, and they give the planet its multi-hued appearance. The highest features are red, due to phosphorus from lower levels carried aloft on rising currents. Ammonia crystals make up the high-altitude white clouds. The lower clouds that make up the belts are largely ammonium hydrosulphide; other substances present include methane, ethane, water, phosphine, hydrogen cyanide, carbon monoxide, and germanium tetrahydride. There is probably a complex cycle of chemical reactions, driven by ultraviolet radiation from the Sun, and also by high-energy lightning and auroral discharges.

The temperature at the cloud-tops is around 125 K, and the pressure is about 0.5 bar. Pressure and temperature increase quite rapidly with depth. Further down, where the temperature and pressure are similar to those at the Earth's surface, there are probably clouds of water-ice crystals and water vapour. At 1000 km (600 miles) below the cloud-tops the atmosphere gives way to an ocean of liquid molecular hydrogen. The temperature here is about 1000 K. At a depth of 20,000 to 25,000 km (12,500 to 15,000 miles), under a pressure of 3 million bars, the hydrogen becomes so compressed that it behaves as a metal. At the centre of Jupiter there is probably an iron–silicate core perhaps five times the Earth's mass, surrounded by an ice mantle of about twenty earth masses. The core temperature is estimated to be 30,000 K. Jupiter emits about 1.7 times as much heat as it receives from the Sun.

The deep metallic hydrogen 'mantle' gives Jupiter a powerful magnetic field and a very extensive **magnetosphere**. The magnetic axis is tilted by nearly 10° with respect to the rotational axis, and the magnetic field is nearly 20,000 times as strong as the Earth's. The magnetosphere is huge and has many unique features. It is several times the size of the Sun, and the magnetotail can stretch further than the orbit of Saturn. The magnetic field traps a large quantity of plasma (charged particles). Low-energy plasma is contained in a vast *magnetodisk* in Jupiter's equatorial plane, while high-energy plasma is funnelled into radiation belts which are ten thousand times as intense as the Earth's **Van Allen Belts**. The magnetosphere interacts with the innermost Galilean satellite, **Io**, to produce a *plasma torus* and a *flux tube*. Most of Jupiter's trapped plasma comes from Io's volcanoes. Some of the sodium from these volcanoes is ejected by Jupiter's magnetosphere and forms a vast 'nebula' with a radius of roughly 35 million km (20 million miles). The magnetosphere is the source of the planet's powerful radio emissions. Some of this is **synchrotron radiation** from high-energy electrons in the radiation belts.

Jupiter has 16 known **satellites**. The four major ones – Io, Europa, Ganymede and Callisto – are often referred to as the **Galilean satellites**. Within the orbit of Io there are four small satellites. These eight together form a compact group orbiting close to Jupiter and in the planet's equatorial plane. An intermediate group of four small bodies orbit at distances between 11 and 12 million km, while the four remaining satellites, also small bodies, orbit at between 21 and 24 million km from the planet – the greatest distances of any known satellites from their planet.

Jupiter also possesses a modest ring system (SEE **ring, planetary**), the brightest component of which is about 50,000 km (30,000 miles) above the cloud-tops.

K

KAO ABBREVIATION FOR **Kuiper Airborne Observatory**

Kapteyn's Star A red subdwarf, 12.7 light years away in the southern constellation Pictor, which has the second-largest proper motion known: 8.7 arc seconds per year. It is of magnitude 8.9, type M1, and is over 250 times less luminous than the Sun. It was discovered by the Dutch astronomer Jacobus Kapteyn (1851–1922).

K corona SEE **corona**

Keck Telescopes A pair of identical telescopes at the W.M. Keck Observatory on Mauna Kea, Hawaii. Both have mirrors 10 m (396 inches) in diameter and consisting of 36 hexagonal segments. Keck I was completed in 1992; its twin, Keck II, is due for completion in 1996. They are run by the University of California and Caltech.

Kellner eyepiece An eyepiece with large **eye relief**, which makes it widely used for binoculars as well as telescopes. It is essentially a **Ramsden eyepiece** with one element replaced by an achromatic doublet (SEE **achromatic lens**). (Invented in 1849 by German optician Carl Kellner, 1826–55.)

Kepler, Johann (1571–1630) German mathematician and astronomer. He supported the Sun-centred Solar System put forward by **Copernicus**. In 1600 he went to Prague, becoming assistant to **Tycho Brahe**, on whose death in 1601 he succeeded as Imperial Mathematician to the Holy Roman Emperor, Rudolf II, with the task of completing tables of planetary motion begun by Tycho. From Tycho's accurate observations he concluded in *Astronomia nova* (1609) that Mars moved in an elliptical orbit, and went on to establish the first of his laws of planetary motion (SEE **Kepler's laws**). Erroneous reasoning led him nevertheless to the second of these laws, and his desire to match celestial and musical harmony led to the third law, published in *Harmonices mundi* (1619). The *Rudolphine Tables*, based on Tycho's observations and Kepler's laws, appeared in 1627 and remained the most accurate until the 18th century. Other works of his include *De stella nova*, on the supernova of 1604 (SEE **Kepler's Star**), and *Dioptrice* (1611), on optics and the theory of the telescope.

Kepler's laws The three fundamental laws governing the motions of the planets around the Sun (and other celestial bodies in closed orbits), first worked out by **Johann Kepler** between 1609 and 1619, and based on observations made by Tycho Brahe.

Law 1 deals with the shape of a planetary orbit: the orbit of each planet is an **ellipse** with the Sun at one of the foci.

Law 2 explains the varying speed of planetary motion such that the planet moves faster the nearer it is to the Sun: the line joining the planet to the Sun (the *radius vector*) sweeps out equal areas in equal times. This is known as the *law of areas*.

Law 3 relates the size of the orbit to the period of revolution: the square of the period of revolution (P) is directly proportional to the cube of the mean distance of the planet from the Sun (a). In practice this means that P^2/a^3 is approximately constant for all bodies orbiting the Sun.

Kepler's Star A **supernova** that appeared in October 1604 in the constellation Ophiuchus, reaching magnitude −3 and remaining visible to the naked eye until March 1606. It was studied by **Johann Kepler**. The smooth light curve closely resembles that of a Type I supernova. The remnant is a radio source, and some faint nebulosity is visible optically. It was the last supernova seen in our Galaxy.

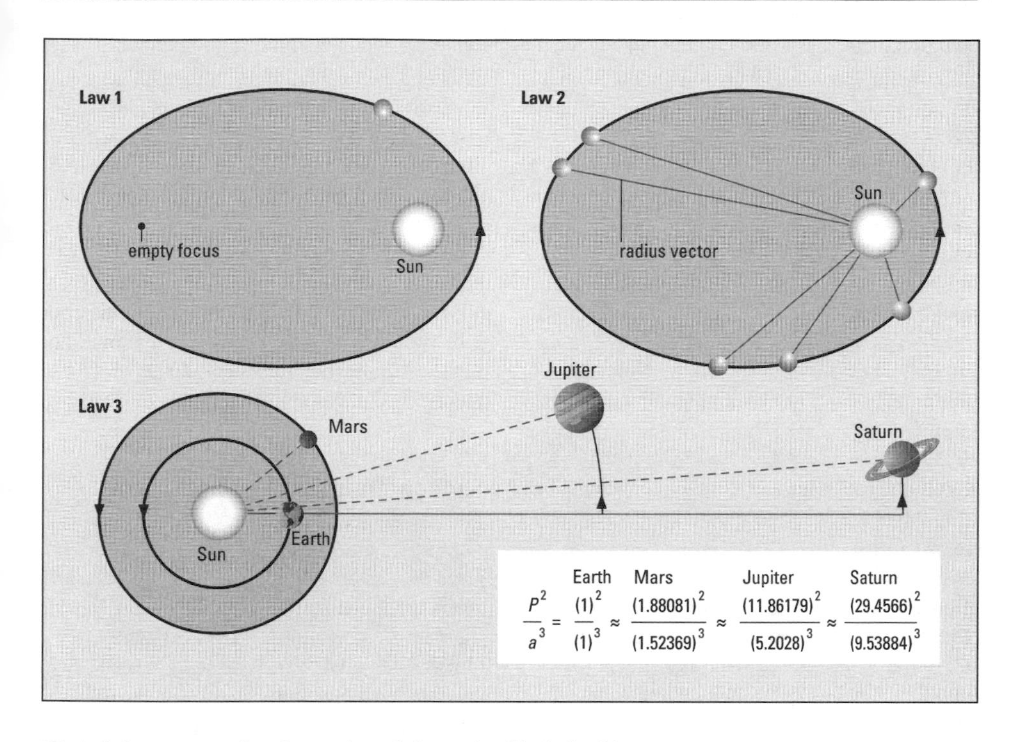

Kepler's laws: governing the motion of planets in elliptical orbits

Kirkwood gaps Regions in the main **asteroid** belt where, because of **resonances** with Jupiter, no stable orbit is possible, and from which asteroids are therefore absent (or nearly absent). The gaps are at distances at which the revolution period of an orbiting body would be **commensurable** with Jupiter's, and Jupiter would perturb the body. There are pronounced gaps at ⅓, ⅖ and ½ of Jupiter's period, for example. The gaps in the ring system of **Saturn** are kept clear by a similar perturbing effect of its satellites, in particular Mimas, whose period is twice that of a body at the distance of the Cassini Division. The American astronomer Daniel Kirkwood (1814–95) explained how the gaps came about.

Kitt Peak National Observatory An observatory near Tucson, Arizona, part of the US National Optical Astronomy Observatories, containing the world's largest collection of telescopes. Its largest instrument is the 4 m (158-inch) Mayall Telescope, opened in 1973. It is also the site of the McMath Solar Telescope, the world's largest solar telescope; a 3.5 m (138-inch) reflector opened in 1994; and the 2.3 m (90-inch) Steward Reflector of the University of Arizona.

Kohoutek, Comet A **comet** discovered in 1973 by the Czech astronomer Luboš Kohoutek (1935–). When first detected it was nearly as far from the Sun as Jupiter, and was predicted to become as bright as the Moon at perihelion, but failed to live up to these expectations. Its initial brightness was probably because it was making its first trip inwards from the **Oort Cloud**, with a full complement of **volatile** material, which it began to shed early on.

K star A star of spectral type K, characterized by numerous absorption lines and the presence of molecular bands of CH and CN in its spectrum. Main-sequence K stars have surface temperatures ranging from 3550 to 4900 K. Giants are about 400 K cooler than this, and supergiants 300 K cooler still. At their lowest point on the main sequence, K stars are about half a solar mass, rising to 0.8 of a solar mass at their highest. As K-type giants and supergiants may be either evolved old stars of about one solar mass or evolved stars of higher masses, a wide range is represented. However, most of the K-type giants are stars of a few solar masses. Only a small fraction of K stars are variable: the coolest Cepheids have K-type spectra, as do the hottest of the Mira stars. A few per cent of K-type giants show spectral peculiarities with an overabundance of carbon and certain **heavy elements**. Bright examples of K stars are Pollux, Arcturus, one component of Alpha Centauri, and Epsilon Eridani.

Kuiper, Gerard Peter (1905–73) Dutch–American astronomer. After World War II he embarked upon an observational programme which revitalized lunar and planetary astronomy. He discovered the satellites Miranda (of Uranus) in 1948, and Nereid (of Neptune) in 1949, and found methane in the atmospheres of Uranus, Neptune and Titan, and carbon dioxide in the atmosphere of Mars. He was involved with many US missions to the Moon and planets, in particular with the Ranger and Mariner probes. He advocated carrying infrared telescopes on high-flying aircraft; the **Kuiper Airborne Observatory** is named in his honour.

Kuiper Airborne Observatory (KAO) A Lockheed C-141 transport aircraft containing a 0.9 m (36-inch) telescope designed for infrared studies at high altitude. The KAO is operated by NASA from the Ames Research Center at Moffett Field, near San José, California, and entered service in 1975.

Kuiper Belt As originally suggested by **Gerard Kuiper**, a region beyond the orbit of Neptune, extending to about 1000 AU, and the source of short-period **comets**. Since 1992 a number of reddish objects with diameters of around 250 km (150 miles) have been discovered orbiting the Sun at average distances approaching or beyond that of Pluto. They have been given asteroidal designations, but are now thought to mark the inner edge of a Kuiper Belt of **planetesimals**, the largest known members of which are Pluto and its satellite, Charon, extending out to perhaps 100 AU. Objects such as **Chiron** may have been perturbed inwards from the Belt; Neptune's moon **Triton** may have originated there too.

L

Lacaille, Nicolas Louis de (1713–62) French astronomer. He spent the years 1750 to 1754 surveying the southern-hemisphere skies from the Cape of Good Hope, and introduced 14 new southern constellations. In 1761 he made an accurate measurement of the Moon's distance that took into account the Earth's oblateness; this made possible a more accurate method of determining terrestrial longitude.

Lacerta A small, obscure northern constellation between Andromeda and Cygnus, representing a lizard, introduced by **Hevelius**. Alpha, its brightest star, is magnitude 3.8.

Lagoon Nebula A bright nebula in Sagittarius, M8 or NGC 6523, surrounding the open star cluster NGC 6530. The nebula is cut by a dark cloud which gives it its name. The nebula and cluster lie at a distance of 5000 light years.

Lagrange(-Tournier), Joseph Louis de (1736–1813) French mathematician, born in Italy. He made many contributions to celestial mechanics, studying the Moon's **libration**, the motion of Jupiter's satellites, and perturbations and stability in the Solar System. From his investigation in 1772 of the **three-body problem** he found special solutions of the equations which suggested the existence of the equilibrium positions now known as **Lagrangian points**.

Lagrangian points One of the points at which a celestial body can remain in a position of equilibrium with respect to two much more massive bodies orbiting each other. At these points the forces acting on the smaller body cancel out. There are five Lagrangian points, in the orbital plane of the two large bodies (SEE the diagram). Bodies at points L_1, L_2 and L_3, on the line joining M_1 and M_2, are less stable than bodies at points L_4 and L_5, which form equilateral triangles with M_1 and M_2. The **Trojan asteroids** lie in stable orbits near the L_4 and L_5 Lagrangian points of Jupiter's orbit around the Sun. (After **Joseph Louis de Lagrange**.)

Lalande, (Joseph) Jérôme Le Français de (1732–1807) French astronomer. His main achievement was a catalogue of over 47,000 stars. He did much to popularize astronomy in France. Lalande observed Neptune in 1795, 51 years before it was located by Johann Galle, without realizing it was a planet.

La Palma observatory SEE **Roque de los Muchachos Observatory**

Laplace, Pierre Simon de (1749–1827) French mathematician and astronomer. He used Isaac Newton's law of **gravitation** to interpret the perturbations of planets and satellites, the shape and rotation of Saturn's rings, and the stability of the Solar System. In 1796 he put forward his *nebular hypothesis* of the origin of the Solar System, which,

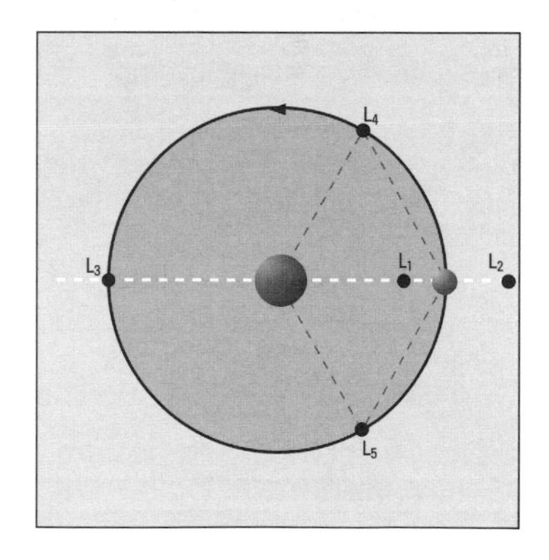

Lagrangian points: stable positions in an orbit

although wrong in detail, is broadly in agreement with the presently accepted view. His five-volume *Traité de mécanique céleste* published over the period 1799–1825 summarized 18th-century advances in celestial mechanics.

Large Magellanic Cloud SEE **Magellanic Clouds**

Larissa One of the small inner **satellites** of Neptune, first detected in 1981 by Harold Reitsema and colleagues when it occulted a star, and confirmed by the 1989 visit of Voyager 2.

Las Campanas Observatory An observatory near La Serena, Chile, run by the Carnegie Institution of Washington. Its main telescope is the 2.5 m (100-inch) Du Pont Telescope, in operation since 1977.

Lassell, William (1799–1880) English brewer and amateur astronomer. From the observatory he built, he discovered several planetary satellites: Neptune's **Triton**, only 17 days after the discovery of Neptune itself, Saturn's **Hyperion**, and Uranus's **Ariel** and **Umbriel**. He also discovered 600 nebulae. Lassell was the first to use the equatorial system of mounting for reflecting telescopes.

last contact SEE **fourth contact**

last quarter One of the Moon's **phases** when it has almost reached western quadrature and is half illuminated as seen from the Earth.

late-type star A cool star of spectral type K, M, C, or S. The name was given when it was believed that the spectral types represented an evolutionary sequence, hot stars being the youngest and cool stars the oldest. This is now known not to be the case. COMPARE **early-type star**

latitude A coordinate on the celestial sphere, or on the surface of a celestial body. *Celestial* (or *ecliptic*) *latitude* is the angular distance of a celestial body from the ecliptic; it is measured at right angles to the ecliptic, from 0° at the ecliptic to 90° at the ecliptic poles, positive to the north and negative to the south. *Galactic latitude* is the angular distance of a body from the galactic equator, measured from 0° to 90° at right angles to the galactic equator. *Heliocentric latitude* is the celestial latitude of a body as seen by an observer at the centre of the Sun, rather than the centre of the Earth. *Heliographic latitude* is the latitude of a point on the Sun's surface, north or south of the Sun's equator.

leap second The adjustment of one second in the radio time signal at midnight on 30 June or 31 December. By international agreement the rate of radio time signals has been maintained since January 1972 in agreement with the International Atomic Time scale. However, the Earth's rotational period is slowing by about 0.003 of a second per day, so the radio time signals gradually diverge from Coordinated **Universal Time** (UTC), which is based on the Earth's daily rotation. To correct this divergence, the radio time seconds are retarded by the addition of one leap second as necessary.

Leavitt, Henrietta Swan (1868–1921) American astronomer. She specialized in variable stars, noticing that the brighter of the **Cepheid variables** tended to have the longer periods. From studies of Cepheids in the Small Magellanic Cloud she formulated the **period–luminosity law**, which was important in establishing the distance scale for other galaxies.

Leda The smallest of the four small **satellites** making up Jupiter's intermediate group, discovered by Charles Kowal in 1974.

Lemaître, Georges Edouard (1894–1966) Belgian priest and astrophysicist. In 1927 he argued that an expanding Universe would have originated in the radioactive explosion of a 'primeval atom' or 'cosmic egg' into which all mass and energy was concentrated. This was the first statement of what would beome known as the **Big Bang** theory.

lens An optical component of an instrument, made of a transparent material (such as glass) which, by the material's property of refraction, modifies an incoming beam of light, usually to aid the formation of an image of the light source. Single lenses have either two curved surfaces, or one curved and one plane surface. Lenses are used in various combinations in astronomical instruments. SEE **aberration, eyepiece, objective, telescope**

lenticular galaxy A type of **galaxy** intermediate in form between spiral and elliptical types. Lenticular galaxies are designated S0, which indicates that they have the flattened form of spirals but not spiral arms. They appear lens-shaped, hence their name.

Leo A large and easily identified constellation of the zodiac, in mythology representing the lion killed by Hercules. Its brightest star is first-magnitude **Regulus**, part of the curved figure of stars, known as the Sickle, that outlines the lion's head and chest. Gamma (Algieba) is a fine, wide binary, magnitudes 2.2 and 3.5, one K0-type giant and one G7-type giant, period 620 years. The Mira variable R Leonis can reach magnitude 4.4 (minimum 11.3, period 310 days). There are five galaxies in Leo on Messier's list: M65, M66, M95, M96 and M105.

Leo Minor A small constellation between Leo and Ursa Major. It has no star above magnitude 3.8.

Leonids A **meteor shower** that occurs in November. Its radiant is in the constellation Leo, in the 'Sickle' asterism. The Leonids are associated with Comet Tempel–Tuttle (SEE **meteor stream**). The shower is periodic, providing modest displays in most years, but producing magnificent meteor storms at intervals of 33 years – rates of 60,000 meteors per hour were recorded in 1966 on the night of 16–17 November.

Lepus A small constellation representing a hare hiding under the feet of Orion. Its brightest star is Alpha (Arneb), magnitude 2.6. It contains the very red Mira variable R Leporis (range 5.5 to 11.7, period 427 days) and the globular cluster M79.

Le Verrier, Urbain Jean Joseph (1811–77) French astronomer. Originally a chemist, he turned to celestial mechanics and made a study of planetary motions and perturbations. In 1845 he predicted that a planet which he prematurely named **Vulcan** lay within the orbit of Mercury, but despite some claimed sightings it was never found. More successful was his prediction the next year, independently of **John Couch Adams**, that an unknown planet (Neptune) was responsible for discrepancies between the calculated and observed orbital motion of Uranus.

Libra An obscure constellation of the zodiac, representing the scales of justice, but formerly seen as the claws of the scorpion, Scorpius, which it adjoins. The brightest star, Beta, is magnitude 2.6. Delta is an eclipsing binary of the Algol type, range 4.9 to 5.9, period 2.33 days.

libration The small oscillation of a celestial body about its mean position. The term is used most frequently in connection with the Moon. As a result of libration it is possible to see, at different times, 59% of the Moon's surface. However, because the areas that pass into and out of view are close to the limb and therefore extremely foreshortened, in practice libration has less of an effect on the features that can be clearly made out on the Moon's disk than this figure might suggest.

Physical libration results from slight irregularities in the Moon's motion produced by

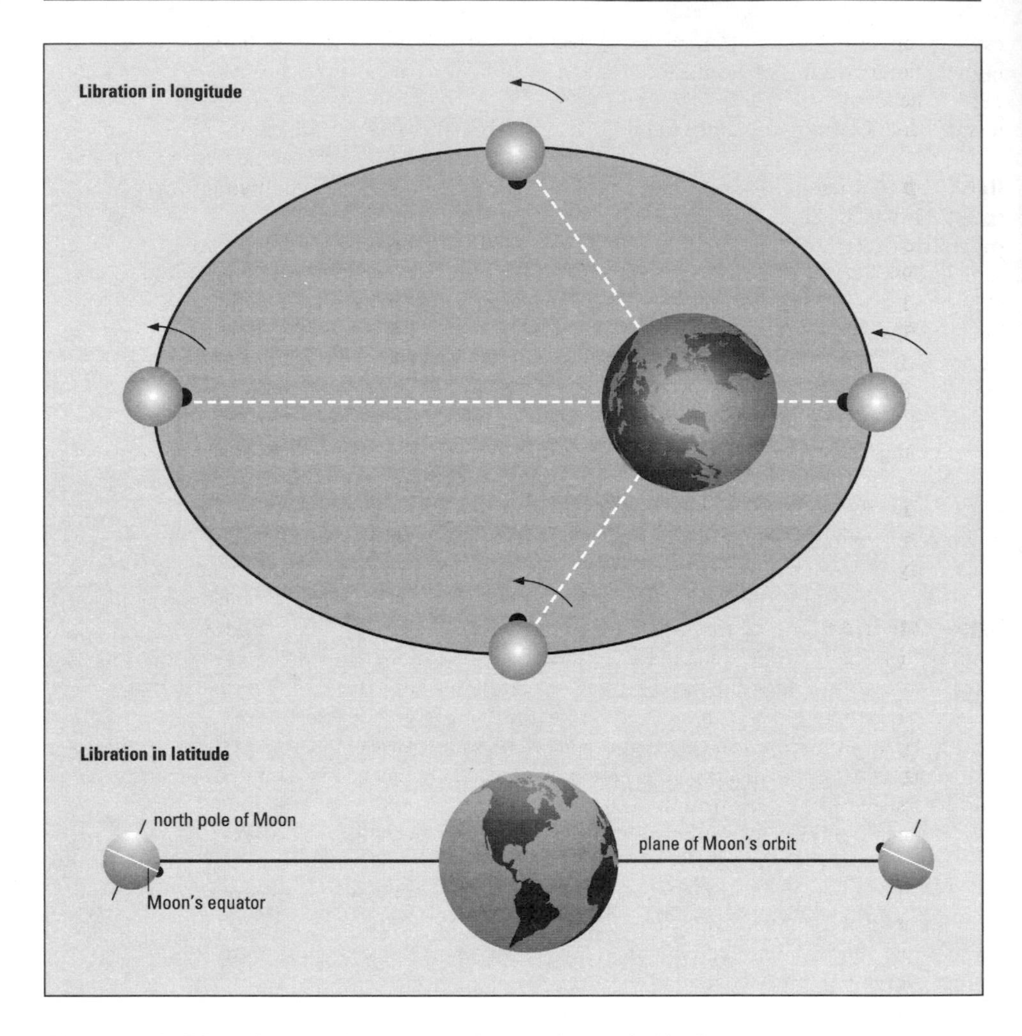

Libration: the Moon does not always present the same face to the Earth

irregularities in its shape. Much more obvious is *geometrical libration*, which results from the Earth-based observer seeing the Moon from different directions at different times. There are three types of geometrical libration. *Libration in longitude* arises from a combination of the Moon's synchronous rotation and elliptical orbit (SEE the diagram above). As a result, at times a little more of the lunar surface is visible at the eastern or western limb than when the Moon is at its mean position. *Libration in latitude* arises because the Moon's equator is tilted slightly from its orbital plane, so that the two poles tilt alternately towards and away from the Earth (SEE the diagram). A smaller effect is *diurnal libration*, by which the Earth's rotation allows us to see more of the Moon's surface at its western limb when it is rising, and more at the eastern limb when it is setting.

Lick Observatory The observatory of the University of California, on Mount Hamilton near San José, California. Its oldest telescope is a 0.91 m (36-inch) refractor, in operation since the observatory opened in 1888. The largest telescope is the 3 m (120-inch) Shane Telescope, completed in 1959.

light, speed of (symbol *c*) The constant speed at which light and other **electromagnetic radiation** travels in a vacuum. Its value is 299,792,458.0 m/s. The first reasonably good estimate of its value was made by **Ole Römer** in 1675. **Albert Einstein** showed with his Special Theory of Relativity that the speed of light in a vacuum is always the same, whatever the relative speed of the observer and the source of the light.

light curve A graph of the change in brightness with time of a variable star or other body showing brightness variations.

light pollution The detrimental effect of artificial lighting – mainly streetlights, but also, for example, floodlights on public buildings and filling-station forecourts, and security lights – on astronomical observation and imaging. The problem has become much worse in recent years with the proliferation of such lighting and the emergence of new types of streetlamp. For amateurs, light pollution reduces the **limiting magnitude** (1) and makes deep-sky objects and comets, for example, difficult or impossible to see, but the development of special **filters** and of **CCDs** and related technology is providing solutions. For professional astronomers it can impede or distort scientific observation, particularly where the area surrounding a once-isolated observatory has become built up.

light-time The time required for light to travel from any celestial object to the Earth. For example, light takes 8.3 minutes to reach the Earth from the Sun.

light-year (symbol l.y.) A unit of distance measurement used in astronomy, equal to the distance travelled by light in a vacuum in one tropical year. 1 l.y. = 9.4607×10^{12} km = 0.3066 parsecs = 63,240 AU.

limb The edge of a celestial body which is visible as a disk, such as the Sun, the Moon, or a planet. SEE ALSO **following**, **preceding**

limb darkening The decrease in the brightness at visible wavelengths of a celestial body towards its limb, giving it a vignetted appearance. The term is used most often in connection with the Sun, where, towards the limb, the observer is looking at the deeper, brighter levels of the Sun through a greater thickness of its atmosphere. The giant planets also show limb darkening.

limiting magnitude (1) The faintest magnitude of objects visible or recordable by photography or electronic means on a given occasion. The limiting magnitude depends on sky conditions, instrument aperture, and sensitivity of the eye or recording apparatus.

limiting magnitude (2) The faintest magnitude of objects recorded in a star atlas.

line of apsides SEE **apse**

line spectrum SEE **spectrum**

local arm SEE **Orion Arm**

Local Group The small group of about 30 galaxies that includes our Galaxy, the two Magellanic Clouds, the Andromeda Galaxy, and the Triangulum spiral. They are distributed over a roughly ellipsoidal space, about 5 million light years across. The members are gravitationally bound so that, unlike more distant galaxies, they are not receding from us or from each other.

local time Time as measured at a given place on Earth, based on the mean solar day. Local time changes with longitude. A difference of 15° in longitude produces a difference of one hour in local time. To avoid the inconvenience of a multiplicity of local times, the Earth has been divided into 24 **time zones**, each of which has a standard time one hour different from the adjacent zone. In addition, *daylight saving time* (summer time) one hour in advance of standard time is kept in some countries for part of the year.

longitude A coordinate on the celestial sphere, or on the surface of a celestial body. *Celestial* (or *ecliptic*) *longitude* is measured eastwards from the vernal equinox (the First Point of Aries) to the great circle passing through the pole of the ecliptic and the object to be measured. It is reckoned in degrees from 0 to 360°. *Galactic longitude* is the angle measured along the galactic equator from the direction of the galactic centre to the point at which the galactic equator intersects the galactic circle of longitude of the object to be measured. *Heliocentric longitude* is the celestial longitude of a body as seen from the centre of the Sun. *Heliographic longitude* is the longitude of a point on the surface of the Sun. It is measured from the solar meridian that passed through the ascending node of the Sun's equator on the ecliptic on 1 January 1854 at Greenwich mean noon, from 0 to 360° in the direction of the Sun's rotation.

long-period comet SEE **comet**

long-period variables Pulsating red giants or supergiants with periods ranging from about 100 to 1000 days, usually of spectral type M or C. They are also known as **Mira** stars after the first of the type to be discovered. Each star has an average period which can vary by about 15%, so they are to some extent irregular. Their amplitudes range from 2.5 to 11 magnitudes, but the amplitude can also change appreciably from one cycle to the next. Long-period variables fall into three main divisions. Those that have a rise steeper than the fall tend to have wide minima and short, sharp maxima; as the asymmetry becomes greater, the period lengthens. Stars with symmetrical curves have the shortest periods. A third group show humps on their curves, or have double maxima and have long and short periods.

Lovell, (Alfred Charles) Bernard (1913–) English radio astronomer. From 1951 to 1981 he was director of the **Jodrell Bank** experimental station for radio astronomy near Manchester, and oversaw the construction there of the world's first large steerable radio telescope. He used the telescope to pioneer the exploration of radio emission from space.

Lowell, Percival (1855–1916) American astronomer. In 1894 he built an observatory (now the **Lowell Observatory**) at Flagstaff, Arizona. From there he observed Mars, producing intricate maps of the so-called **canals**, which he ascribed to the activities of intelligent beings. He went on to study the orbits of Uranus and Neptune, and calculated that their orbital irregularities were caused by an undiscovered **Planet X**. His predictions led to the discovery of the planet Pluto in 1930 by **Clyde Tombaugh**. It was at Lowell's urging that Vesto M. Slipher carried out the observations that led to the discovery of galactic **redshifts** in 1916.

Lowell Observatory An observatory at Flagstaff, Arizona, founded by Percival Lowell in 1894. Its telescopes include a 0.61 m (24-inch) refractor installed by Lowell in 1896. At a nearby dark-sky site on Anderson Mesa is the 1.8 m (72-inch) Perkins Reflector of the Ohio State and Ohio Wesleyan universities. A Planetary Research Center funded by NASA is located at Lowell Observatory.

lower culmination SEE **culmination**

low-surface-brightness galaxy (LSB galaxy) A galaxy that is very faint at optical wavelengths because it contains few stars. LSB galaxies of many types are known, from dwarf irregular galaxies to giant spirals. It is possible that, though difficult to detect, they are very common, and they might account for the **missing mass** of the Universe.

luminosity The absolute brightness of a star, given by the amount of energy radiated from its entire surface per second. It is expressed in watts (joules per second), or in terms of the Sun's luminosity. *Bolometric luminosity* is a measure of the star's total energy output, at all wavelengths. **Absolute magnitude** is an indication of luminosity at visual wavelengths.

luminosity class SEE **Morgan–Keenan classification**

luminosity function A measure of the distribution of stars in space according to their **absolute magnitude**. It is usually expressed as the number of stars per cubic parsec for a given range of absolute magnitude. The value of the luminosity function in the solar neighbourhood rises to a maximum at absolute magnitudes of 14–15, then falls away for fainter magnitudes. This implies that faint M-type dwarfs of about one ten-thousandth the Sun's luminosity are the most abundant type of star locally. Luminosity functions can also be plotted for other categories, such as particular types of star, star clusters, galaxies, or clusters of galaxies.

Luna The name of a series of lunar probes launched by the former Soviet Union. The first three were originally named *Lunik*. Luna 1, launched in January 1959, missed the Moon, but Luna 2 became the first man-made object to hit the Moon, on 13 September 1959. Luna 3, launched on 4 October 1959, obtained the first photographs of the farside of the Moon. Luna 9 achieved the first soft landing on the Moon on 3 February 1966 and transmitted some photographs from the surface. Luna 10, in March 1966, was the first craft to enter lunar orbit. In September 1970 Luna 16 made the first automatic return of a soil sample to Earth. Luna 17 carried the first unmanned lunar roving vehicle, **Lunokhod 1**. The series ended with Luna 24, another sample-and-return mission, in August 1976.

lunar eclipse An **eclipse** of the Moon by the Earth. Since the Moon's orbital plane is inclined to the plane of the ecliptic, a lunar eclipse can happen only when the Moon is at opposition (i.e. at full Moon) and at the same time is at or near one of its nodes (SEE **ecliptic limits**). The Sun, Earth and

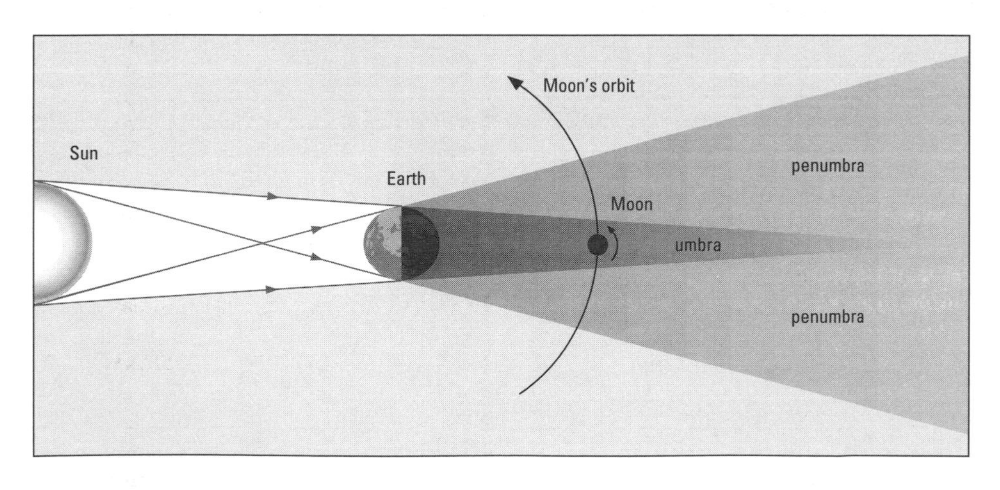

Lunar eclipse (not to scale): may be total or partial

Moon are then very nearly in a straight line. Lunar eclipses are visible from anywhere on Earth where the Moon is above the local horizon. The Moon does not become completely invisible during an eclipse because it is partially lit by sunlight refracted by the Earth's atmosphere, and it appears reddish because the blue component is removed from the sunlight by **scattering**.

A *total lunar eclipse* occurs when the Moon is entirely within the umbra, the dark central part of the Earth's shadow (SEE the diagram on page 117). The eclipse is *partial* when the Moon is partly within the umbra, and *penumbral* when the Moon passes through the penumbra but completely misses the umbra. The overall duration of a total lunar eclipse, from **first contact** to **fourth contact**, can last for 3½ hours; totality, from **second contact** to **third contact**, for 1¾ hours. There are two or three partial or total lunar eclipses each year.

Lunar Orbiter A series of US probes put into orbit around the Moon to photograph its surface in detail far superior to anything obtainable from Earth. The photographs taken by the probes were processed on board and telemetered back to Earth. There were five probes in the series, all successful, from Lunar Orbiter 1 in August 1966 to Lunar Orbiter 5 in August 1967.

Lunar Rover (Lunar Roving Vehicle) One of the battery-powered cars used on the Moon by the crews of Apollos 15, 16 and 17 (SEE **Apollo programme**). They enabled the astronauts to travel further afield in search of rock samples.

lunation The time interval between two identical phases of the Moon, for example from one new Moon to the next new Moon. This corresponds to the Moon's **synodic period** or the **synodic month** of 29.53059 days.

Lunik SEE **Luna**

luni-solar precession The main component of **precession**, resulting from the gravitational effects of the Moon and Sun on the Earth. It amounts to about 50.4 arc seconds per year.

Lunokhod One of two lunar roving vehicles landed on the Moon by the former Soviet Union. All operations, such as travelling across the lunar surface, surveying the area, testing the mechanical properties of the surface and making chemical analyses of it, were controlled from Earth. Lunokhod 1 was placed on the Moon by Luna 17 in November 1970; Lunokhod 2 followed in January 1973.

Lupus A constellation in the southern hemisphere, adjoining Centaurus, representing a wolf. Its brightest star is Alpha, magnitude 2.3, but it has no objects of particular interest.

Lyman series The series of lines in the ultraviolet region of the **hydrogen spectrum** analogous to the **Balmer series** in the visible part. The first Lyman line, Lα, is at 121.6 nm, and was first detected in the solar spectrum by photography from a rocket. The second, Lβ, is at 102.6 nm; Lγ is at 97.2 nm. In the spectra of **quasars**, Lyman lines are redshifted into the visible region. (After American physicist Theodore Lyman, 1874–1954.)

Lynx A dim northern constellation between Auriga and Ursa Major, introduced by **Hevelius**. Its leading star is Alpha, magnitude 3.2, but the only other star above magnitude 4 is 38 Lyncis, at magnitude 3.9.

Lyot, Bernard Ferdinand (1897–1952) French astrophysicist. He studied the Sun, and invented the **coronagraph** for observing the corona, and the **Lyot filter**. He also pioneered astronomical polarimetry, the measurement of the degree of polarization of light from celestial bodies, and invented a number of devices for the purpose.

Lyot filter A **filter** for observing the Sun at a particular wavelength, developed by **Bernard Lyot** in the 1930s. Sunlight is directed through a succession of alternating layers of quartz plates and polaroid sheets, and each separate passage reinforces some wavelengths and cancels others. The result is a number of very narrow (0.1 nm) and widely spaced wavebands which are easy to isolate.

Lyra A small but prominent constellation. In mythology, it represents the harp given to Orpheus by Apollo. Its brightest star is **Vega**. Lyra contains the famous 'Double Double', the quadruple star Epsilon; the two main components, magnitudes 4.6 and 4.7, can be separated with the naked eye and each is again double. Beta (Sheliak) is an eclipsing variable, range 3.3 to 4.4, period 12.94 days; it is the prototype of the **Beta Lyrae stars**. Between Beta and Gamma (Sulafat), magnitude 3.2, lies the **Ring Nebula**, M57, a planetary nebula.

Lyrids A **meteor shower** that occurs in April, and has its radiant on the border of the constellations Lyra and Hercules, south-west of the star Vega. The Lyrids are associated with Comet Thatcher (1861 I).

Lysithea One of the four small **satellites** making up Jupiter's intermediate group, discovered by Seth Nicholson in 1938.

McDonald Observatory The astronomical observatory of the University of Texas, on Mount Locke, Texas. Its largest telescope is a 2.7 m (107-inch) reflector opened in 1969.

Maffei galaxies Two nearby galaxies so close to the plane of the Milky Way that at optical wavelengths they are almost completely obscured by galactic dust. Maffei I is an S0 (lenticular) galaxy over 3 million light years away, on the edge of the Local Group. Maffei II is an Sb spiral but, at a distance of 15 million light years, it is unrelated to Maffei I and the Local Group. They were discovered as infrared sources in 1968 by the Italian astronomer Paolo Maffei.

Magellan A NASA space probe to Venus, launched in May 1989. It went into orbit around Venus in August 1990 and made a detailed radar map of 98% of the planet, revealing impact craters, volcanic mountains, lava flows and other features. In 1993 Magellan's highly elliptical orbit was circularized by repeated passes through the top of the planet's atmosphere, a technique known as *aerobraking*. Tracking of the probe in its new orbit allowed astronomers to map the gravitational field of Venus. Following a further lowering of the orbit in 1994, Magellan burnt up in the atmosphere.

Magellanic Clouds The two nearest galaxies to our own, visible to the naked eye like isolated patches of the Milky Way in the southern sky. The Large Magellanic Cloud (LMC) lies in the constellations Dorado and Mensa and is about 6° across. The Small Magellanic Cloud (SMC) lies in the constellation Tucana and is about 3° across. They are named after the Portuguese explorer Ferdinand Magellan (1480–1521), who described them. Their distances are 169,000 light years and 190,000 light years, with an uncertainty of about 10%. Although usually classified as irregular galaxies, both Clouds are somewhat bar-shaped with faint outer structure that can be interpreted as rudimentary spiral arms. The mass of the LMC is about 10^{10} solar masses, and that of the SMC about 2×10^9 solar masses.

At the north-east end of the LMC lies the **Tarantula Nebula**, one of the largest and brightest groupings of hot stars and bright nebulosity known in any galaxy. Near here occurred **Supernova 1987A**, the first naked-eye supernova since 1604. The irregular distribution of star clusters and diffuse nebulae in the LMC contrasts with the more uniform stellar distribution in the SMC. Young stars in the LMC lie in a thin disk, which is seen nearly face-on (about 27° to the plane of the sky). There is no evidence for a spherical halo of the kind that surrounds our Galaxy.

The SMC is greatly extended in the line of sight, with a depth of about 60,000 light years, five times the dimension it presents to us. It appears to have been distorted by tidal forces, possibly the result of a close passage of the SMC and LMC some 200 million years ago.

The *Magellanic Stream* is a vast cloud of neutral hydrogen (SEE **HI region**) stretching between the Clouds and extending towards the Galaxy. It was probably torn out from the Clouds by gravitational interaction with the Galaxy.

magnetohydrodynamics (MHD) The study of the behaviour of plasma (a mixture of ions and electrons) in a magnetic field. A moving plasma is in effect an electric current, which interacts with the magnetic field to produce a force which modifies the motion of the plasma. The subject was pioneered by the Swedish physicist Hannes Alfvén (1908–), who predicted that waves (*Alfvén waves*) would propagate through a plasma, and he applied MHD to a variety of problems in astrophysics. Plasmas occur in many astronomical contexts, so MHD has proved fruitful in the study of solar phenomena such as sunspots, the interaction of the solar wind with planetary **magnetospheres**, star formation and supernovae.

magnetosphere The region of space surrounding a planet in which the planet's magnetic field predominates over the **solar wind**, and controls the behaviour of plasma (charged particles) trapped within it. The boundary of the magnetosphere is called the *magnetopause*, outside which is a turbulent magnetic region called the *magnetosheath*. Where the steady outward flow of the solar wind is first interrupted there is a **bow shock**. Downwind from the planet, the solar wind draws the magnetosphere out into a long, tapering *magnetotail*.

Mercury, the Earth, and the giant planets have magnetospheres. The Earth's contains the **Van Allen Belts** of charged particles. Jupiter's magnetosphere is many times larger than the Sun, and contains a vast disk of plasma, and **Io**'s plasma torus.

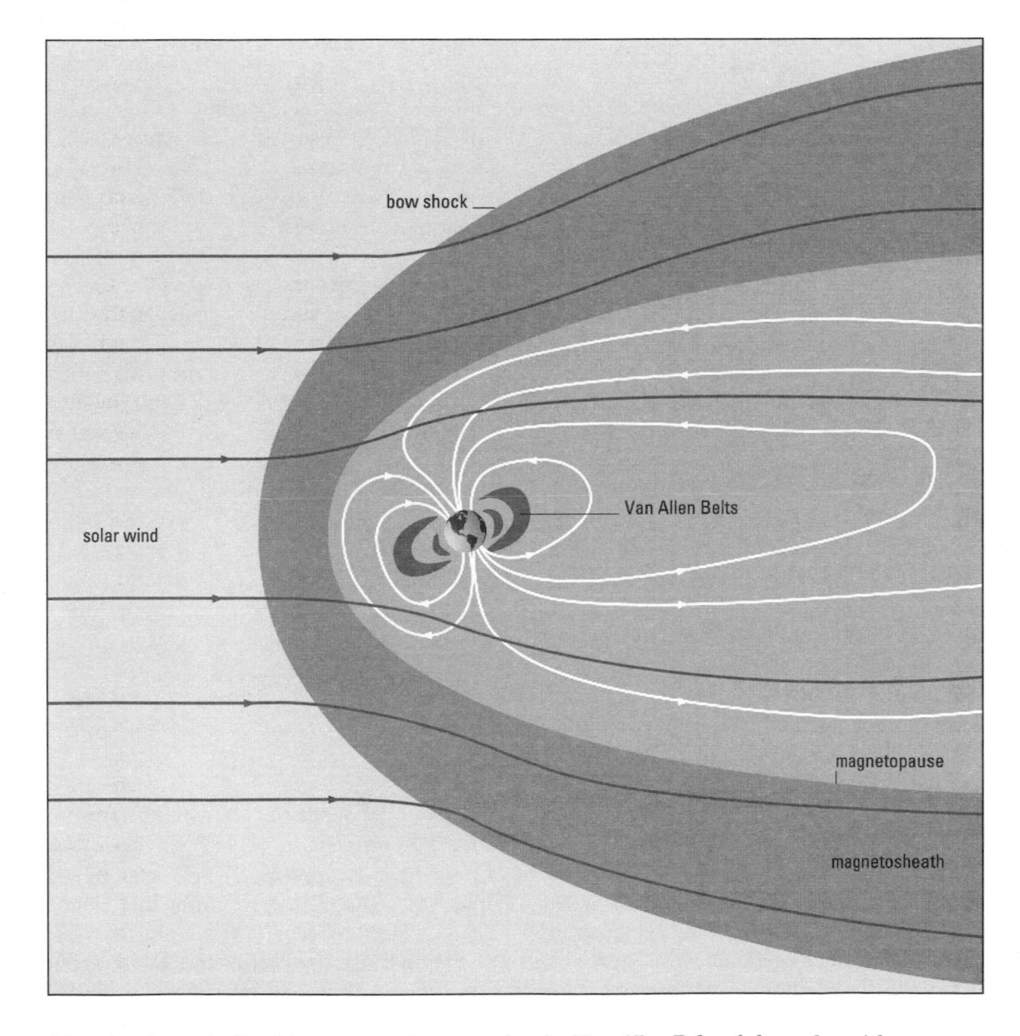

Magnetosphere: the Earth's magnetosphere contains the Van Allen Belts of charged particles

magnitude A measure of the brightness of celestial objects. In the 2nd century BC the Greek astronomer **Hipparchus of Nicea** classified the stars into six magnitudes, first magnitude being the brightest and sixth magnitude the faintest visible to the naked eye. In the 19th century it was found that a star of first magnitude was about 100 times as bright as one of sixth magnitude. It was decided to adopt a scale of magnitudes in which this ratio is exactly 100 : 1. Hence, 100 stars of magnitude 6 taken together are as bright as a single star of magnitude 1. The ratio of brightness between one magnitude and the next is thus the fifth root of 100, which is 2.512. Thus, differences of 2.5, 5 and 10 magnitudes correspond to brightness ratios of 10, 100 and 10,000 respectively. Very bright bodies, Sirius, Venus, and the Sun, have negative magnitudes. The faintest objects detectable from ground-based telescopes are about magnitude 30.

Apparent magnitude is the brightness of an object as seen from Earth. **Absolute magnitude** is the true brightness of an object, taking into account its distance from Earth. The difference between the apparent and absolute magnitudes is the *distance modulus*. SEE ALSO **bolometric magnitude, photoelectric magnitude, photographic magnitude, photovisual magnitude, visual magnitude**

main sequence The region on the **Hertzsprung–Russell diagram** where most stars lie, including the Sun. It runs diagonally from hot, bright stars at the upper left down to cool, faint stars at the bottom right. The position of a star on the main sequence depends on its mass, the most massive stars being the brightest. The upper limit is about 60 solar masses at stars of spectral types O and B. The lower limit is at 0.08 of a solar mass with stars of spectral type M. Stars spend most of their lives on the main sequence, producing energy from the fusion of hydrogen to helium in their cores. The *zero-age main sequence* is where stars lie when they first start to burn hydrogen. As stars use up their internal hydrogen they move away from the main sequence. The more massive a star, the sooner it evolves off the main sequence. All stars on the main sequence are termed **dwarfs**, irrespective of whether they are larger or smaller than the Sun.

major axis The longest diameter of an elliptical orbit. Half the major axis (the **semimajor axis**) is the average distance of the orbiting object from its primary.

major planet Any of the nine largest planets of the Solar System – Mercury, Venus, Earth, Mars, Jupiter, Saturn, Uranus, Neptune, and Pluto – as opposed to the asteroids, which are also known as *minor planets*. Some astronomers feel that Pluto, because of its small size, does not merit being classified as a major planet.

Maksutov telescope A modification of the **Schmidt camera**, devised independently by Bouwers (1940) and Maksutov (1944). The Schmidt's aspherical correcting plate is replaced by a concave lens which is designed so as to cancel out the chromatic and spherical aberrations – SEE **aberration** (2) – of the primary concave spherical mirror. A secondary mirror directs the light path through a hole in the primary to a Cassegrain focus, although various other configurations have been used. The Maksutov is a compact instrument that has been popular with amateur observers, but it is now losing out to the **Schmidt–Cassegrain telescope**. (After Dmitrii Dmitrievich Maksutov, 1896–1964.)

many-body problem (*n*-body problem) The problem in celestial mechanics of finding how a number (n) of bodies move under the influence of their mutual gravitational attraction only. It is an extension of the **three-body problem**. The many-body problem has no general solutions. It requires a vast amount of computation to obtain good approximate solutions, but with modern computers this is now easily accomplished. Analysing how the planets of the Solar System perturb one another, and how space probes such as the Voyager craft move between them, are examples of the many-body problem.

mare (plural **maria**) An extensive dark plain on the Moon, composed of basalt (solidified lava). Most of the maria were formed 3 to 4 billion years ago, after large meteorites struck the Moon and weakened the crust, and lava later erupted from below. They range from the vast Oceanus Procellarum, covering much of the north-western quadrant of the Moon, down to the areas less than 100 km (60 miles) across. Nearly all are on the Moon's nearside, where the crust is thinner than on the farside. Maria is Latin for 'seas', which is what early observers thought they were. Similar plains on Mars were also known as maria, but since detailed mapping by spacecraft these are now known as *planitiae.*

Mariner A series of US space probes to the planets. Mariner 2 flew past Venus in 1962, discovering that the planet was hot. Mariner 4 flew past Mars in July 1965, photographing craters on its surface. Mariner 5 passed Venus in October 1967, making measurements of the planet's atmosphere. Mariners 6 and 7 obtained further photographs of Mars in 1969. Mariner 9 became the first probe to enter planetary orbit when it went into orbit around Mars in November 1971. It made a year-long photographic reconnaissance of the planet's surface, and obtained close views of the two moons, Phobos and Deimos. Mariner 10, the last of the series, was the first two-planet mission, passing Venus in February 1974 and then encountering Mercury three times, in March and September 1974 and March 1975.

mantle SEE **differentiation**

Mariner Valley A series of enormous, interconnected canyons on Mars, stretching for 4500 km (2800 miles) just south of the equator. Individual canyons extend for up to 200 km (125 miles) and are 7 km (4½ miles) deep. Mariner Valley is a fault structure whose formation is probably linked to that of the volcanic Tharsis region at its western end. The name commemorates the **Mariner** probes to Mars; the official version is Valles Marineris.

Markarian galaxies Galaxies catalogued by the Armenian astronomer Beniamin Markarian, distinguished by their strong ultraviolet emission. They include **Seyfert galaxies, N galaxies**, and **quasars**.

Mars The fourth major planet from the Sun, and the second-smallest of the inner planets. Mars appears red to the naked eye, and is also known as the Red Planet. When viewed through a telescope it appears as a small orange-red disk with lighter and darker markings, and white patches often visible at one pole or the other. Because of its markedly eccentric orbit, its distance

MARS: DATA	
Globe	
Diameter (equatorial)	6787 km
Diameter (polar)	6752 km
Density	3.94 g/cm^3
Mass	6.42×10^{23} kg
Sidereal period of axial rotation	24h 37m 23s
Escape velocity	5.0 km/s
Albedo	0.15
Inclination of equator to orbit	25° 11′
Surface temperature (average)	220 K
Surface gravity (Earth = 1)	0.38
Orbit	
Semimajor axis	1.524 AU = 227.9×10^6 km
Eccentricity	0.0934
Inclination to ecliptic	1° 51′
Sidereal period of revolution	686.980d
Mean orbital velocity	24.13 km/s
Number of satellites	2

from Earth at opposition varies from 56 to 101 million km (35 to 63 million miles), with corresponding variations in apparent size and brightness of 14 to 25 arc seconds and magnitude −1.0 to −2.8. Favourable oppositions occur at intervals of 15 to 17 years.

Mars shows some variation in **phase**, appearing slightly gibbous at quadrature. Its axial tilt is similar to the Earth's, so it passes through a similar cycle of **seasons**, and seasonal variations in its appearance may be observed through a telescope. The polar caps vary in size; and the darker markings may vary in shape and extent, which was once thought to indicate seasonal changes in vegetation but is now known to be caused by seasonal dust storms covering and uncovering darker surface rocks. For a long time the planet was imagined to be an abode of intelligent life, and some observers reported seeing linear surface markings which they believed were artificial **canals**, but no trace of life has been found there. Much information has been returned by various **Mars probes**. The main data for Mars are given in the table.

The surface of Mars reveals a long and complex history of geological activity. Although similar processes have operated on the Earth and Mars, the results on Mars have been dramatically different, mainly because of its smaller size and gravity, but also because of the absence of plate **tectonics**. The major difference in terrain is between the largely smooth, lowland volcanic plains of the northern hemisphere and the heavily cratered uplands of the south. There is also a prominent bulge, 5000 km (3000 miles) across, in the volcanic Tharsis region. The biggest volcanic structure on Mars – and in the whole Solar System – is **Olympus Mons**. Nearby, on the Tharsis Ridge, are three other extinct volcanoes: Arsia Mons, Ascraeus Mons, and Pavonis Mons. All four are hundreds of kilometres across and all rise to the same height, 27 km (17 miles). On the Earth, plate tectonics can eventually move a volcano away from the underlying source of lava, forming a volcanic chain, but on Mars such volcanoes stayed in one place and grew to a great size.

Stretching east from Tharsis for 4500 km (2800 miles) is the network of giant canyons known collectively as the **Mariner Valley**. While some Martian canyons and similar features are geological faults, others, in the Mariner Valley and elsewhere, are clearly channels in which water once flowed across the Martian surface in the form of rivers. Some features have even been interpreted as dried-up sea-beds. At some time in the past Mars must have had a very different climate, with a denser atmosphere. Perhaps all the water is now frozen into the subsurface permafrost layer; more likely is that the planet's low gravity has simply not been enough to hang on to all of its air and water, which have slowly been escaping.

Two large impact basins, Hellas (1800 km/1500 miles) and Argyre (900 km/550 miles) dominate the southern hemisphere. Here there are two terrain types: an ancient surface almost saturated with craters and crossed by many channels, and smoother, less heavily cratered *intercrater plains*. Many Martian craters show signs of erosion by the dust storms that blow across the surface. In places the dust forms dune fields that shift with the winds. Craters at higher latitudes have a smoothed appearance, probably because ground ice allows the crater walls to undergo 'creep'.

The variable polar ice caps appear to be composed of solid carbon dioxide with underlying caps of water-ice. The northern water cap is exposed in the summer when the carbon dioxide warms and becomes part of the atmosphere, but the southern cap always retains a carbon dioxide covering.

Before the space age the nomenclature for surface features was that introduced in the 19th century by **Giovanni Schiaparelli**, who produced a map bearing names from classical literature and of geographical locations on Earth. However, these *albedo features* visible from Earth represent differences in surface brightness, and they do not all correspond to topographical features visible in images returned by probes. The modern nomenclature borrows the old names where appropriate; otherwise, large craters are named after astronomers and other people associated with Mars, small craters after towns and villages on Earth, and the various types of channel after the name for Mars in various languages or after rivers on Earth.

The atmosphere consists mainly of 95% carbon dioxide, 2½% nitrogen, and 1½% argon, with smaller quantities of oxygen, carbon monoxide, and water vapour. Although the atmospheric pressure at the surface of the planet is less than 1% of that at the surface of the Earth, a variety of meteorological phenomena occur. Winds of up to 300 km/h (200 mile/h) drive seasonal dust storms that may last for weeks and lift dust particles to heights of 40 km (25 miles), sometimes obscuring the whole planet. Water vapour forms into a variety of cloud types, and produces morning mists and ground frosts. The surface temperature on Mars varies between extremes of 130 K and 290 K.

The crustal rocks of Mars are iron-rich, giving the planet its red colour; in composition they may resemble minerals like haematite and olivine. Much of the surface may be covered with a hard, ferrous clay-like deposit. Basalts cover the volcanic plains. From the planet's density, its interior structure probably consists of a mantle overlying a core. However, the extremely weak magnetic field suggests that if there is an iron or nickel–iron core it is not hot and fluid. As for its **magnetosphere**, Mars does possess an ionosphere, with which the solar wind interacts to produce a weak bow shock.

Mars has two tiny **satellites** in very close orbits: Phobos and Deimos.

Mars Environmental Survey (MESUR) A US project to place a series of landers on Mars. The first part of the project, MESUR Pathfinder, is due for launch at the end of 1996 and will land an automatic roving vehicle on Mars in July 1997. In the second part of the project, MESUR Network, several landers will be sent to Mars in 2000 and 2001.

Mars probes Between 1962 and 1973 the former Soviet Union launched seven probes called *Mars* towards that planet, plus one probe in the **Zond** series. Mars 5 in 1973 orbited the planet and sent back pictures, but the other Soviet probes to Mars were either failures or produced little of value. By contrast, the United States had success with several probes to Mars in its **Mariner** series, as well as the two **Viking** craft. In 1988 the Soviet Union launched two probes called *Phobos* to study Mars and its largest moon. One failed en route, while the second failed shortly after entering orbit around Mars. A US probe, *Mars Observer*, launched in 1992, also failed shortly before entering orbit around the planet in August 1993. Future US and Russian missions will land automatic rovers on the surface of Mars.

mascon A high-density region below the Moon's surface; the word is derived from 'mass concentration'. Mascons cause localized increases the Moon's gravitational field, which is how they were first detected by Lunar Orbiter 5 in 1967. The dozen or so mascons that are known have diameters of 50–200 km (30–125 miles) and lie about 50 km (30 miles) below the surface. They were probably formed when impacts from large meteorites blasted away the lighter crust, and the denser, underlying mantle material bulged upwards.

maser (1) A cloud of molecular **interstellar matter** in which **electromagnetic radiation** emitted by high-energy molecules at a particular wavelength in the microwave region stimulates other molecules to emit radiation at the same wavelength. The word is an acronym of 'microwave amplification by stimulated emission of radiation'. A maser therefore amplifies natural microwave emissions, just as a laser does for light.

maser (2) An amplifier of extremely low noise used in radio-astronomy laboratories and in satellite-tracking stations.

Maskelyne, Nevil (1732–1811) English astronomer. He was appointed the fifth Astronomer Royal in 1765. Building on a method of his for determining longitude at sea by observing the Moon, in 1767 he published the first *Nautical Almanac*, an annual book for the same purpose. In 1774 he measured the mass of the Scottish mountain of Schiehallion by seeing how far it caused a plumb-line to deviate, from which he calculated the average density of the Earth to be 4.5 g/cm^3 (the true value is 5.52 g/cm^3).

mass–luminosity relation A relationship between luminosity and mass for stars on the **main sequence**, derived theoretically by **Arthur Eddington** in 1924. It is expressed approximately as

$$L \propto M^x,$$

where L is the luminosity, M is the mass, and the exponent x varies between 3 and 3.5 according to the mass. The relationship is not obeyed by white dwarfs, or by giants or supergiants. The relationship enables the mass of a star to be determined if its absolute magnitude is known. The theoretical upper limit (SEE **Eddington limit**) of stellar mass is about 100 solar masses, but very few stars exceed 30 solar masses. The lower limit of a star's mass is 0.08 of a solar mass.

Mauna Kea Observatory A major observatory in Hawaii, at an altitude of 4200 m (13,800 ft). Its main telescopes include the twin 10 m (396-inch) **Keck Telescopes**; the 3.6 m (142-inch) Canada-France-Hawaii Telescope; the 3.8 m (150-inch) UK Infrared Telescope; and the 15 m (49-ft) James Clerk Maxwell Telescope, owned by the UK and used for **millimetre-wave astronomy**. Mauna Kea has also been chosen as the site for the Japanese 8.3 m (326-inch) Subaru Telescope due for completion in 1999, and for one of the twin 8.1 m (319-inch) **Gemini Telescopes**.

Maunder minimum The period 1645–1715, when few sunspots were observed, as identified in the 1890s from historical records by British solar physicist Edward Walter Maunder (1851–1928). This was not simply the result of a lack of data from those years: recent studies, especially of tree rings from the period, reveal a genuine lull in solar activity. Also, the Maunder minimum correlates well with the long succession of below-average temperatures during the period, known as the *Little Ice Age*. It was caused by a number of much longer-term fluctuations acting together to damp down the 11-year **solar cycle**.

mean anomaly SEE **anomaly**

mean solar time Time based on the motion of the **mean Sun**, which travels at a uniform rate along the celestial equator.

mean Sun A fictitious Sun moving along the celestial equator at a constant rate equal to the average rate of the true Sun along the ecliptic. It completes its annual course with respect to the vernal equinox in exactly the same time as the true Sun. The true Sun travels at a variable rate along the ecliptic, and cannot therefore give a uniform measure of time.

Mensa A far southern constellation, introduced by **Lacaille**, representing the Table Mountain at the Cape of Good Hope where he worked. Mensa has no star above the 5th magnitude, but part of the Large Magellanic Cloud extends into it.

Mercury The innermost planet of the Solar System, and the smallest of the four inner planets. Mercury is difficult to observe because it is never more than 28° from the Sun, and its small angular diameter (8 arc seconds at greatest elongation) makes identification of surface detail very difficult, but through a telescope light and dark areas can be made out. As an inferior planet, Mercury

MERCURY: DATA	
Globe	
Diameter	4878 km
Density	5.43 g/cm^3
Mass (Earth = 1)	0.0553
Volume (Earth = 1)	0.0562
Sidereal period of axial rotation	58d 15h 30m
Escape velocity	4.3 km/s
Albedo	0.06
Inclination of equator to orbit	0°
Surface temperature	100–800 K
Surface gravity (Earth = 1)	0.38
Orbit	
Semimajor axis	0.387 AU = 57.91 $\times$ 10^6 km
Eccentricity	0.206
Inclination to ecliptic	7° 00′
Sidereal period of revolution	87.969d
Mean orbital velocity	47.89 km/s
Satellites	0

passes through **phases**. It reaches magnitude −0.7 at brightest. A **transit** of Mercury across the Sun's disk occurs at intervals of either 7 or 13 years. The planet has no known satellite. The main data for Mercury are given in the table.

Mercury rotates three times on its axis for every two revolutions of the Sun. As a result, the planet has a 'hot pole' in each hemisphere – areas that spend longer facing the Sun and get heated more. Also, the **synodic period** of 116 days is very close to twice the rotation period. Thus, at greatest elongation west, say, Mercury presents to Earth almost the same face as at the previous greatest elongation west. This led earlier observers into thinking that Mercury had **synchronous rotation**. The true rotation period was established in 1965 by **radar astronomy**.

Very little was known about Mercury's surface until the Mariner 10 probe (the only mission to Mercury so far) made three close approaches to the planet in 1974 and 1975 and returned pictures of nearly half the surface. These showed a heavily cratered, lunar-like world. The largest impact feature, at 1300 km (800 miles) across, is the Caloris Basin, similar to the Moon's Imbrium Basin; the next-largest is the 625 km (390-mile) diameter crater Beethoven. Like the Moon, Mercury has **ray** craters. Unlike the Moon, it has *lobate scarps*: cliffs up to 3 km (2 miles) high and hundreds of kilometres long, and sometimes cutting across craters. They may be compression faults formed when the planet's crust cooled and contracted. The *intercrater plains* appear to be volcanic, like the lunar maria, but unlike the maria they contain many small craters. There are also valleys and ridges.

Most craters on Mercury are named after writers, artists, and composers; plains are named after the word for Mercury in different languages; scarps take the names of famous explorers' ships; valleys are named after radar-astronomy stations; and the names of ridges commemorate astronomers with a strong association with the planet.

The surface temperature at noon on the equator can reach over 800 K at perihelion, while on the night side it can fall to 90 K. Radar mapping of Mercury's polar regions in 1991 and 1992 revealed what may be water-ice on the floors of craters permanently in shadow, where the temperature is estimated never to rise above 112 K. There is a very tenuous atmosphere, mainly of helium and sodium, deriving mainly from the **solar wind**, but also from particles sputtered (knocked off) from the surface by solar radiation, and possibly from material slowly outgassing from the interior.

The presence of a weak magnetic field and the planet's high density for its size suggest that Mercury has a very large iron-rich core, proportionally far bigger than for any other Solar System body. This could be because a small body forming so close to the Sun would have trouble hanging on to lighter elements. Alternatively, Mercury could have suffered a collision with another body which blasted off most of its outer layers.

Mercury's orbital velocity is greater than that of any other planet. Consequently, the advance of its perihelion is that much more pronounced. (The advance of a planet's perihelion results from the slow rotation of the orbit's **semimajor axis** in the direction of the planet's revolution around the Sun.) The gravitational attraction of the other planets accounts for all but 43 arc seconds per century in Mercury's case. In the 19th century the discrepancy was attributed to an intra-Mercurial planet named **Vulcan**, but the answer was provided by **Einstein's** General Theory of Relativity.

meridian SEE **celestial meridian**

MERLIN (Multi-Element Radio-Linked Interferometer Network) An array of six radio telescopes linked to Jodrell Bank. Five of the telescopes are of 25 m (82 ft) diameter at distances up to 127 km (204 miles) from Jodrell Bank; a sixth dish, of 32 m (105 ft) diameter at Cambridge, gives a maximum baseline of 230 km (370 miles).

Messier, Charles Joseph (1730–1817) French astronomer. He kept a record of various fuzzy extended objects that were in fixed positions in the northern sky so that he should not confuse them with comets, of which he was a successful hunter, discovering over twenty. But it is for the catalogue of **Messier objects** – what we now know to be galaxies, clusters and nebulae – that he is remembered today.

Messier object Any of the nebulae, star clusters and galaxies, numbering over 100, which appear in the list begun by **Charles Messier**. Messier drew up the list so that he and other comet-hunters would not confuse these permanent, fuzzy-looking objects with comets. The first edition of the catalogue was published in 1774, with supplements in 1780 and 1781. Not all the objects were discovered by Messier himself. Objects in the catalogue are given the prefix M, and are still widely known by their *Messier numbers*. Some of the best-known Messier objects are listed in the table above.

SOME OF THE BEST-KNOWN MESSIER OBJECTS

Messier number	Popular name	Description
1	Crab Nebula	Remnant of supernova in Taurus
8	Lagoon Nebula	Emission nebula in Sagittarius
11	Wild Duck	Open star cluster in Scutum
13	—	Globular cluster in Hercules
17	Omega, Swan or Horseshoe Nebula	Emission nebula in Sagittarius
20	Trifid Nebula	Emission nebula in Sagittarius
27	Dumbbell Nebula	Planetary nebula in Vulpecula
31	Andromeda Galaxy	Spiral galaxy in Andromeda
42	Orion Nebula	Emission nebula in Orion
44	Praesepe	Open star (or Beehive) cluster in Cancer
45	Pleiades	Open star cluster in Taurus
51	Whirlpool Galaxy	Spiral galaxy in Canes Venatici
57	Ring Nebula	Planetary nebula in Lyra
64	Black Eye Galaxy	Spiral galaxy in Coma Berenices
97	Owl Nebula	Planetary nebula in Ursa Major
104	Sombrero Galaxy	Edge-on spiral galaxy in Virgo

MESUR ABBREVIATION FOR **Mars Environmental Survey**

metals (heavy elements) In astronomy, all elements heavier than helium. Population I stars (SEE **stellar populations**), which are young, have relatively high metal contents (up to 2%), while Population II stars, which formed first, have low metal contents (as little as 0.01%). The metal content of a star affects its structure and evolution.

meteor The brief streak of light in the night sky caused by a **meteoroid** entering the Earth's upper atmosphere at high speed from space. Meteors are popularly known as *shooting stars*. They occur at altitudes of about 100 km (60 miles). The typical meteor lasts for a few tenths of a second to a second or two, depending on the meteoroid's impact speed, which can vary from about 11 to 70 km/s (7 to 45 mile/s).

Most meteor-producing meteoroids are low-density (0.3 g/cm^3) dust particles that originated in comets. They are rapidly eroded away by **ablation** and leave a trail of ionized and excited atmospheric atoms that is visible as the meteor. Most such meteors have a magnitude in the range 0–4. Larger, denser meteoroids that originated in asteroids produce brighter meteors, the brightest of which may rival the full Moon or, in rare cases, even the Sun (SEE **fireball, bolide**). Ionization trails are detectable by radar (SEE **radar astronomy**), enabling meteors to be studied in the daytime.

A few meteors per hour may be seen on any clear, moonless night at any time of year. But at certain times of the year there are **meteor showers**, when meteors are more numerous than usual. Showers occur when the Earth passes through a **meteor stream** – dust particles spread around the orbit of a comet. Most of the meteors appearing during the year are *sporadic meteors*, not associated with cometary orbits.

Meteor Crater (Arizona Meteor Crater, Barringer Crater) An enormous circular crater, about 1200 m (4000 ft) in diameter and 180 m (600 ft) deep, near Flagstaff in the Arizona desert. It is the best-known of the Earth's impact craters, and the first to be identified as such. The impacting body is estimated to have been a nickel–iron meteorite, about 40 m (130 ft) across and weighing a quarter of a million tonnes, which fell around 50,000 years ago. Quantities of meteoritic nickel–iron have been recovered from the plain around the crater, but most of the meteorite was probably vaporized on impact. The crater was discovered in 1891.

meteorite That part of a large **meteoroid** that survives passage through the Earth's atmosphere and reaches the ground. Hundreds of tonnes of meteoroidal matter is swept up by the Earth each day. Most burns up in the atmosphere to produce **meteors**, but about 10% reaches the surface as meteorites and **micrometeorites**.

Meteorites whose fall is observed are known as *falls*. Their atmospheric passage produces a **fireball** or **bolide**, and analysis of the trajectories of some falls reveals them to have been in orbits like those of **near-Earth asteroids**. Over 99% of meteorite falls are unobserved. Meteorites discovered on the ground are known as *finds*. Many recent finds come from Antarctica, recovered from the ice surface.

There are three main classes of meteorite: **stony meteorites** are composed mainly of silicates of iron, calcium, aluminium, magnesium, and sodium; **iron meteorites** are composed of iron and nickel, and when polished and etched show characteristic **Widmanstätten patterns**; **stony-iron meteorites** have an intermediate composition. Stony meteorites, which make up 93% of all falls, are further subdivided into **chondrites** and **achondrites**. Iron meteorites account for just 6% of falls, but over half of all finds, iron being much easier than stony material to pick out on the ground. Just 1% of falls are stony-irons. The exterior of all meteorites is a *fusion crust* of material, melted by friction during passage through the atmosphere and then resolidified.

The largest known single meteorite weighs about 60 tonnes and lies where it fell at Hoba, in Namibia. Stony bodies weighing several tonnes often fragment before impact; the fall at Allende, Mexico, in 1969 deposited 5 tonnes of material over 500 km^2 (200 mile^2). Meteoroids weighing more than 100 tonnes that do not break up are not decelerated as much as lighter bodies, and produce impact craters such as **Meteor Crater**.

Most meteorites are fragments of asteroids, but a few are of lunar origin, and the **SNC meteorites** are believed to be of Martian origin. Analysis of meteorites can shed light on the geological histories of the parent bodies.

meteoroid A small particle or body following an Earth-crossing orbit, with the potential to become a **meteor** or **meteorite**. Meteoroids are of cometary or asteroidal origin. Those of cometary origin are dust particles given off by **comets** and spread around the comet's orbit, and are responsible for **meteor showers**. Those of asteroidal origin are fragments produced by collisions in the main **asteroid** belt (or they may be small asteroids in their own right), and may penetrate the Earth's atmosphere to land as **meteorites**.

meteor shower The appearance of **meteors** from the same point in the sky, the **radiant**, around the same time each year. Nearly all showers are named after the constellation in which their radiant lies. There are a dozen or so major showers and many minor ones. Meteors belonging to a shower are called *shower meteors* to distinguish them from **sporadic meteors**. During major showers there is a build-up of activity to a *maximum*, when 10–100 shower meteors may be visible each hour from any one location (under ideal conditions; SEE **zenithal hourly rate**), and then a falling away. In a typical shower, activity lasts for around two weeks in all.

A shower occurs when the Earth passes through a **meteor stream**. In an *annual shower* the meteor rates vary little from one year to the next because the meteoroids in the stream are spread evenly around the orbit. With streams in which the meteoroids are bunched together in a swarm, meteor numbers are low except when the Earth intersects the swarm. Such *periodic showers* may produce a *meteor storm* (SEE **Leonids**). SEE ALSO the names of individual meteor showers.

MAJOR METEOR SHOWERS

Shower	Dates	ZHR*
Quadrantids	1–6 Jan	100
Lyrids	19–25 Apr	10–15
Eta Aquarids	24 Apr–20 May	50
Delta Aquarids	15 Jul–20 Aug	20–25
Perseids	25 Jul–20 Aug	80
Orionids	15 Oct–2 Nov	30
Leonids	15–20 Nov	100+
Geminids	7–15 Dec	100

**Approximate zenithal hourly rate.*

meteor stream A large quantity of **meteoroids** circling the Sun in a common orbit. Meteor streams are associated with **comets**, and consist of particles shed by comets at perihelion passage over the course of their active lifetimes. A **meteor shower** occurs when the Earth intersects a meteor stream. In a young meteor stream, cometary debris is bunched together in a *meteor swarm*, which produces a periodic meteor shower. As the stream evolves material is spread more evenly around the orbit. With time, the orbits of individual meteoroids move further away from the primary orbit to produce a torus-shaped stream; the showers such streams produce are low-activity showers lasting a number of weeks. Eventually, a stream will be indistinguishable from the 'background' concentration of dust that pervades the inner Solar System (SEE **interplanetary matter**).

Metis One of the small inner **satellites** of Jupiter discovered in 1979 during the Voyager missions. It orbits within Jupiter's main ring.

Metonic cycle A 19-year cycle of **lunations**, after which the phases of the Moon begin to repeat themselves on the same days of the year. This is because 19 **tropical years** are equal to 235 **synodic months**, to within 2 hours. The discovery is often attributed to the 5th-century BC astronomer Meton of Athens. The cycle forms the basis of the Greek and Jewish calendars. SEE ALSO **saros**

MHD ABBREVIATION FOR **magnetohydrodynamics**

micrometeorite A micrometre-sized **meteorite**. Micrometeorites are continually settling on the Earth's surface. The *micrometeoroids* that give rise to them are too small to be burnt up in the atmosphere; they are quickly decelerated and drift gently downwards. The quantity that falls greatly exceeds the total of all other meteorites. There are three main types: spheres of high density, irregular compacted particles, and fluffy non-compacted particles of lower density.

Microscopium A small southern constellation representing a microscope, introduced by **Lacaille**. Its brightest star is only of magnitude 4.7.

microwave background radiation SEE **cosmic microwave background**

midnight Sun The name given to the Sun when it is visible at midnight from inside the Arctic or Antarctic Circle. At the North Pole the Sun is continuously visible for the 6-month period for which it is north of the celestial equator, and continuously below the horizon for the 6 months it spends south of the celestial equator. A similar situation holds for the South Pole.

Milky Way The faint band of light visible on clear dark nights encircling the sky along the line of the galactic equator. It is the combined light of an enormous number of stars, in places obscured by clouds of interstellar gas and dust such as the **Coalsack**. It is in fact the disk of our **Galaxy**, viewed from our vantage point within it.

millimetre-wave astronomy Observations of **electromagnetic radiation** at wavelengths from about 1 to 10 mm, at the shortest end of the radio spectrum. Millimetre waves are given out by complicated molecules in space, which are found in dense clouds where stars are forming. Waves with a wavelength shorter than 1 mm are known as *submillimetre waves* and bridge the gap between radio and infrared astronomy.

millisecond pulsar A **pulsar** that flashes every few thousandths of a second, the first of which was discovered in 1982. Unlike normal pulsars, millisecond pulsars are not slowing down, and they have weak magnetic fields. They are thought to be old pulsars that have been speeded up by the influx of gas from another star.

Mills Cross One of the earliest types of radio array, designed by the Australian radio astronomer Bernard Mills (1920–). In a Mills Cross, the aerials are arranged in two lines at right angles, acting as an **interferometer** to improve resolution.

Mimas The smallest of Saturn's regular **satellites**, with a diameter of 392 km (244 miles), and the innermost, orbiting near the inner edge of the tenuous E Ring. It was discovered by William Herschel in 1789. Mimas is heavily cratered; the largest crater, named Herschel, is 130 km (80 miles) across.

minor axis The shortest diameter of an elliptical orbit.

minor planet An alternative term for an **asteroid**, which tends to be preferred by professional astronomers.

minute of arc SEE **arc minute, second**

Mira The star Omicron Ceti, a red giant that is the prototype of the **long-period variables**, also known as *Mira stars*. It was the first star discovered to vary in a periodic manner. David Fabricius, a Dutch astronomer, first noticed Mira at 3rd magnitude on 13 August 1596. A few months later it had vanished, but he saw it again in February 1609. In 1660 it was shown to vary in an approximate period of 11 months.

Mira's mean period is 332 days, but this is subject to irregularities. Its mean range is magnitude 3.5 to 9.1, but maxima as bright as 2.0 and minima as faint as 10.1 have been observed; its light curve is shown in the diagram. Its spectrum varies from M5 to M9. Mira has a faint companion, now known as VZ Ceti, a variable with a magnitude range of 9.5 to 12.0, possibly a white dwarf with an **accretion disk**. Mira lies about 200 light years away.

Miranda The smallest and innermost of the five main **satellites** of Uranus, discovered in 1948 by **Gerard Kuiper**. Its diameter is 472 km (293 miles) and it has the lowest density of the five, at 1.35 g/cm^3. Miranda displays one of the most complex and baffling surfaces ever seen. A rolling, cratered terrain containing both old and new craters is interrupted by well-defined regions called *coronae* containing parallel or concentric light and dark bands. One of these, Inverness Corona, contains a bright, tick-shaped feature known as the Chevron, and faults running from it lead to cliffs 20 km (12 miles) high. To explain these features, it has been suggested that after its formation Miranda was broken up by impact and then reassembled itself.

mirror An optical component of an instrument (which may be plane, spherical, ellipsoidal, paraboloidal or hyperboloidal) which redirects an incoming beam of light and may also modify it so as to aid the formation of an image of the light source. Mirrors are used in many astronomical instruments, in particular the **reflecting telescope**. They are made of glass or quartz, ground and polished to achieve the required accuracy of curvature, and coated with a thin layer of a highly reflective material. Aluminium is the usual coating material as it is cheap and easily deposited on the glass. The primary mirrors of reflecting telescopes were once made of a copper–tin alloy called *speculum*, and this term was sometimes used to refer to the mirror itself. A variety of sophisticated and exotic techniques, such as honeycomb

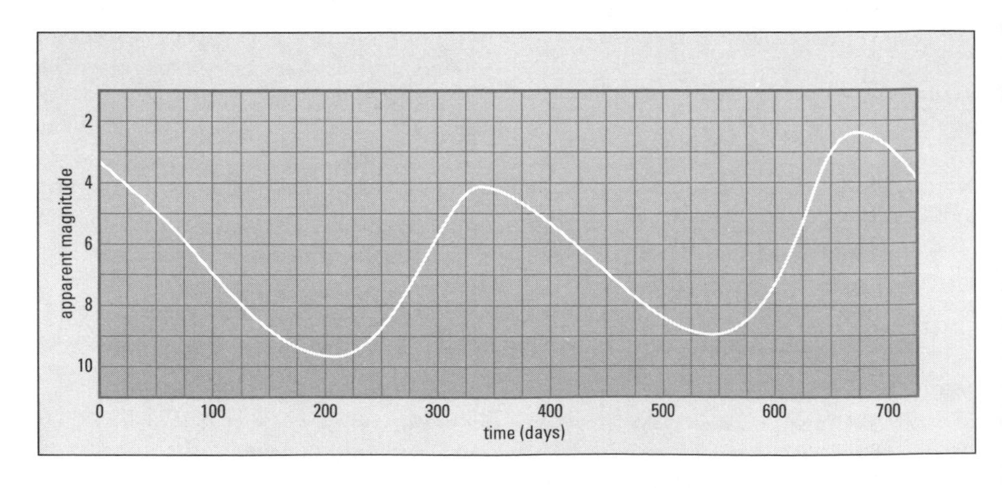

Mira: light curve, showing variation of maximum and minimum brightness

structures and the use of ceramic materials, are now being used to fashion primary mirrors with diameters of 10 m (33 ft) or more for professional reflecting telescopes.

missing mass Invisible or low-luminosity matter that surrounds individual galaxies and pervades clusters of galaxies. Analysis of the rotation of our Galaxy strongly suggests that the galactic halo is embedded in an enormous *corona* that is roughly 600,000 light years in diameter and contains at least 10^{12} solar masses of low-luminosity matter. Other galaxies exhibit similar rotation curves, so they too are presumed to be surrounded by large, massive coronae. In addition, clusters of galaxies must contain a substantial amount of non-luminous matter otherwise there would not be enough gravity to hold the clusters together.

The issue of the missing mass also arises on the cosmological scale. By measuring the rate at which the cosmic expansion is slowing down (the so-called **deceleration parameter**), astronomers can deduce the average density of matter in the Universe. Estimates of the cosmic deceleration indicate that the average density of matter is near the **critical density** of 5×10^{-30} g/cm^3, which is equivalent to about three hydrogen atoms per cubic metre of space. But the average density of matter that astronomers actually see in space is about 3×10^{-31} g/cm^3. Thus the observed matter is less than one-tenth that needed to account for the behaviour of the Universe. SEE ALSO **dark matter**

Mizar A double star in Ursa Major, also known as Zeta Ursae Majoris, of magnitude 2.3. Its wide companion Alcor, magnitude 4.0, can be seen with the naked eye. Mizar and Alcor are too far apart to be a genuine binary. However, with a small telescope a 4th-magnitude companion can be seen closer to Mizar, and this does form a very long-period binary with Mizar. In addition, Mizar, Alcor, and the companion to Mizar are all **spectroscopic binaries**.

MK system ABBREVIATION FOR **Morgan–Keenan classification**

MMT ABBREVIATION FOR **Multiple-Mirror Telescope**

mock Sun SEE **parhelion**

Monoceros A constellation, representing a unicorn, which adjoins Orion. Its brightest star, Alpha, is only of magnitude 3.9, but it is crossed by the Milky Way and contains some famous clusters and nebulae, including the **Rosette Nebula** and the open cluster M50. Also in Monoceros is the massive binary known as **Plaskett's Star**.

month The period of revolution of the Moon around the Earth. Various months, defined according to the choice of reference point, are listed in the table. The *calendar month* is one of the 12 divisions of the Gregorian calendar (SEE **calendar**), and is approximately equal in length to one synodic month. SEE ALSO the entries for each type of month listed in the table below.

TYPES OF MONTH	
anomalistic month	27.55455 mean solar days
draconic month	27.21222 mean solar days
sidereal month	27.32166 mean solar days
tropical month	27.32158 mean solar days

moon (with lower-case 'm') Another word for **satellite**.

Moon (with capital 'M') The Earth's only natural **satellite**. Apart from the Sun it is the brightest object in the sky because of its proximity, being at a mean distance of only 384,000 km (239,000 miles). Like the Sun, its apparent diameter is about half a degree. With an actual diameter of 3476 km (2160 miles) and a density of 3.34 g/cm^3, the Moon has 0.0123 of the Earth's mass and 0.0203 of its volume. The Earth and Moon revolve around their common centre of gravity, the **barycentre**. Although it is so bright (the full Moon has a visual magnitude of −12.7) its surface rocks are dark, and the Moon's albedo is only 0.07. It is a cratered world, in many ways a typical Solar System satellite. It is the only other world whose surface features we can easily see from the Earth, and the only other world on which humans have walked.

As the Moon orbits the Earth, it is seen to go through a sequence of **phases** as the proportion of the illuminated hemisphere visible to us changes. One complete sequence, from one new Moon, say, to the next, is called a *lunation*. At the crescent phase, the dark side of the Moon is seen to be faintly illuminated by *earthshine* – sunlight reflected from the Earth. At new Moon or full Moon, **eclipses** can occur. An observer on Earth always sees the same side of the Moon because the Moon has **synchronous rotation**: its orbital period (the **sidereal month**) around the Earth is the same as its axial rotation period. The visible side is called the *nearside*, and the side invisible from Earth is the *farside*. In fact, the face the Moon presents to us does vary slightly because of a number of effects known collectively as **libration**.

The Moon has been studied by many space probes: the **Luna** and **Zond** series launched by the former Soviet Union, and the US series **Ranger**, **Surveyor** and **Lunar Orbiter**. The manned **Apollo** missions and some of the later Luna probes returned samples of lunar material to the Earth for study. Most recently the **Clementine** probe gathered much valuable new information.

The surface features may be broadly divided into the darker maria (SEE **mare**), which are low-lying volcanic plains, and the brighter highland regions (sometimes called *terrae*), which are found predominantly in the southern part of the Moon's nearside and over the entire farside. There are **impact features** of all sizes. The largest are called *basins*, produced during the early history of the Moon when bombardment by impacting objects was at its heaviest. On the nearside, where the crust is thinner, some basins were subsequently filled with upwelling lava to produce the maria, which in turn became cratered. The greatest lunar depression, discovered by Clementine, is centred on the crater Aitken near the south pole, and is about 2500 km (1500 miles) in diameter – the largest such structure in the Solar System. The floor of Aitken is 13 km (8 miles) below its rim, the deepest point on the lunar surface. The smaller basins are similar to the largest craters, formerly called *walled plains*, which have flat floors and are surrounded by a ring of mountains, like the 298 km (185-mile) diameter crater Baily. Other features are mountain peaks and ranges, valleys, elongated depressions known as **rilles**, **wrinkle ridges**, low hills called *domes*, and bright **rays** of ejecta radiating from the sites of the more recent cratering impacts.

Early lunar cartographers assumed that the darker lunar features were expanses of water, and named them after fanciful oceans (oceanus), seas (mare), bays (sinus), lakes (lacus) and swamps (palus). Other features are named after famous people, principally astronomers and other scientists. The old names are still in use on lunar maps.

The origin of the Moon is uncertain. A current theory, known as the *giant impactor* or *big splash* theory, is that a Mars-sized body collided with the newly formed Earth, and debris from the impact accreted (SEE **accretion**) to form the Moon. One piece of evidence for this is the correspondence between certain radioactive isotopes in lunar and terrestrial rocks. Subsequent large impacts produced the Moon's basins, and smaller impacts the craters. A volcanic origin for the craters has long been dismissed by most astronomers, but Clementine photographed what appears to be a volcanic crater near the large farside crater Schrödinger. Mild *moonquakes* occur at depths of roughly 700 km (450 miles), and **transient lunar phenomena** appear to be associated with them; otherwise, the Moon is now geologically inactive.

Seismic measurements at Apollo landing sites yielded some information on the Moon's internal structure. The lunar crust is about 100 km (60 miles) thick in the highland regions, but only a few tens of kilometres under the mare basins. Under the largest basins the underlying mantle has bulged upwards to form **mascons**. An iron-rich core a few hundred kilometres across might exist, with a temperature at the centre of about 1500 K. Although there is now no significant overall magnetic field, some areas of the surface show a degree of magnetization, so the Moon could have had a stronger magnetic field in the past.

The chemical composition of material brought back from the Moon has been found to consist mainly of silica, iron oxide, aluminium oxide, calcium oxide, titanium dioxide, and magnesium oxide. Lunar rocks are igneous. The maria have a basaltic composition, while the highlands consist largely of anorthosite. As a result of the constant bombardment, crustal rocks are in the form of **breccias**, perhaps to a depth of several kilometres, and the surface is covered with a **regolith**. The Moon has only the most tenuous of atmospheres. Apollo instruments detected traces of gases such as helium, neon, and argon. The atmosphere is probably made up of **solar wind** particles retained temporarily, and atoms sputtered (knocked off) from the surface by the solar wind. Consequently the surface temperature variation is extreme: from 100 to 400 K. Results obtained by the Clementine probe suggest that small quantities of water-ice might exist on the permanently shadowed floors of craters near the Moon's poles.

Morgan–Keenan classification The categorization of a star's spectrum from its absorption features. The system was devised by William W. Morgan, Philip C. Keenan, and Edith Kellman of Yerkes Observatory for their *Atlas of Stellar Spectra* published in 1943; hence it is also known as the *MK system*, *MKK system*, or *Yerkes system*. It established formal rules for describing a star by spectral type (O, B, A, F, G, K, M, in sequence from hot to cool surface temperature) and luminosity class (I = supergiant, III = giant, V = dwarf; with interpolations Ia, Iab, Ib, II and IV). SEE ALSO **spectral classification**

morning star Name sometimes given to the planet Venus when it is visible in the east before sunrise.

mounting SEE **telescope, mounting of**

Mount Palomar Observatory SEE **Palomar Observatory**

Mount Stromlo Observatory An observatory near Canberra, operated by the Australian National University. It was founded in 1924 as a solar observatory, but changed focus to stellar astronomy after World War II. Its largest telescope is a 1.8 m (74-inch) reflector. It also operates telescopes at **Siding Spring Observatory**.

Mount Wilson Observatory An observatory in the San Gabriel Mountains near Pasadena, California, site of the 2.5 m (100-inch) Hooker reflector, opened in 1917, with which **Edwin Hubble** discovered the expansion of the Universe. The Mount Wilson Observatory was founded in 1904 by **George Ellery Hale**, originally for solar astronomy; a 1.5 m (60-inch) reflector was opened in 1908. The 2.5 m reflector was temporarily closed between 1985 and 1993 when the observatory's owners, the Carnegie Institution of Washington, withdrew support. The Observatory is now operated by the Mount Wilson Institute.

moving cluster A group of stars, often widely scattered, that share the same motion through the Galaxy. The stars in a moving cluster all follow nearly parallel paths. However, because of a perspective effect, they seem to be moving towards, or away from, a point known as the *convergent point*. This allows the distance of the cluster to be determined. First, the position of the convergent point must be established. Some clusters are very scattered, for example the Ursa Major moving cluster, which includes five stars of the Plough (Beta, Gamma, Delta, Epsilon and Zeta). Several other bright stars show the same motion through the Galaxy as these five Ursa Major stars. They include Sirius, Delta Leonis, Beta Aurigae, Beta Eridani, and Alpha Coronae Borealis. In total about 100 members of the group are known, scattered widely across the sky. Their separation in space is larger than in a normal cluster. Our Sun is moving through the outskirts of the group, which is why its members appear so widely spread across the sky.

M star A star of spectral type M, characterized by absorption bands of titanium oxide molecules. M-type dwarf stars, also known as red dwarfs, define the lower end of the **main sequence**. Proxima Centauri and Barnard's Star are examples. The surface temperatures of M-type dwarfs range from 3500 to 2500 K; their masses range from 0.08 to 0.5 solar masses, and their radii from 0.1 to 0.6 times that of the Sun. The most massive M-type dwarfs are nearly one-tenth as luminous as the Sun, while those at the bottom of the main sequence have less than one-thousandth of the Sun's luminosity. Most of their radiation is emitted in the infrared, so their visual luminosity is even fainter, from one twenty-fifth to one ten-thousandth that of the Sun. Many M-type dwarfs are **flare stars**; none are visible to the naked eye.

M-type giants and supergiants are all evolved stars with masses much greater than those of M-type dwarfs. Their surface temperatures are typically a few hundred degrees cooler than dwarfs of the same spectral type. The brightest M-type supergiants are up to a million times as luminous as the Sun. The largest of them are several thousand times the Sun's diameter – as large as the orbit of Saturn or Uranus. Even the coolest M-type giants are as large as the Earth's orbit. Such distended stars are unstable and therefore variable; examples are **Betelgeuse** and **Mira**. The high luminosity and low surface gravity of M-type giants and supergiants causes a **stellar wind** of gas in which the mass loss can be as high as one solar mass every 100,000 years.

Mullard Radio Astronomy Observatory The radio astronomy observatory of the University of Cambridge, situated at Lords Bridge near Cambridge. It was here that the technique of **aperture synthesis** was developed. The Observatory's main instrument is the Five-Kilometre Telescope (opened in 1972), consisting of four fixed and four movable dishes 13 m (43 ft) in diameter, arranged in an east–west line. The One-Mile Telescope (opened in 1964) consists of three dishes 18 m (60 ft) in diameter, one movable along 0.8 km (0.5 mile) of track. This same track is shared by two movable dishes of the Half-Mile Telescope (opened in 1968), which consists of a total of four 9 m (30-ft) aerials.

Müller, Johann (also known as **Regiomontanus**) (1436–76) German astronomer and mathematician. In 1463 he completed the translation into Latin of the ***Almagest*** begun by his tutor, **Georg von Purbach**. In 1471 he set up a printing press and became one of the first publishers of astronomical and scientific books. There is some evidence that he had the idea of a heliocentric universe, and that this reached and therefore influenced **Nicholas Copernicus**.

Multi-Element Radio-Linked Interferometer Network SEE **MERLIN**

Multiple-Mirror Telescope (MMT) An innovative reflecting telescope, using six 1.8 m (72-inch) mirrors arranged in a hexagon on an altazimuth mount. The six images are brought together at a common focus, giving the light-gathering power of a 4.5 m (176-inch) telescope. The MMT was opened in 1979 on Mt Hopkins, near Tucson, Arizona. The six mirrors are due to be replaced by a single 6.5 m (255-inch) mirror in 1995, but the instrument will still be known as the MMT.

multiple star A gravitationally connected group of more than two stars. **Castor**, for example, has six components. In such systems there is a recognizable hierarchy or order. In Castor, for instance, the two main pairs rotate about a common centre of gravity while a pair of cool red dwarfs rotates about the same centre but at a larger distance. There is no clear distinction between a multiple star with many members and a small cluster. A good example is the **Trapezium** in the Orion Nebula, which consists of four young stars enmeshed in nebulosity with a further five possible members nearby.

Musca A small but distinctive southern constellation adjoining Crux, representing a fly. Its brightest star, Alpha, is magnitude 2.7. The globular cluster NGC 4833 is visible with binoculars.

N

nadir The point on the **celestial sphere** vertically below the observer. It is diametrically opposite the zenith.

Nagler eyepiece A telescope eyepiece with a very wide field of view, 80° or more, making it suitable for activities such as searching for comets. (After its designer, Albert Nagler, 1935– .)

Naiad One of the small inner **satellites** of Neptune discovered in 1989 during the Voyager 2 mission.

naked-eye A term applied to observations made by the eye alone, without the aid of any optical instrument, or to celestial objects thus visible.

NASA ABBREVIATION FOR **National Aeronautics and Space Administration**

Nasmyth focus A focus of a refelcting telescope on an **altazimuth** mount, situated to one side on the altitude axis. The light beam is directed there by an extra flat mirror angled at 45°. Heavy instruments positioned at a Nasmyth focus do not have to be moved bodily as the telescope tracks an object. It is named after Scottish engineer James Nasmyth (1808–90), who incorporated it in his 6-inch (150 mm) reflector. Its use has been revived with the present generation of large, altazimuthally mounted professional instruments. SEE ALSO **coudé**.

National Aeronautics and Space Administration (NASA) The US government agency for the peaceful exploration of space, formed in 1958. Its headquarters are in Washington, DC, and it operates various field stations. These include the Goddard Space Flight Center at Greenbelt, Maryland, which handles space science research and Earth satellite tracking; the Jet Propulsion Laboratory in Pasadena, California, which is mission control for NASA space probes and manages NASA's deep-space tracking stations around the world; the Ames Research Center, San Francisco, for aeronautics and planetary science; the John F. Kennedy Space Center at Cape Canaveral, Florida, the main US launch site; and the Lyndon B. Johnson Space Center in Houston, Texas, mission control for manned space missions.

National Radio Astronomy Observatory (NRAO) The major US radio observatory, with its headquarters at Green Bank, West Virginia. Its instruments include a 42.7 m (140 ft) radio dish and an interferometer consisting of three 26 m (85 ft) dishes plus one of 13.7 m (45 ft). A 90 m (300 ft) dish at Green Bank collapsed in 1988 and is being replaced by a new one of 100 m (330 ft) diameter, due for completion in 1995. NRAO also operates a 12 m (39 ft) dish for millimetre-wave studies at Kitt Peak, Arizona, and the **Very Large Array** in New Mexico.

nautical twilight SEE **twilight**

***n*-body problem** SEE **many-body problem**

neap tide SEE **tides**

near-Earth asteroid (NEA) A small **Amor asteroid**, **Apollo asteroid** or **Aten asteroid** whose orbit brings it relatively close to the Earth. The closest approach so far detected was by asteroid 1994 XM_1, about 10 m (33 ft) across, which in 1994 passed just 100,000 km (60,000 miles) away. Objects this size striking the Earth would burn up in the atmosphere; larger objects would penetrate further, causing a powerful low-level airburst like the **Tun-**

guska Event, or producing a crater like **Meteor Crater**.

Near Earth Asteroid Rendezvous (NEAR) A NASA space probe to the asteroid **Eros**, due for launch in February 1996. It should arrive at the end of 1998 and go into orbit around Eros. On the way, NEAR will fly past a small member of the main asteroid belt, 2968 Iliya.

nebula (plural **nebulae**) A region of interstellar gas and dust. The word nebula is Latin for 'cloud'. There are three main types of nebula. **Emission nebulae** are bright diffuse nebulae which emit light and other radiation as a result of ionization (removal of electrons) and excitation of the gas atoms by ultraviolet radiation. The source of the ultraviolet is usually one or more hot stars. When gas ions recombine with free electrons, and forbidden transitions (SEE **forbidden lines**) occur in excited atoms, radiation is emitted, giving rise to emission lines in the spectrum. Examples of emission nebulae are **H II regions** such as the Orion Nebula, and **planetary nebulae** such as the Ring Nebula in Lyra. **Supernova remnants** are a form of emission nebula in which the gas is made to glow not by the ultraviolet radiation of the star within, but by frictional heating as it collides with surrounding interstellar gas; an example is the Cygnus Loop. The Crab Nebula is a type of supernova remnant that shines by the mechanism of **synchrotron radiation**. In contrast, the brightness of **reflection nebulae** results from the scattering by dust particles of light from nearby stars. **Dark nebulae** are not luminous: interstellar gas and dust absorbs light from background stars, producing apparently dark patches in the sky. This third class includes a group known as **Bok globules**, which are dense absorption nebulae of nearly spherical shape.

Neptune The eighth major planet from the Sun, and the smallest and most distant of the four giant planets. With a mean magnitude of 7.8, Neptune is invisible to the naked eye. Its average apparent size is 2.2 arc seconds. Through a telescope it appears as a small greenish blue disk on which very few details can be distinguished. In its size, mass, atmosphere and colour, Neptune resembles Uranus. Like all the giant planets, it possesses a ring system and a retinue of satellites. The main data for Neptune are given in the table below.

Neptune was first identified by **Johann Galle**, assisted by Heinrich D'Arrest, on 23 September 1846, close to a position predicted by **Urbain Le Verrier**. Le Verrier and, independently, **John Couch Adams**, had calculated the mass, orbit, and position of an unseen planet to account for perturbations in the orbital motion of Uranus.

The fly-by of the Voyager 2 probe in 1989 provided most of our current knowledge of the planet. The upper atmosphere is about 85% molecular hydrogen and 15% helium. Its predominant blue colour is due to a trace

NEPTUNE: DATA

Globe	
Diameter (equatorial)	49,528 km
Diameter (polar)	48,686 km
Density	1.64 g/cm^3
Mass (Earth = 1)	17.14
Volume (Earth = 1)	57.67
Sidereal period of axial rotation	19h 10m
Escape velocity	23.3 km/s
Albedo	0.84
Inclination of equator to orbit	29° 34′
Temperature at cloud-tops	55 K
Surface gravity (Earth = 1)	0.98
Orbit	
Semimajor axis	30.06 AU = 4497 × 10^6 km
Eccentricity	0.0097
Inclination to ecliptic	1° 46′
Sidereal period of revolution	164.79y
Mean orbital velocity	5.43 km/s
Satellites	8

of methane, which strongly absorbs red light. Several different atmospheric features were visible at the time of the Voyager encounter. There were faint bands parallel to the equator, and spots, the most prominent of which was the oval Great Dark Spot (GDS), about 12,500 km (8000 miles) long and 7500 km (4500 miles) wide. It is about the same size relative to Neptune as the **Great Red Spot** (GRS) is to Jupiter, and like the GRS it is a giant anticyclone. White, cirrus-type clouds of methane crystals cast shadows on the main cloud deck some 50 km (30 miles) below. Neptune's atmosphere has **differential rotation**, some features taking 19 hours or more for one rotation. The atmosphere thus has retrograde rotation with respect to the interior, which rotates in just over 16 hours. There are also the highest wind speeds recorded on any planet – over 2000 km/h (1250 mile/h) in places.

The GDS was not visible on images obtained in 1993 and 1994 from ground-based telescopes and the Hubble Space Telescope, which also revealed a bright band not seen by Voyager. Neptune's dynamic and changing atmosphere probably has to do with its internal heat source – it radiates more than twice as much heat as it receives from the Sun. The amount of white cloud also varies with the **solar cycle**, which is hard to explain for such a distant planet.

Below the visible atmosphere the composition and structure are less certain. A chemical cycle, driven by ultraviolet radiation from the Sun, might operate in which methane is converted into other hydrocarbons and back again. At deeper levels there may be a mantle of water and ammonia about six times the mass of the Earth, surrounding an iron–silicate core of about one Earth mass. Like Uranus, Neptune has a magnetic field in which the axis is markedly tilted (by 47°) to the axis of rotation and is displaced (by about 10,000 km/6000 miles) from the planet's centre. In Neptune's case it could be that the mantle is at least partially fluid and the ammonia–water mixture is ionized, and therefore electrically conducting, and that this is the source of the magnetic field.

Of Neptune's eight known **satellites, Triton** and **Nereid** were discovered before the Voyager 2 fly-by, and another (**Larissa**) was suspected from an **occultation** measurement. Triton's orbital motion is **retrograde**, and Nereid's is the most eccentric of any Solar System satellite. The six small satellites imaged for the first time by Voyager are all in close, regular orbits.

Neptune's ring system was detected from Earth during an **occultation** in 1984, but the measurements suggested a number of incomplete 'ring arcs'. Voyager revealed four continuous rings: two narrow ones, and two faint broad ones. The outermost and brightest, the Adams Ring, contains 'clumps' of material that appeared as arcs in the occultation data. SEE ALSO **ring, planetary**

Nereid Neptune's outermost **satellite**, discovered by **Gerard Kuiper** in 1949. It has the most eccentric orbit of all known satellites, taking it from 1.4 million km (850,000 miles) out to 9.7 million km (6 million miles). It is almost certainly a captured body.

neutron star An extremely small, dense star that consists mostly of neutrons. Neutron stars are observed as **pulsars** and are also thought to be one of the components in **X-ray binaries**. They are formed when a massive star explodes as a supernova, blasting off its outer layers and compressing the core so that its component protons and electrons merge into neutrons. Neutron stars have masses comparable to that of the Sun, but diameters of only about 20 km (12 miles) and average densities of about 10^{15} g/cm^3. A neutron star's gravity is so great that objects would weigh 10^{11} times more at its surface than on the surface of the Earth, and it also produces a significant **gravitational redshift** (z about 0.2). The maximum mass for a neutron star is about 3 solar masses, beyond which it would collapse further, into a **black hole**.

Being so small and having such strong gravity, neutron stars can spin rapidly without disintegrating – up to several hundred times per second. They retain the magnetic field of the original star but compressed into a much smaller volume, so the magnetic fields of neutron stars are about a million million times stronger than that of the Earth. A neutron star's structure is complicated because of the great variation in pressure from the surface to the centre. The crust is made of iron about 10,000 times denser than terrestrial iron and a million times stronger than steel. Beneath this is a liquid of neutrons, and at the very centre may be a core of exotic nuclear particles.

Newcomb, Simon (1835–1909) Canadian-born American mathematical astronomer. He calculated highly accurate values for astronomical constants, such as the solar **parallax**, and prepared extremely accurate tables of the motions of the Moon and the planets. It was Newcomb who calculated the discrepancy in the advance of **Mercury**'s perihelion that was to provide a proof of **Einstein**'s Special Theory of Relativity.

new Moon The **phase** of the Moon when at conjunction. The dark side of the Moon then faces the Earth, and the illuminated side is invisble to us.

New Technology Telescope A 3.6 m (142-inch) reflector at the **European Southern Observatory** in Chile, completed in 1989. It uses a system of **active optics** to keep the main mirror in shape.

Newton, Isaac (1642–1727) English physicist and mathematician. His achievements were so important for astronomy that the phrase 'Newtonian revolution' is often used. Most of his theories – on light, gravitation, and calculus, for example – he developed in basic form while in his early twenties. In 1668 he built the first reflecting telescope in order to overcome the **chromatic aberration** inherent in the lenses used in refracting telescopes. In 1684, encouraged by **Edmond Halley**, he began work on his *Philosophiae naturalis principia mathematica*. Published in 1687, and usually referred to as the *Principia*, this was one of the most epoch-making books in the history of science. In it he put forward his law of **gravitation** and laws of motion (SEE **Newton's laws of motion**), and showed that the orbit of a body moving under an **inverse-square law** of gravitation would be an ellipse. He went on to demonstrate how gravitation explained the second and third of **Kepler's laws** and also many outstanding problems in astronomy. In his *Opticks* (1704) he set out his theories of light, including how white light is made up of the colours of the spectrum. Newton's mathematical achievements were many, but his invention of calculus (independently of Gottfried Leibniz) was the greatest.

Newtonian telescope The simplest form of **reflecting telescope**. The first, built by

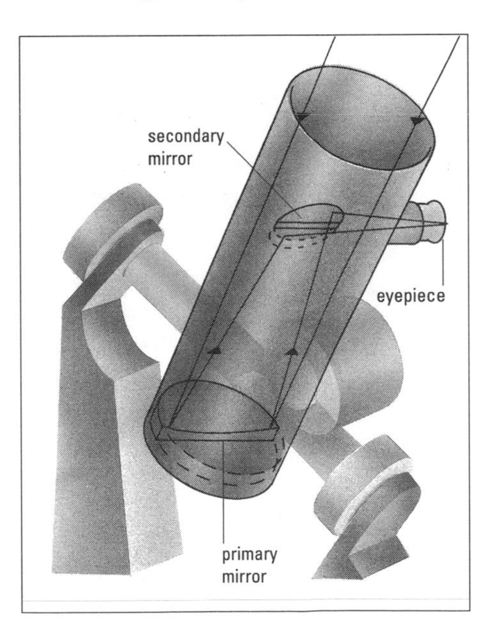

Newtonian telescope: light path

Isaac Newton in 1668, had a primary concave spherical mirror of speculum metal (an alloy of copper and tin) which reflected the light up the tube of the telescope. A flat speculum mirror near the top of the tube, angled at 45°, directed the light to a focus – the *Newtonian focus* – to one side of the tube where the image of the object being observed was viewed through a small planoconvex eyepiece. This is still the basic design, although speculum has given way to glass, and for all but the smallest apertures the primary mirror is now a **paraboloid**. The Newtonian has long been a favourite with amateur astronomers.

Newton's law of gravitation SEE **gravitation**

Newton's laws of motion The fundamental laws of dynamics, set out by Isaac Newton in his *Principia*:

1. *Law of inertia*: Every body continues in its state of rest, or of uniform motion in a straight line, until acted upon by some outside force.
2. *Law of acceleration*: The acceleration of a body is directly proportional to the force acting upon it, and is in the direction in which the force is acting.
3. *Law of action and reaction*: To any action there is an equal and opposite reaction.

N galaxy A type of active galaxy with a small, bright nucleus, intermediate in its properties between a **Seyfert galaxy** and a **quasar**. Like quasars, N galaxies are often radio sources and show similar broad emission lines and variability. Their redshifts are somewhat lower than quasars, since they are closer and less luminous, making it easier to see the stars in their outer parts.

NGC Abbreviation for the *New General Catalogue of Nebulae and Clusters of Stars*, published in 1888 by **J.L.E. Dreyer**, and incorporating previous catalogues by **William** and **John Herschel**. The NGC contains nearly 8000 objects – including galaxies, although their real nature was not then known – and another 5000 are listed in two subsequent ***Index Catalogues*** (IC). Objects are still referred to by their NGC or IC numbers.

noctilucent cloud (NLC) A cloud at very high altitude (average 82 km/51 miles) in the Earth's atmosphere, visible in the summer in twilight hours (noctilucent means 'night-shining') from latitudes between 50° and 70°. NLCs have a pearly or blue colour and resemble delicately interwoven cirrus cloud. They are made up of water droplets that have been carried aloft and then condensed and frozen on to particles. It is not known exactly what these particles are; they may be ions produced by solar radiation, or they may be **micrometeorites**, or even products of industrial pollution. NLCs have become more common in recent years.

node The point at which one orbit cuts another. Specifically, a node is one of the two points in the orbit of a planet or comet at which it cuts the ecliptic, or at which the orbit of a satellite cuts that of its primary. The *ascending node* is the node at which a celestial body passes from south to north; the *descending node* is the node at which a celestial body passes from north to south. The *line of nodes* joins the ascending and descending nodes. The *regression of the nodes* is the backward motion of the Moon's nodes around its orbit. They move slowly westward so that the nodal line makes one complete revolution of the orbit in 18.6 years. The regression is caused by the Sun's gravitational attraction.

non-gravitational force Any force acting on a celestial body which is not gravitational, in particular one which produces similar effects to gravity. The term is most commonly used in the context of comets which, on their approach to perihelion, lose volatile material through one or a small number of vents in their crust, producing a jet or jets in a particular direction. This can alter their course in much the same way as the gravitational **perturbations** to which they are prone. The **Poynting–Robertson effect** results from another type of non-gravitational force. SEE ALSO **magnetohydrodynamics**

Norma A small southern constellation between Ara and Lupus, representing a surveyor's level. It contains no star above magnitude 4.0.

North America Nebula A large emission nebula in the constellation Cygnus. It is so called because its shape resembles the continent of North America. It is also known as NGC 7000. An associated region of nebulosity, IC 5070, is the *Pelican Nebula.*

Northern Cross A popular name sometimes given to the cross-shaped arrangement of the brightest stars in the constellation Cygnus.

North Polar Sequence A list of stars with accurately measured magnitudes within 2° of the north celestial pole. It includes stars from magnitude 2 down to magnitude 20. Comparison with the North Polar Sequence enables the magnitudes of other stars to be determined.

nova (plural **novae**) An existing star, usually quite faint, whose brightness suddenly increases by 10 magnitudes or more before slowly returning to its pre-nova state. A typical nova rises to maximum in a few days, and declines thereafter in brightness by a factor of about ten in 40 days, although cases of both slower and faster declines are known. The diagram below shows the light curve of the brightest nova (as opposed to **supernova**) of modern times, Nova Aquilae in 1918. About 40 novae are estimated to explode in the Galaxy each year, although we see only a small percentage of them.

Novae occur in close binary systems where one component is a white dwarf and the other is a giant. Matter drains from the surface of the giant, where gravity is weak, to the surface of the white dwarf, where gravity is strong. Nuclear reactions involving the infalling material cause a violent explosion on the surface of the white dwarf. The energy of a typical explosion is about the same as the Sun emits in 10,000 years. Clouds of gas are ejected at speeds around 1500 km/s (900 mile/s) during the outburst.

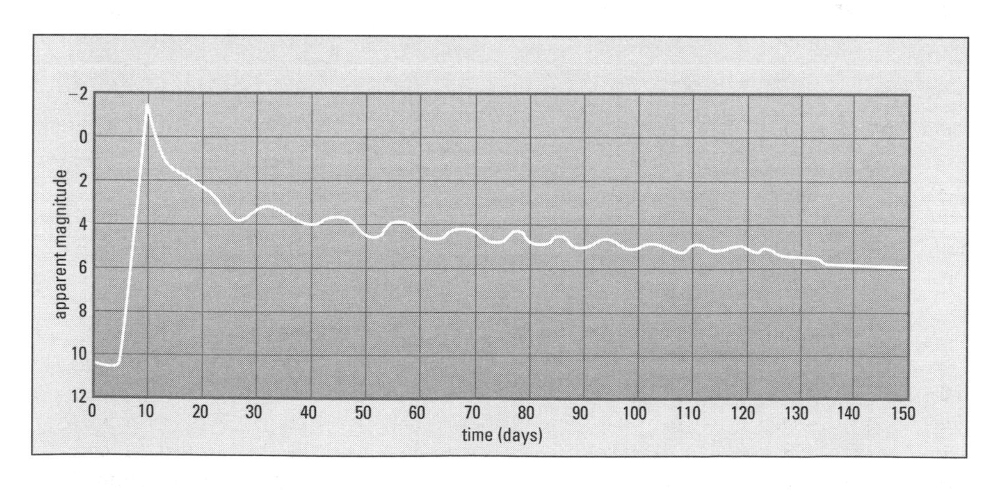

Nova: light curve, showing steep rise in brightness, and gradual decline

The total amount of material lost is estimated to be about one ten-thousandth the mass of the white dwarf. Some stars, known as **recurrent novae**, undergo explosions repeatedly. Given time, all novae probably recur. SEE ALSO **dwarf nova**

Nubecula Major, Nubecula Minor Obsolete names for the Large and Small **Magellanic Clouds**.

nucleosynthesis The production of chemical elements from other chemical elements via naturally occurring nuclear reactions. These reactions are of two types: *fusion*, in which heavier elements are built up from lighter ones, and *fission*, in which heavier elements are broken down into lighter ones. Fusion created most of the Universe's helium from hydrogen in the first few minutes following the **Big Bang**, and fusion processes such as the **carbon–nitrogen cycle** operate inside stars to create helium and heavier elements which are carried into intersteller space by stellar winds. Both fusion and fission occur in the extreme conditions of novae and supernovae to produce many different elements which are then ejected into the surrounding space.

nucleus (1) The solid part of a **comet**.

nucleus (2) The central region of a **galaxy**.

nutation The periodic oscillation in the precessional motion of the Earth's axis of rotation (SEE **precession**). It was discovered by **James Bradley** in 1747. It is in effect a slight 'nodding' of the Earth's axis, caused by the combined effect of the gravitational attractions of the Sun and Moon upon the Earth. The effect is constantly varying as the relative positions of the three bodies are continuously changing. *Lunar nutation*, which causes the Earth's axis to describe an ellipse, has a period of 18.6 years. *Solar nutation* has a period of 0.5 of a tropical year. The *fortnightly nutation* has a period of 15 days.

OB association A region of space in which recent star formation has produced high-mass O and B stars. Such associations are less tightly bound than a normal star cluster and are slowly dispersing. Since O and B stars are very luminous, OB associations are recognizable at great distances, even through intervening dust. Tens or hundreds of these massive stars can occur in a single association. Fainter, less massive stars may also exist, but are harder to see at great distances. The Orion Nebula is part of an OB association that is relatively close to us, so young, low-mass stars are also detectable within it. OB associations are a feature of the gas-rich spiral and irregular galaxies, but not of elliptical galaxies, in which no young stars are found. COMPARE **T association**

Oberon One of the five main **satellites** of Uranus, discovered in 1787 by William Herschel. It is 1524 km (947 miles) in diameter and consists largely of a mixture of rock and water-ice, with a density of 1.58 g/cm^3. Oberon has many craters, some with **rays** and some with dark floors, and mountains, one of which appears to be 20 km (12 miles) high. The high density of craters is evidence of a world unaltered by geological activity.

objective The main light-gathering element of an optical instrument. In a refracting telescope it is the lens or lens system (called the *object glass* in older texts) that collects and focuses light from a celestial body to form an image of it.

objective prism A large, narrow-angle prism mounted across the aperture of a telescope so that the image of each star in the field of view is transformed into a low-resolution spectrum. From photographs taken by this means it is possible to rapidly assign spectral classifications to the stars in a given area.

oblateness (ellipticity) A measure of how far a **spheroid** deviates from a true sphere, given by the ratio of the difference between the equatorial and polar radii to the equatorial radius. The oblateness of a celestial body is an indication of the speed at which it is rotating.

oblate spheroid SEE **spheroid**

obliquity of the ecliptic The angle between the plane of the ecliptic and the celestial equator. Alternatively, it may be defined as the angle between the axis of rotation of a planet and the pole of its orbit. The Earth's obliquity is currently about 23° 26′, and is decreasing by 0.47′ per year. The angle will start to increase again in about 1500 years; the maximum range is 21° 55′ to 28° 18′, in a period of 40,000 years. The obliquity is responsible for the seasons. It represents the greatest angular distance that the Sun can lie north and south of the equator.

observatory A structure that is built and equipped for making celestial observations and measurements, acting principally as a housing for a telescope. The best location for astronomical purposes is on a mountain or high plateau where there are many cloudless or near-cloudless nights. The most favourable locations include Hawaii, Chile, western parts of the USA, and the Canary Islands.

Professional optical and infrared telescopes are installed in buildings with a rotating roof, usually dome-shaped, although square buildings are now becoming fashionable, and in some cases the entire building may rotate. The dome has a slit, closed by sliding shutters when not in use, through which observations are made. In addition there are control rooms, laboratories, photographic darkrooms, workshops, living accommodation for personnel, and other facilities. There are over 200 professional optical observatories throughout the world.

Sites for radio telescopes are less critical than for optical observatories because radio waves are not affected so much by the atmosphere. Radio telescopes consisting of an array of antennae do, however, require a considerable area of land.

The latest observatories are those being launched into Earth orbit, entirely above the absorbing and distorting effects of the Earth's atmosphere. In addition to optical and radio studies these craft carry out measurements in the infrared, ultraviolet, X-ray and gamma-ray regions of the spectrum which do not penetrate the Earth's atmosphere and so cannot be studied from ground-based observatories. SEE ALSO the names of individual observatories.

occultation The temporary cutting off of the light from one celestial body as another, nearer one passes in front of it. For example, the Moon may pass in front of a star; a planet may do the same and may also occult its own satellites, as in the case of Jupiter. The rings of **Uranus** were first detected when the planet occulted a star in 1977. Strictly speaking, a solar eclipse is an occultation of the Sun by the Moon. Accurate timings of lunar occultations are valuable for accurately determining the Moon's position and can also provide information about the occulted body. *Grazing occultations*, in which the star appears to skim the Moon's limb, are of particular interest.

Octans A faint constellation, notable only because it contains the south celestial pole. The nearest naked-eye star to the pole is Sigma Octantis, magnitude 5.4. Octans was introduced by **Lacaille** and represents an octant, the forerunner of the sextant. The brightest star in the constellation is Nu, magnitude 3.8.

ocular Another word for **eyepiece**.

Olbers, Heinrich Wilhelm Matthäus (1758–1840) German doctor and amateur astronomer. He is best known for having posed the question now known as **Olbers' paradox**. He made the second and fourth asteroid discoveries, of Pallas (1802) and Vesta (1807). Olbers invented a method for calculating cometary orbits, and discovered several comets. He also proposed the theory that the tails of comets are highly rarified matter expelled from the head of the comet by pressure of some kind from the Sun, thus anticipating the concept of **radiation pressure**.

Olbers' paradox The paradox that arises in attempting to answer the question, 'Why is the sky dark at night?' If space is infinite and uniformly filled with galaxies, then an observer would end up looking at the surface of a star or galaxy at every point in the sky, so the entire night sky should be bright. In fact, space is not infinite. The Universe is believed to have been born in the **Big Bang** about 15,000 million years ago, and so light from the most distant objects has not yet reached us. In addition, the light from distant galaxies is weakened by the **redshift** caused by the expansion of the Universe. The paradox is named after **Heinrich Olbers**, who drew attention to it in 1823.

Olympus Mons An extinct volcano on Mars, and the largest known volcano in the Solar System, rising to 27 km (17 miles) above the planet's mean surface level. It is about 700 km (450 miles) across, including the surrounding area of eroded volcanic rock. The caldera (summit crater) is over 80 km (50 miles) across.

Omega Centauri The brightest globular cluster in the sky, visible to the naked eye as a fuzzy star of magnitude 3.7 in the constellation Centaurus. It is 17,000 light years away, one of the nearer globular clusters. It contains several hundred thousand stars in a volume 200 light years across.

Oort, Jan Hendrik (1900–92) Dutch astronomer. He carried out research on the structure and dynamics of stellar systems, especially the Galaxy, whose rotation he discovered in 1927. He was also a pioneer of radio astronomy, in particular with respect to the 21 cm radiation (SEE **twenty-one centimetre line**) of interstellar hydrogen, collaborating with **Hendrik van de Hulst** to map the Galaxy at this wavelength. In 1950 he proposed the existence of what has come to be called the **Oort Cloud** of comets.

Oort Cloud (Oort–Öpik Cloud) The region of space surrounding the Solar System in which **comets** are thought to reside. It is spherical, extending halfway to the nearest star, with the greatest concentration of comets in a torus-shaped region perhaps 10,000 to 20,000 AU from the Sun. The size and structure of the Oort Cloud have been worked out from statistical studies of comets' orbits, although there is no direct evidence for its existence. The idea was put forward by Ernst Julius Öpik (1893–1985) in 1932, and first worked out in detail by **Jan Oort** in 1950. SEE ALSO **Kuiper Belt**

open cluster A group of young stars in the spiral arms of our Galaxy, containing from a few tens of stars to a few thousand. Open clusters are also known as *galactic clusters*. They are usually several light years across. Examples are the **Hyades** and **Pleiades** in Taurus, and **Praesepe** in Cancer. Related to open clusters are *associations*, loose groupings of stars of common origin. Often one of more open clusters are found in the central parts of associations. SEE ALSO **star cluster**, **OB association**, **T association**

Ophelia One of the small inner **satellites** of Uranus discovered in 1986 during the Voyager 2 mission. It and Cordelia act as **shepherd moons** to the planet's Epsilon Ring.

Ophiuchus A large constellation spanning the celestial equator, commemorating Aesculapius, a great healer. Its brightest star, Alpha (Rasalhague), is magnitude 2.1 and lies near Alpha Herculis. RS Ophiuchi, a **recurrent nova**, flared up to naked-eye brightness in 1898, 1933, 1958, 1967 and 1985. Rho Ophiuchi is a celebrated multiple star, and 70 Ophiuchi is a well-known binary, magnitudes 4.2 and 6.0, orbit 88 years. The constellation contains several fairly bright globular clusters: M9, M10, M12, M14, M19, M62 and M107. It also contains **Barnard's Star**.

opposition The alignment of the Earth and Sun with another body in the Solar System so that the body lies exactly opposite the Sun in the sky, or the time when this occurs. The term is used mostly for alignments with the major planets, although it is also applied to asteroids. Only a **superior planet** can come to opposition (SEE the diagram at **elongation**). Opposition is the most favourable time to observe a body, for it is then closer to the Earth than at any other point in its orbit.

optical double Two stars that appear to be close together, but are not components of a binary system. They appear close together because they happen to be in almost the same line of sight, but are in fact separated by great distances.

optical interferometer An instrument for studying a celestial object by **interferometry** at optical wavelengths. The first was developed by Albert Michelson in the 1920s. It consisted of mirrors at either end of a steel beam placed across the aperture of the 2.5 m (100-inch) telescope at Mount Wilson Observatory, and with it he measured the diameters of a few large stars by examining the interference pattern formed in the eyepiece. Present-day systems working on the same principle use separate telescopes linked either electronically or by laser beams, and

photometers in place of the human eye. Another type is the *speckle interferometer*, which uses an image enhancer or a CCD to capture a succession of very-short-exposure, high-magnification pictures. These individual pictures, which 'freeze' the effects of atmospheric turbulence, are made up of a large number of 'speckles'. Computer image-processing combines these pictures and yields an image from which the effects of the atmosphere are electronically subtracted. Other types, such as the **Fabry–Pérot interferometer**, exploit interference to capture high-resolution spectra.

orbit The path of a celestial body in a gravitational field. The path is usually a closed one about the focus of the system to which it belongs, as with those of the planets about the Sun, or the components of a binary system about their common centre of mass. Most celestial orbits are elliptical, although the eccentricity can vary greatly. It is rare for an orbit to be parabolic or hyperbolic. To define the size, shape, and orientation of the orbit, seven quantities must be determined by observation. These are known as *orbital elements*. In the case of a planetary orbit, only six are needed. They are as follows:

a the semimajor axis, in AU,
e the eccentricity,
i the inclination of the orbital plane to the plane of the ecliptic,
Ω the longitude of the ascending node,
ω the longitude of perihelion,
T the epoch (the time of perihelion passage).

In the case of the orbit of a binary star system a seventh element is needed, if the mass is not known; this is the period (P), or alternatively the mean motion (n).

In general, the same considerations apply in determining the orbit of a satellite as in the case of determining the orbit of a planet, except that the inclination is usually referred to the equatorial plane of the primary planet, instead of to the plane of the ecliptic. SEE ALSO **ellipse, parabola, hyperbola**

orbital element One of the quantities needed to define an orbit, or the position of a body in its orbit at a given time. SEE **orbit**

Orion Perhaps the most splendid of all the constellations, representing a hunter. Its brightest stars are **Betelgeuse** and **Rigel**. Gamma (Bellatrix) is a giant of type B2, magnitude 1.6, 360 light years away. Orion is so distinctive as to be unmistakable. A line of three stars – Zeta (Alnitak), magnitude 1.8; Epsilon (Alnilam), magnitude 1.7; and Delta (Mintaka), magnitude 2.2 – make up Orion's Belt. The celestial equator passes closely north of Mintaka. From the belt hangs Orion's Sword, marked by the **Orion Nebula**. Within the nebula lies Theta, a multiple star known as the **Trapezium**. The **Horsehead Nebula** is a dark nebula near Zeta Orionis.

Orion Arm A spiral arm of the **Galaxy**, about 30,000 light years from the galactic centre, also known as the *local arm*. The Sun is located on the inner side of a part of this arm known as the *Orion Spur*.

Orionids A **meteor shower** that occurs in October. Its radiant is near the border of the constellations Orion and Gemini, close to the star Gamma Geminorum. Like the Eta Aquarids, the Orionids are associated with Halley's Comet.

Orion Nebula A gaseous emission nebula, visible to the naked eye as a diffuse glow marking Orion's Sword. It lies about 1500 light years away. It is an **H II region** lit up by a group of new-born stars called the Trapezium, probably less than one million years old. Because they are very hot, about 50,000 K, these stars emit most of their energy in the ultraviolet, which makes the nebula glow. The Trapezium stars have destroyed most of the volatile dust in the nebula, etching out a roughly spherical hollow that is still growing. The visible Orion Nebula is therefore a hole in a much larger

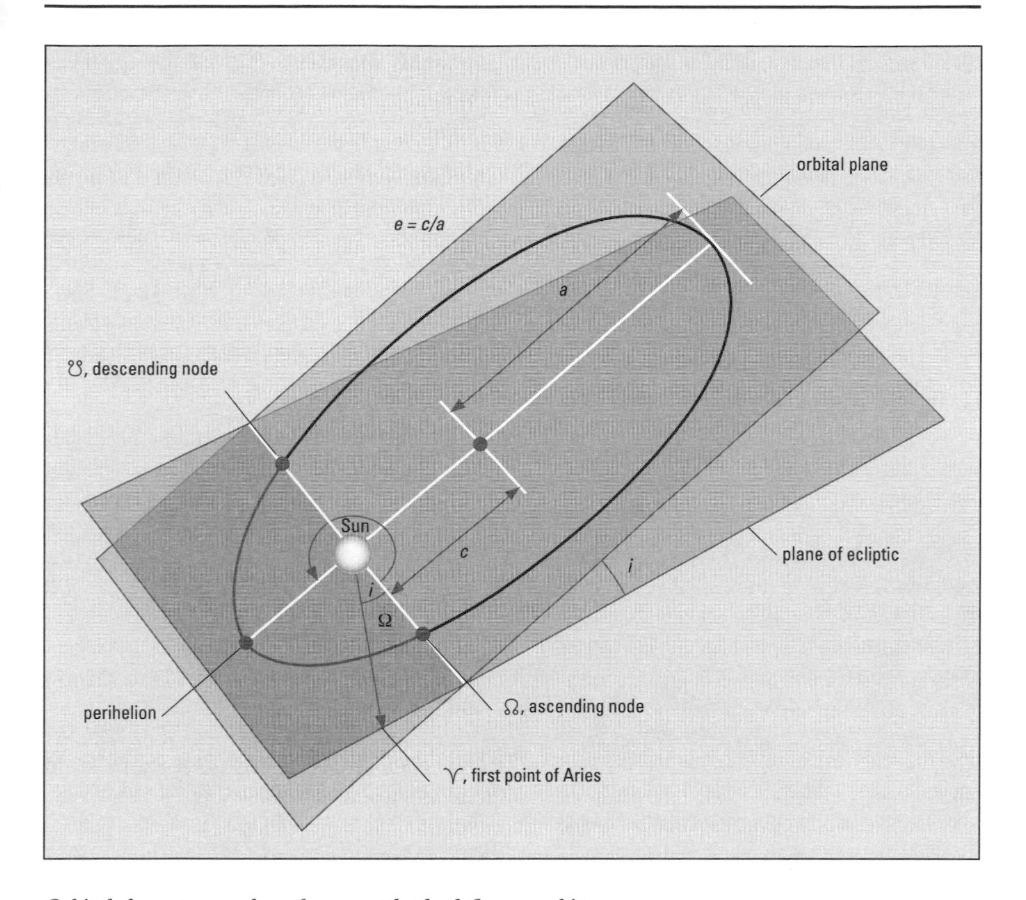

Orbital elements: together, these completely define an orbit

dark nebula. Within the densest portions of the dark cloud, star formation is still under way. The most recent stars to be formed lie unseen within it. These include an infrared source called the *Becklin–Neugebauer Object*, a B star with a surface temperature of 20,000 K only a few tens of thousands of years old, discovered by Eric Becklin and Gerald Neugebauer in 1967.

orrery A mechanical model of the Solar System. Orreries varied from simple ones with just the Earth, Moon, and Sun to highly complicated representations of the whole Solar System, where the planets with their satellites not only revolve in their orbits but also rotate on their axes. The name is from Charles Boyle, fourth Earl of Orrery (1676–1731), who commissioned one from John Rowley in 1712, although this was not the first to be made.

orthoscopic eyepiece A standard telescope lens with good definition and **eye relief**. It is a two-element lens, one of which is a cemented triple achromatic (SEE **achromatic lens**).

oscillating universe A variant of the **Big Bang** theory in which it is suggested that the Universe passes through successive cycles of expansion and contraction (or collapse). At the end of the collapse phase, with the Universe packed into a small volume of great density, it is possible that a 'bounce' would occur. The Universe would thus oscillate between Big Bang and 'Big Crunch' episodes and so be infinite in age. However, for this to happen the density of the Universe would have to be above a certain value (the **critical density**), which is not thought to be the case.

osculating orbit The path that an orbiting body, such as a planet or comet, would follow if it were not subject to any perturbations, but were governed solely by the **inverse-square law** of attraction by the Sun or other central body. The orbital elements, in this case, are called *osculating elements*, and are used in calculating perturbations.

O star A star of spectral type O, characterized by absorption lines of ionized helium in its spectrum. O stars are the hottest and brightest stars on the main sequence, with surface temperatures in the range 30,000 to 50,000 K and luminosities from 50,000 up to a million times that of the Sun. They are also the heaviest stars, ranging from 20 to 50 solar masses. O stars are rare – there are only about 300,000 in the Galaxy at present – because they remain on the main sequence for less than a million years. However, there has been time for 10,000 generations of O stars during the estimated life of the Galaxy. As they return most of their mass to space via stellar winds and supernova explosions, they have been an important source of heavy elements (SEE **metals**). Together with the hotter B stars, they form **OB associations**.

Most O and B stars form in binary systems, many with components of similar mass, an example being **Plaskett's Star**. When one star explodes as a supernova, it may release its companion from orbit at high velocity. As many as 20% of O stars may be such runaways. If the binary survives the supernova explosion intact, the result will be a neutron star or black hole orbiting through the stellar wind from its companion, producing an X-ray binary such as **Cygnus X-1**. Examples of O stars are Zeta Puppis, Delta Orionis, and Zeta Orionis.

outer planets The planets Jupiter, Saturn, Uranus, Neptune and Pluto, which orbit beyond the main asteroid belt.

Owl Nebula A planetary nebula, M97, in Ursa Major, of 11th magnitude and apparent diameter 3 arc minutes. It takes its name from two dark patches, like eyes, that give it the appearance of an owl's face. It lies about 1500 light years away and has a true diameter of 1.5 light years.

P

Pallas Asteroid no. 2, the second to be discovered, by **Heinrich Olbers** in 1802. It is the second-largest, with a diameter of 580 × 470 km (360 × 290 miles).

Palomar Observatory An observatory located on Palomar Mountain, near San Diego, California, site of the 5 m (200-inch) Hale reflector, opened in 1948, and a 1.2 m (48-inch) Schmidt telescope. The Schmidt was used for the *Palomar Sky Survey*, an all-sky photographic survey. Palomar Observatory was founded by **George Ellery Hale** and is now run by the California Institute of Technology (Caltech).

Pan A small **satellite** of Saturn, identified in 1991 by Mark Showalter from images obtained during the Voyager missions. It orbits within the **Encke Division** in Saturn's rings.

Pandora One of the small inner **satellites** of Saturn discovered in 1980 during the Voyager missions. Pandora and Prometheus are the two **shepherd satellites** of Saturn's F Ring.

parabola An open curve. In mathematics the parabola is a type of conic section, so called because it is one of the intersections of a plane with a cone, and is formed when a plane cuts a cone parallel to its slope. The **eccentricity** of a parabola is 1. In astronomy, it is a type of open **orbit**, perhaps most familiar in the context of cometary orbits.

Half of all comets observed have an orbital eccentricity of approximately 1. Because these comets are observable over only a short arc of their orbits near perihelion, it is not possible to determine their orbital elements sufficiently accurately to distinguish between exactly parabolic orbits and extremely elongated elliptical orbits.

paraboloid The figure obtained by rotating a **parabola** about its axis. It is the shape most often used for the primary mirror of reflecting telescopes, because a paraboloidal reflector, unlike a spherical one, brings an incoming beam of light to a single focus and is thus free from spherical **aberration** (2).

parallax (symbol π) The angular distance by which a celestial object appears to be displaced with respect to more distant objects, when viewed from opposite ends of a baseline, used as a measure of the object's distance. If the parallax is determined by direct means, using the principle of triangulation and a known baseline, it is known as **trigonometric parallax**. If the parallax is deduced from examination of a star's spectrum, it is known as **spectroscopic parallax**.

In trigonometric parallax, the choice of baseline is made according to the remoteness of the object. For fairly close objects, such as members of the Solar System, the Earth's equatorial radius is used; this gives a *geocentric* (or *diurnal*) *parallax*. For more remote objects the radius of the Earth's orbit – the **astronomical unit** (AU) – is used; this gives the *annual* (or *heliocentric*) *parallax*. The parallax of an object is therefore the angular size of the radius of the Earth or of its orbit as it would appear from that object.

The *solar parallax* is the Sun's geocentric parallax, defined as the angular size of the Earth's equatorial radius from a distance of 1 AU. It has been historically very important in establishing the scale of the Solar System. SEE ALSO **distance modulus**

parhelion (mock Sun, sundog) A round patch of light at the same altitude as the Sun. Parhelia are images of the Sun produced when sunlight is refracted by ice crystals in the upper atmosphere. They often occur in pairs, one either side of the Sun.

Parkes Observatory The National Radio

Astronomy Observatory of Australia, in New South Wales. Its main instrument is a 64 m (210 ft) dish aerial, completed in 1961, which is now part of the **Australia Telescope**.

parsec (symbol pc) The distance at which a star would have a **parallax** of one second of arc; equivalent to 3.2616 light years, 206,265 astronomical units, or 30.857×10^{12} km.

Parsons, William Algernon SEE **Rosse, Third Earl of**

partial eclipse SEE **lunar eclipse, solar eclipse**

Pasiphae One of Jupiter's four small outermost **satellites**, discovered in 1908 by P. J. Melotte. It is in a retrograde orbit – SEE **retrograde motion** (1) – and may well be a captured asteroid.

Pavo A southern constellation, representing a peacock; its brightest star, Alpha, magnitude 1.9, is called Peacock. Kappa is a **Cepheid variable** with a magnitude range from 3.9 to 4.8 in a period of 9.1 days.

Pegasus A conspicuous constellation of the northern sky, representing the flying horse of Greek mythology. Its most famous feature is the *Square of Pegasus*, made up of four stars, although one of these stars is actually part of Andromeda. Beta (Scheat) is a red giant irregular variable with a small magnitude range (2.3 to 2.7). Pegasus is not a rich constellation but it does contain a major globular cluster, M15, which at magnitude 6 is an easy binocular object.

Pelican Nebula SEE **North America Nebula**

penumbra (1) The outer part of the shadow cast by a celestial body illuminated by an extended source such as the Sun. An observer in the penumbral region of a shadow sees only part of the source obscured, as, for example, in a partial **solar eclipse**. SEE ALSO **lunar eclipse**. COMPARE **umbra** (1)

penumbra (2) The lighter zone surrounding the dark central part of a sunspot. COMPARE **umbra** (2)

penumbral eclipse SEE **lunar eclipse**

Penzias, Arno Allan (1933–) German–American astrophysicist. In 1964 he was working with **Robert Wilson**, using a small radio telescope to measure noise that might interfere with satellite communications. They detected a mysterious signal that, after consultation with Robert Dicke and James Peebles, they realized was the **cosmic microwave background** – radiation left over from the Big Bang, as had been predicted by **George Gamow**. Penzias and Wilson shared the 1978 Nobel Prize for Physics for what was the most important discovery in modern cosmology.

peri- A prefix referring to the point in an object's orbit at which it comes closest to its primary, as in *perigee* and *perihelion*.

periastron The point of nearest approach of the components of a binary star in their orbit.

perigee The point of nearest approach to Earth by the Moon or an artificial satellite.

perihelion The point in the orbit of a planet or comet at which it is nearest to the Sun.

periodic comet A common term for a short-period **comet** – one whose period is less than 200 years.

period–luminosity law A relationship between the period of light variation and the mean absolute magnitude (luminosity) of a **Cepheid variable** star: the absolute magnitude increases as the period increases. Once the period of a Cepheid is known, its absolute magnitude can be deduced from the period–luminosity law. Furthermore, the absolute magnitude taken in conjunction with the apparent magnitude then enables the distance of the Cepheid to be calculated. It was discovered by **Henrietta Leavitt** in 1912.

Perseids A **meteor shower** that occurs at the end of July and the beginning of August. Its radiant starts in the constellation Cassiopeia, and moves through the north of the constellation Perseus. The shower is associated with Comet **Swift–Tuttle**. The Perseids are one of the strongest, best known, and most long lived of all meteor showers. In the late Middle Ages they were known as the *tears of St Lawrence*, who was martyred in AD 258, on 10 August, close to the shower's maximum.

Perseus A prominent northern constellation, representing the Greek hero who rescued Andromeda. Alpha Persei (Algenib or Mirfak) is magnitude 1.8. Beta is the prototype eclipsing binary, **Algol**. Rho is a red giant semiregular variable with a magnitude range from 3.3 to 4.0 and a very rough period of around 50 days. Perseus is a rich constellation, crossed by the Milky Way; there are several open clusters, including the famous **Double Cluster** and M34.

personal equation A correction factor applied to visual observations made by a particular person. An individual observer may record timings of, for example, transits that are consistently early or late. Such systematic errors cannot be measured directly, but in any programme of work carried out by several observers a personal equation can be found by carrying out a statistical analysis of all the results. It can then be applied to improve the accuracy of future observations.

perturbation An irregularity in the orbital motion of a celestial body, brought about by the gravitational attraction of other bodies. Attractions by planets and satellites cause significant displacements of bodies from where they would be in a purely elliptical orbit. *Periodic perturbations* are small oscillations with periods similar to the orbital period of the perturbed body. *Secular perturbations* are slow continuous changes in orbital motion, for example the slow rotation of the orbit of a planet (SEE **perihelion**), and in the orbit's eccentricity and inclination, that arise from commensurabilities (SEE **commensurable**). It is such perturbations that produce **precession**. Perturbations are responsible for dramatic changes in the orbits of **comets**, changing long-period orbits into short-period ones. SEE ALSO **non-gravitational force**

Phaethon Asteroid no. 3200, discovered in 1983 by Simon Green and John Davies, using the **Infrared Astronomical Satellite**. It is a 6.5 km (4 mile) diameter **Apollo asteroid** with a perihelion within the orbit of Mercury. Its orbit coincides with the **Geminid** meteor shower, so it seems likely that it is an extinct comet.

phase The proportion of the illuminated hemisphere of a body in the Solar System (in particular the Moon or an inferior planet) as seen from Earth. The phase of a body changes as it, the Sun, and the Earth change their relative positions. All the phases of the Moon (SEE the diagram on page 154) – new, crescent, half, gibbous, and full – are observable with the naked eye. The inferior planets, Mercury and Venus, show phases from a slender crescent to a fully illuminated disk when observed through a telescope. Of the superior planets, Mars can show quite a marked gibbous phase, and Jupiter a very slight gibbous phase, but with Saturn and the planets beyond no phase is ever discernible. Phase is sometimes expressed as the percentage of the visible disk that is illuminated.

Phobos (1) The larger of the two **satellites** of Mars, discovered in 1877 by **Asaph Hall**. It is a dark, irregular body, measuring 27 ×

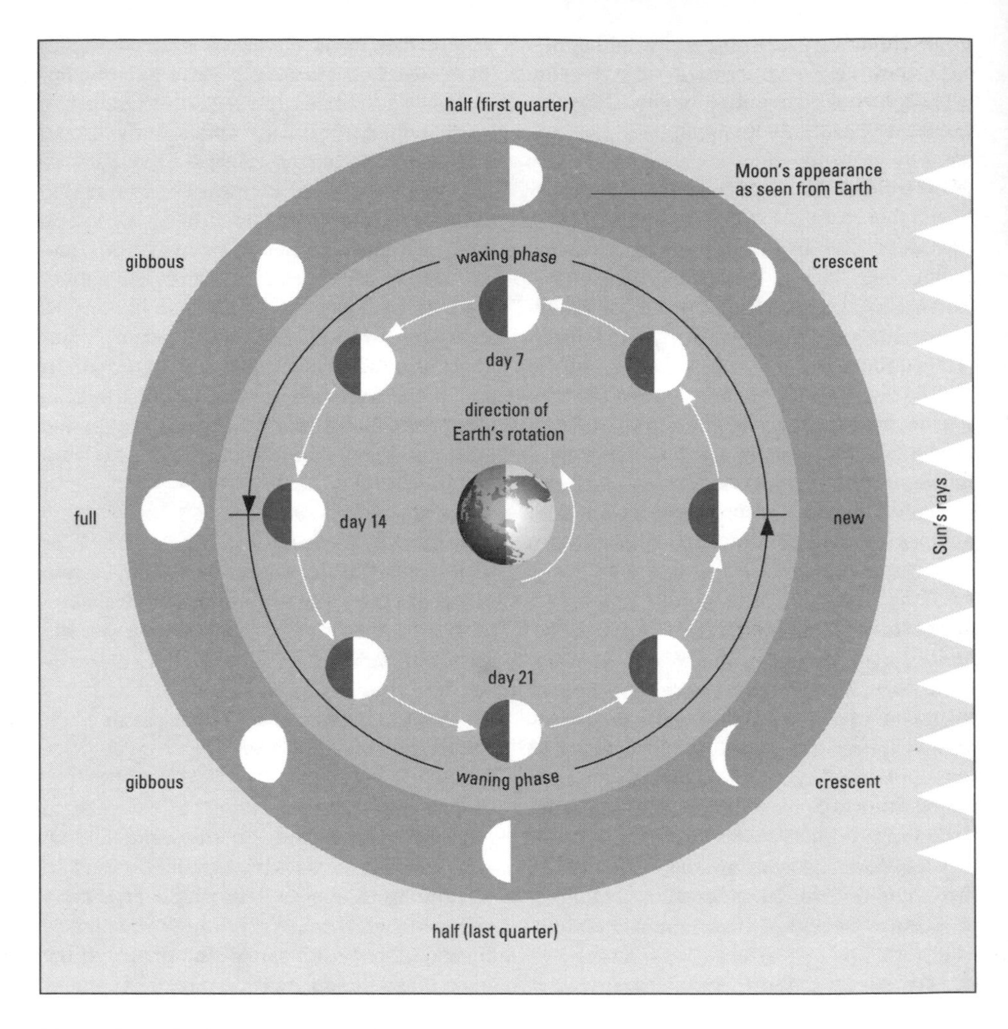

Phase: the cycle of changes in the Moon's appearance

22 × 19 km (17 × 14 × 12 miles), and may well be a captured asteroid. It has two large craters, named Stickney and Hall.

Phobos (2) SEE **Mars probes**

Phoebe The outermost **satellite** of Saturn, discovered in 1898 by William Pickering. Phoebe is quite small (diameter 220 km/ 135 miles), moves in a distant, inclined and retrograde orbit, and has a non-**synchronous rotation** period of 9.4 hours, all of which suggests that it is a captured body. Its similarities with **Chiron** indicate a possible common origin.

Phoenix A southern constellation representing a phoenix. Alpha (Ankaa), magnitude 2.4, is the brightest star. Zeta is an eclipsing binary with a range from magnitude 3.9 to 4.4 in a period of 1.67 days.

photoelectric magnitude The **apparent magnitude** of an object measured by a

photoelectric photometer attached to the telescope. The measurement of brightness is made through a number of coloured filters in turn. SEE ALSO **UBV system**

photoelectric photometer An instrument that measures the brightness of a celestial object by generating an electrical signal that is directly proportional to the intensity of the light from that object falling on it. There are various types of detector. A *photomultiplier tube* is a device in which electrons given off when light falls upon a special screen liberate more electrons to produce a measurable electric current. A **CCD** can be used instead (strictly speaking, it is not a photoelectric device, but it performs the same function). The output from the detector is fed into a computer. The brightness thus measured is an object's **photoelectric magnitude**. An accuracy of 0.01 magnitude is possible, even with fairly basic equipment, which is far better than can be achieved by photographic methods. The photoelectric photometer is therefore used for accurate tracking of the variations of variable stars, and also for measuring the brightness variations of asteroids, and thus their rotation periods. The light falling on the detector can be confined to a particular part of the visible spectrum, or infrared or ultraviolet wavelengths can be selected.

photographic magnitude (symbol m_{pg}) The **apparent magnitude** of an object as measured on a traditional photographic emulsion which is more sensitive to blue light than the human eye is. Hence the photographic magnitude differs significantly from the visual magnitude.

photographic zenith tube (PZT) A fixed telescope directed at the local zenith and used to photograph stars as they pass across its field of view. The photographs are used to determine the zenith distance and time of transit of **fundamental stars**. These measurements yield highly accurate values of the stars' positions, from which can be calculated the *clock error* – a very small non-uniformity in the Earth's rotation. The photographic zenith tube has a much greater degree of accuracy than was possible with traditional **transit instruments**.

photometer SEE **photoelectric photometer**

photosphere The visible surface of the Sun. It is a layer of highly luminous gas about 500 km (300 miles) thick and with a temperature of about 6000 K, falling to 4000 K at its upper level, where it meets the **chromosphere**. As this is the region of the Sun from which all its output of visible light is emitted, the photosphere is the source of the Sun's visible spectrum. The lower, hotter gases produce the continuous emission spectrum, while the higher, cooler gases absorb certain wavelengths and so give rise to the **Fraunhofer lines**. This absorption contributes to **limb darkening**. Sunspots and other visible features of the Sun are situated in the photosphere.

photovisual magnitude (symbol m_{pg}) The **apparent magnitude** of an object measured photographically using emulsions and filters that mimic the spectral response of the human eye.

Piazzi, Giuseppe (1746–1826) Italian astronomer and monk. He discovered the first asteroid, Ceres, on 1 January 1801, during a lengthy programme of taking accurate measurements of star positions (published as a catalogue in two parts, in 1803 and 1814). Piazzi proposed the name 'planetoid', but 'asteroid' has prevailed.

Pic du Midi Observatory An observatory in the French Pyrenees, at an altitude of nearly 2900 m (9400 ft), noted for its solar and planetary observations. Its main instrument is a 2 m (78-inch) reflector.

Pickering, Edward Charles (1846–1919) American astrophysicist, who was director of the Harvard College Observatory. Most of his work was on stellar photometry and the spectral classification of stars. He built a meridian photometer and began a programme of accurate stellar magnitude determinations, published in the *Revised Harvard Photometry*. Pickering discovered spectroscopic binary systems, and invented a method of obtaining the spectra of several stars on one photographic plate. The spectroscopic work he instituted culminated in the *Harvard system* of **spectral classification** and the ***Henry Draper Catalogue*** of stellar spectra. This work was largely carried out by a team of women, originally hired as assistants but who grew to be able astronomers in their own right. They included **Annie Jump Cannon**, Williamina Fleming (1857–1911), and Antonia C. Maury (1866–1952). Edward's younger brother, **William Henry Pickering** (1858–1938), was a lunar and planetary astronomer who instituted the search for **Pluto**. He also discovered Saturn's satellite Phoebe, in 1898.

Pictor An unremarkable constellation in the southern sky introduced by **Lacaille**, representing a painter's easel. Its leading star is Alpha, magnitude 3.3, and the only other star above magnitude 4 is *Beta Pictoris*, magnitude 3.8, which in 1984 was discovered to have a disk of gas and dust around it, believed to be a planetary system in the process of formation.

Pioneer A series of US space probes launched between 1958 and 1978. The first four were failed Moon probes. Pioneers 5 to 9 were put into orbit around the Sun to study the solar wind and solar flare activity. Pioneer 10, launched in March 1972, was the first probe to travel through the asteroid belt; it passed Jupiter in December 1973, sending back photographs. Pioneer 11 followed it, flying past Jupiter in December 1974 and moving on to Saturn, which it reached in September 1979. Both probes are now on their way out of the Solar System. The last two in the series were the **Pioneer Venus** probes.

Pioneer Venus Two NASA space probes to the planet Venus, launched in May and August 1978. Pioneer Venus 1 went into orbit around the planet in December 1978, making a radar map of the surface and photographing the planet's clouds; it finally burned up in the planet's atmosphere in October 1992. Pioneer Venus 2 dropped five small probes into the atmosphere of Venus in December 1978, which sent back data on the planet's clouds and atmospheric conditions.

Pisces A constellation of the zodiac, representing a pair of fishes. It is very obscure, consisting of a chain of rather faint stars south of Pegasus. Its brightest star, Eta, is magnitude 3.6. Alpha (Alrescha) is a close binary, magnitudes 4.2 and 5.2, with a period of over 900 years.

Piscis Austrinus The constellation of the southern fish. First-magnitude **Fomalhaut** is its only star brighter than magnitude 4.

plage A bright cloud of hot gas in the Sun's **chromosphere**. Plages are intimately connected with **faculae**, which occur below them in the photosphere, and are found in active regions, where the local magnetic field is enhanced.

planet A large non-stellar body in orbit around a star, shining only by reflecting the star's light. In our **Solar System** there are nine **major planets**, as opposed to the thousands of small bodies known as asteroids or minor planets. It is believed that other stars, apart from our Sun, may have planets orbiting them. One such was long thought to be Barnard's Star, irregularities in whose **proper motion** was attributed to the gravitational attraction of one or more planets. A strong candidate is the **millisecond pulsar**

in Virgo, designated PSR 1257+12, variations in whose pulse rate are consistent with three terrestrial-sized planets in orbit around it. SEE ALSO **Planet X**

planetarium (plural **planetaria**) A structure in which a representation of the stars and planets as visible in the night sky is projected on to the inside of a dome for an audience seated below. They vary in size and sophistication from small portable projectors used in inflatable 3 m (10 ft) domes to highly complex installations in permanent theatres seating hundreds of people. In all planetaria the positions and motions of celestial bodies can be simulated. The first planetarium, built by the Carl Zeiss company, began operation in Munich in 1923. From then until the early 1980s, planetarium projectors consisted of many individual optical projectors mounted on a complicated rotating frame. In the new generation of projectors, a computer-generated image is projected through a single, stationary fish-eye lens.

planetary nebula An emission nebula formed when a red giant or supergiant sheds its outer layers, leaving a hot core to excite the gas. Their sizes range from about that of our Solar System to a light year or so across. They were so named because the first ones to be discovered gave the impression of planetary disks when viewed in small telescopes; the Ring Nebula (M57) in Lyra is a well-known example. However, only 10% of the 1000-plus now known have a circular shape, and about 70% have two lobes.

The central stars of planetary nebulae are hot and blue, with temperatures of 30,000 to 400,000 K. They form a transitional stage at the end of a star's evolution between the giant stars and the white dwarfs. Ultraviolet light from the central stars make planetary nebulae fluoresce, as in other emission nebulae. In addition to hydrogen and helium, planetary nebulae contain small proportions of oxygen, nitrogen, and neon, which accounts for their various colours as seen on photographs.

Planetary nebulae are thought to result from stellar winds blown out from red giants and supergiants, leaving the red giant's core as the central star (SEE **RV Tauri star**). Over time, this star cools to become a white dwarf. Because red giants are so big the gravity at their surfaces is very low, and storms on the surface can easily throw off material. A star up to about 8 solar masses can lose three-quarters of its mass into space before it turns into a white dwarf of about 1.4 solar masses. The visible nebula is simply the central region of a much larger object, the outer parts of which are not lit up by the central star. All planetary nebulae are expanding, at speeds of around 20 km/s (12 mile/s).

The curious shapes of planetary nebulae are probably linked to the magnetic field and rotation of the stars. The magnetic field lines might control the outflow, channelling it in particular directions, and the rotation of the star could add spiral twists. The double-lobed structure of most planetary nebulae must be connected with the star's direction of spin or magnetic axis. Some planetary nebulae, such as the Eskimo and Saturn nebulae, have a bright central nebula surrounded by fainter extensions, as though the central star has ejected successive shells. Since they are expanding as the central stars fade, planetary nebulae are relatively transitory objects with lifetimes of tens of thousands of years. Several new planetary nebulae form each year in our Galaxy.

planetary precession The smaller of the two components of **precession**. It results from the gravitational effects of other planets upon the Earth. It moves the equinoxes eastwards by about 0.12 arc seconds per year.

planetesimal In theories of the origin of the Solar System (and other planetary systems), a body, between a few millimetres and a few kilometres in size, that condensed out of a cloud of gas and dust (the solar nebula), away from the centre where the Sun was forming. Once they had reached a few kilometres in size, planetesimals' gravitational attraction would allow them to combine with one another, by a process of **accretion**, to form **protoplanets**. The term was first used for the very small planetary bodies supposed to have condensed from matter torn out of the Sun by a passing star in the 'planetesimal theory' proposed by Forest Ray Moulton (1872–1952) and Thomas Chrowder Chamberlin (1843–1928) in the early 1900s.

Planet X The name given to a hypothetical tenth major planet orbiting the Sun beyond **Pluto**, the X representing both 'ten' and 'unknown'. It was originally the name given by **Percival Lowell** to the planet whose gravitational **perturbations** he supposed accounted for irregularities in the orbits of Uranus and Neptune. Pluto's mass was calculated accurately in 1978 and found to be far too small to account for these perturbations, reviving interest in Planet X. A trans-Plutonian planet was also invoked to explain how comets and objects like **Chiron** become diverted into the region of the major planets. But the perturbations of Uranus and Neptune are now almost certainly explained away by errors in observing their positions, and the inward diversion of cometary bodies is now believed to be caused by *chaos* in the outer Solar System – in other words, orbits way beyond the major planets lack long-term stability. Several long-term searches, including one by **Clyde Tombaugh** that continued for 13 years after he discovered Pluto, have turned up no sizeable body. The **Kuiper Belt** objects found from 1992 onwards are nowhere near planet-sized. There is no longer any reason to invoke a Planet X, and it is highly unlikely that one exists.

planisphere A simple device for showing the stars on view from a certain latitude at any time and on any date. It consists of two disks held together by a pivot at their centres. On the lower disk is printed a map of the night sky, with the celestial pole at the pivot, and into the upper disk is cut an oval window. The planisphere is set by rotating the disks so that the correct date, marked on the rim of one, is aligned with the correct time, marked on the rim of the other. There is some distortion of constellation shapes, particularly near the rim, but the planisphere is nevertheless a useful and easily portable guide.

Plaskett's Star An exceptionally massive binary star in the constellation Monoceros. In 1922 the Canadian astronomer John Plaskett (1865–1941) showed it to be a spectroscopic binary with a period of 14 days. Each component is a supergiant of over 50 solar masses. It is the most massive binary known.

Platonic year The time taken for the celestial pole to describe a circle around the pole of the ecliptic, as a result of **precession**. Its value is about 25,800 years. (After the Greek philosopher Plato, *c.* 427 – *c.* 347 BC.)

Pleiades A young open cluster in the constellation Taurus, popularly called the *Seven Sisters*. Although only six or seven stars are visible to the naked eye, there are in fact over a thousand embedded in a reflection nebula. The brightest member is Alcyone, a B-type giant over 300 times as luminous as the Sun. Another member of the cluster, Pleione, is a slightly variable **shell star** that throws off shells of gas as a result of its fast rotation. The other named Pleiades are Maia, Atlas, Electra, Merope and Taygete. The cluster lies just over 400 light years away.

plerion SEE **supernova remnant**

Plössl eyepiece A telescope eyepiece with good **eye relief** and a wider field of view (about 40°) than an orthoscopic lens. The most common form has two identical achromatic doublets (SEE **achromatic lens**), but some quite different designs have been called 'Plössls'. (After Austrian optician Simon Plössl, 1794–1868.)

Plough Popular name for the shape formed by the seven main stars in the constellation Ursa Major: Alpha (Dubhe), Beta (Merak), Gamma (Phekda), Delta (Megrez), Epsilon (Alioth), Zeta (**Mizar**), and Eta (Alkaid). In the US, the stars are known as the *Big Dipper*.

Pluto The smallest and outermost planet of the Solar System. It is in a much more inclined and eccentric orbit than are the other major planets, and for a 20-year period around perihelion (as, for example, from 1979 to 1999) it comes within the orbit of Neptune. However, a **resonance** between the two planets' orbits (their periods are in the ratio 3 : 2) means they never get close. Pluto's small size and great distance give it a maximum apparent diameter of 0.1 arc seconds, and it is never brighter than magnitude 14.0. The main data for Pluto are given in the table.

After the discovery of Neptune (SEE **Adams, John Couch**), it seemed that there were still unexplained irregularities in the motion of Uranus. Independently, William H. Pickering (SEE **Pickering, Edward**) and **Percival Lowell** concluded that an undiscovered planet beyond Neptune was responsible, and made calculations of its size, orbit, and position in the sky. The planet was eventually located by **Clyde Tombaugh**, within 5° of Lowell's predicted position, on photographic plates taken in January and February 1930. The discovery was publicly announced on 13 March 1930.

Pluto has yet to be visited by a space probe. Most of our knowledge of it dates from the discovery in 1978 of its satellite, **Charon**, which allowed, among other things, Pluto's mass and unusual axial tilt to be determined. In 1988 Charon's orbit around Pluto was edge-on to the line of sight from Earth. A series of mutual **occultations** of planet and moon from 1985 until 1990 gave information on their relative brightnesses and sizes, and a very rough brightness map of Pluto's surface. The Hubble Space Telescope has enabled these results to be refined further.

Although there is still some uncertainty in their measurements, the Pluto–Charon pair is known to be unique in many ways. Charon is by far the largest satellite with respect to its primary, with between 8 and 16% of Pluto's mass, and from Pluto's surface its apparent size is 4°. Charon has both a **synchronous orbit** and **synchronous rotation**, so it keeps the same face turned towards Pluto, and hangs motionless in Pluto's sky.

The coupling of Pluto and Charon could well have brought about sufficient tidal heating (SEE **tides**) to cause **differentiation** in both bodies, giving them rocky cores and icy mantles. Pluto seems to a have mottled surface with light and dark regions, and signs of polar caps. The surface is covered with icy deposits consisting of 98% nitrogen, with traces of methane, and also probably water, carbon dioxide, and carbon monoxide. For

PLUTO: DATA	
Globe	
Diameter	2300 km
Density	2.0 g/cm^3
Mass (Earth = 1)	0.0021
Volume (Earth = 1)	0.0058
Sidereal period of axial rotation	6.387d (retrograde)
Escape velocity	1.1 km/s
Surface gravity (Earth = 1)	0.03
Albedo	0.9
Inclination of equator to orbit	122° 27′
Surface temperature (average)	45 K
Orbit	
Semimajor axis	39.53 AU = 5914 × 10^6 km
Eccentricity	0.248
Inclination to ecliptic	17° 09′
Sidereal period of revolution	248.54y
Mean orbital velocity	4.74 km/s
Satellites	1

a 60-year period in each orbit, when Pluto's high orbital eccentricity brings it closer to the Sun, the surface warms to above 50 K, sufficient to release a thin, temporary atmosphere of nitrogen. Interaction with the **solar wind** probably draws some of this atmosphere into a comet-like tail. Pluto therefore loses material at each perihelion passage; some of it may be swept up by Charon.

Pluto is certainly not the massive planet whose supposed perturbations of Uranus and Neptune led to Pluto's discovery. The realization of just how small Pluto is has led some astronomers to question its status as a major planet. Pluto and Charon may be alternatively regarded as the largest members of the **Kuiper Belt** so far detected.

Poincaré, (Jules) Henri (1854–1912) French mathematician. In astronomy he studied celestial mechanics in general and the **three-body problem** in particular. He showed that the **many-body problem** can never be solved exactly, and in the process discovered chaotic orbits – ones that lack any long-term stability. Poincaré established many new directions in celestial mechanics, with many innovative mathematical treatments. He also made important early studies of the Special Theory of Relativity.

Pointers The stars Alpha and Beta Ursae Majoris (Dubhe and Merak), which point towards Polaris – the Pole Star and the brightest star in Ursa Minor.

polar caps Deposits of ice in the polar regions of a planet or satellite. The Earth's permanent polar caps are of water-ice, and show only a little seasonal variation. Mars has caps of water-ice and carbon dioxide that show pronounced seasonal variation because of the planet's eccentric orbit. The axial inclination of Neptune's moon **Triton** with respect to the Sun produces a migration of its nitrogen-ice cap from one pole to the other over the course of the planet's 165-year orbital period.

polar distance The angular distance of a celestial body from the celestial pole. It is the complement of **declination** (i.e. polar distance plus declination equals 90°).

Polaris The star Alpha Ursae Minoris, lying within 1° of the north celestial pole. **Precession** is bringing Polaris closer to the pole, and it will be at its closest around AD 2100. Polaris is a supergiant of spectral type F, magnitude 2.0, distance 800 light years, luminosity nearly 10,000 times that of the Sun. It was originally listed as a **Cepheid variable** of small amplitude, but during the 20th century its pulsations slowly died out until by the 1990s it had virtually ceased to vary. It is an optical double, with an 8th-magnitude companion 18″ away.

pole The points on a sphere that are 90° from the equatorial plane. The *celestial poles*, the points where the Earth's axis of rotation intersects the celestial sphere, are 90° north and south of the celestial equator. Similarly, the *ecliptic poles* are 90° from the plane of the ecliptic, and the *galactic poles* are 90° from the galactic plane.

Pole Star The nearest naked-eye star to the celestial pole. In the northern hemisphere it is **Polaris**; in the south it is 5th-magnitude Sigma Octantis. Because of **precession** the positions of the celestial poles, and hence of the pole stars, are continually changing. By AD 14,000 the north pole star will be Vega.

Pollux The star Beta Geminorum, magnitude 1.14, distance 36 light years, luminosity 45 times that of the Sun. It is a giant of type K0.

Population I star A relatively young star containing a high abundance of metals, found especially in the spiral arms of galaxies. The Sun is a Population I star. SEE ALSO **stellar populations**

Population II star An old star containing

a lower abundance of metals than a Population I star. Population II stars, which include RR Lyrae stars and subdwarfs, are found in the halo of our Galaxy, especially in globular clusters, and in the nuclei of spiral galaxies. The oldest of them are sometimes called *halo population stars*. SEE ALSO **stellar populations**

pore A very small and short-lived **sunspot**.

Portia One of the small inner **satellites** of Uranus discovered in 1986 during the Voyager 2 mission.

position angle The orientation in the sky of one celestial body with respect to another, measured from 0 to 360° from north via east. For a binary system it is the angle between the direction of the celestial north pole and the direction of the line joining the stars.

post-nova A **nova** after its outburst and return to its original magnitude.

Poynting–Robertson effect A **non-gravitational force** produced by the action of solar radiation on small particles in the Solar System which causes them to spiral inwards towards the Sun. When particles in orbit around the Sun absorb energy from the Sun and then re-radiate it, they lose kinetic energy and their orbital radius shrinks slightly. The effect is most marked for particles between about a micrometre and a centimetre in size. A similar effect makes particles in planetary ring systems spiral inwards. The effect is named after John Henry Poynting (1852–1914), who first described it in 1903, and Howard P. Robertson (1903–61), who proved its reality in 1937 by derivation from the theory of relativity.

Praesepe A large open cluster in the constellation Cancer, known also as the *Beehive* or *Manger*; its number on Messier's list is M44. It was known to the Greeks as a hazy naked-eye patch, but binoculars and telescopes show several dozen stars. The cluster is believed to be about 650 million years old, and lies 520 light years away.

preceding (abbreviation *p*) Describing the leading edge, feature, or member of an astronomical object or group of objects. For example, the preceding limb of the Moon is the edge facing away from its direction of motion; and the preceding spot (or *p*-spot) of a group of sunspots is the first of the group to be brought into view by the Sun's rotation.

precession The circular motion of the celestial poles around the poles of the ecliptic, and the associated westward movement of the equinoxes with respect to the background stars. The main component of precession is caused by the gravitational pull of the Sun and Moon on the Earth's equatorial bulge. This is called the *luni-solar precession*.

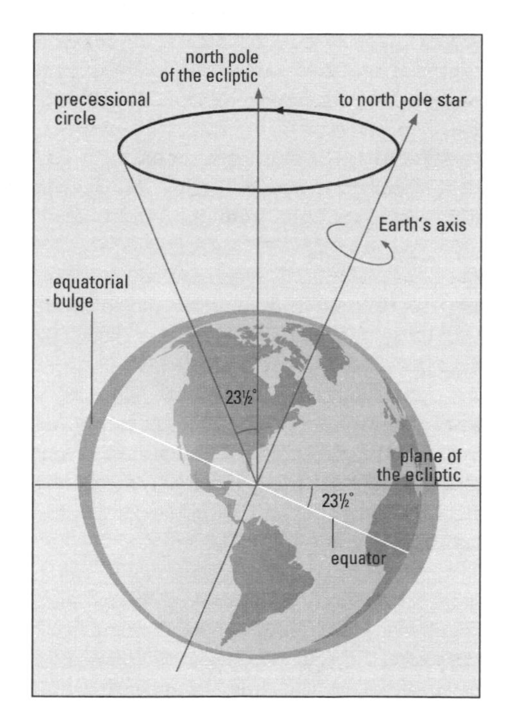

Precession

A smaller, similar effect caused by the gravitational influence of the other planets is called the *planetary precession*, but it acts in the opposite direction. The total effect, called the *general precession*, amounts to about 50.3 arc seconds per year. The celestial pole therefore describes a circle on the sky of radius 23½° (the inclination of the Earth's axis), centred on the pole of the ecliptic, in about 25,800 years. Because of precession, the positions of all stars (including, of course, the pole star) are continuously changing. Precession also has the effect of moving the **equinoxes** around the ecliptic, and is therefore sometimes called the *precession of the equinoxes*. The First Point of Aries, for example, is now in Pisces. Star catalogues give positions for a given **epoch**, e.g. 2000, so these positions must be corrected for precession to the date of observation.

primary (1) The largest of a system of celestial bodies, around which the others orbit. The term is used for the more massive component of a binary or multiple star, and for a planet with respect to its moons.

primary (2) **(primary mirror)** The largest mirror in a **reflecting telescope**, which collects light from the object being observed.

prime focus The point at which the objective lens or primary mirror of a telescope brings light to a focus in the absence of any other optical component, such as a secondary mirror. In large telescopes, special instruments are sometimes mounted at the prime focus. In the **Schmidt camera** the photographic plate is placed at the prime focus.

In *prime focus photography*, a technique used with small telescopes, the lens of a camera is removed and the camera back attached to the telescope via an adapter, in place of the eyepiece; photography does not, despite the name, actually take place at the prime focus.

primeval fireball The explosion which, according to the **Big Bang** theory, gave the present expanding Universe.

primordial black hole SEE **black hole**

Procyon The star Alpha Canis Minoris, the eighth-brightest star in the sky, magnitude 0.38, distance 11.4 light years, luminosity seven times that of the Sun. It is a main-sequence star of type F5 and has a white dwarf companion of 11th magnitude-that orbits it every 41 years.

prograde SEE **direct motion** (1)

prolate spheroid SEE **spheroid**

Prometheus One of the small inner **satellites** of Saturn discovered in 1980 during the Voyager missions. Prometheus and Pandora are the two **shepherd moons** of Saturn's F Ring.

prominence A cloud of matter extending outwards from the Sun's chromosphere, into the corona. Prominences are regions of higher density and lower temperature than the surrounding corona. Ones near the Sun's limb are visible as bright protrusions into the dark corona during total eclipses or with the aid of an instrument such as a **coronagraph** or a special filter. Away from the limb, they are silhouetted against the bright photosphere as dark **filaments**. *Quiescent prominences*, which extend for tens of thousands of kilometres, can last for weeks or months. *Active prominences* are short-lived, high-speed, flamelike eruptions that can reach heights of as much as 700,000 km (450,000 miles) in just 1 hour. All prominences have spectra showing lines of neutral hydrogen, helium, and ionized calcium. Their forms and behaviour are very varied, and their direction of flow is controlled by the Sun's magnetic field.

proper motion (symbol μ) The apparent motion of a star on the celestial sphere, as a result of its movement relative to the Sun. About 300 stars are known to have proper motions greater than 1 arc second per year, but most annual proper motions are smaller than 0.1 arc seconds. **Barnard's Star** has the largest known proper motion, 10.3 arc seconds per year. Proper motion is determined by comparing the star's position on photographic plates exposed on widely separated occasions, usually many years or decades apart, perhaps with the aid of a **comparator**. Considerably improved proper motions for over 100,000 stars have been obtained by the position-measuring satellite **Hipparcos**.

Proteus Neptune's second-largest **satellite**, discovered in 1989 during the Voyager 2 mission. It is a squarish, dark body, heavily cratered, and its main feature is the Southern Hemisphere Depression, 250 km (150 miles) across.

proton–proton reaction The nuclear reaction thought to be the main source of energy for main-sequence stars with masses equal to or less than the Sun's. In more massive stars the main energy source is the **carbon–nitrogen cycle**. The proton–proton reaction is a three-stage process in which hydrogen is converted into helium with the liberation of an enormous amount of energy. In the first stage two protons (hydrogen nuclei) combine to form a deuterium nucleus, releasing of a positron, a neutrino, and radiation. In the second stage the deuterium nucleus combines with a proton to form an isotope of helium (^{3}He), again releasing radiation. In the third stage two ^{3}He nuclei combine to form the normal helium nucleus (^{4}He), releasing two protons and radiation. Overall, four hydrogen nuclei have combined to form one helium nucleus and give out radiation. The reaction requires a temperature of about 10 million K.

protoplanet In theories of the origin of the Solar System (and other planetary systems), a body formed by the accretion of **planetesimals**. Once protoplanets attained the size of, say, the Moon, they could grow no further by accretion. Instead, collisions between them caused some to grow by accumulating fragments. The major planets of our Solar System are the result of this second stage of growth, while the collisional fragments that were not captured survive as bodies ranging in size from asteroids down to **micrometeoroids**.

protostar An embryonic star, formed as dense clumps of gas and dust collapse within molecular clouds. Protostars are the earliest, optically invisible stage of star formation. They are very red and can be observed at infrared wavelengths. The Sun's protostellar stage lasted about 0.1 to 1 million years. SEE ALSO **stellar evolution**

Proxima Centauri The nearest star to the Sun, slightly closer than the nearby star Alpha Centauri. It is magnitude 11.2, and its distance is 4.25 light years. Proxima Centauri is an M-type dwarf 20,000 times fainter than the Sun. It is a **flare star**. It was long thought to be part of the Alpha Centauri system, but some astronomers now believe it to be an unrelated star making a close approach.

Ptolemy (Claudius Ptolemaeus) (2nd century AD) Egyptian astronomer and geographer. His chief astronomical work, known by its title in Arabic translation, the *Almagest*, was largely a compendium of contemporary knowledge, including a star catalogue, drawing heavily on the work of **Hipparchus of Nicaea**. The *Ptolemaic system* is based on the geocentric world system of the ancient Greeks, with the Earth fixed at the centre, and the Moon, Mercury, Venus, the Sun, Mars, Jupiter, and Saturn revolving about it. Beyond these planets lies the sphere of fixed stars. Each of the bodies moves around a small circle called the *epicycle*, the centre of which in turn revolves around the Earth on a larger circle called the *deferent*. Ptolemy added to each orbit two more points, the *eccentric* and the *equant*, equally spaced either side of the Earth. He made the epicycle revolve around the eccentric rather than the Earth, and the planet have uniform motion with respect to the equant. The resulting model reproduced the apparent motions of the planets, including retrograde loops, so well that it remained unchallenged until the revival of the heliocentric theory by **Nicholas Copernicus** in the 16th century. Ptolemy's *Geography* remained a definitive text for a similar length of time.

Puck The largest of the small inner **satellites** of Uranus discovered during the Voyager 2 mission. It was first detected at the end of 1985, shortly before the encounter itself in January 1986. It is cratered and approximately spherical.

Pulkovo Observatory A historically important observatory in Russia, near St Petersburg (Leningrad). Both **F.G.W. Struve** and his son Otto were directors there. It was completely destroyed during the siege of Leningrad in World War II and rebuilt after the war, but because observational conditions there are poor, major Russian telescopes are now sited elsewhere.

pulsar An object emitting radio waves in pulses of great regularity. They were first noticed in 1967 by **Jocelyn Bell**, working at the Mullard Radio Astronomy Observatory, Cambridge. The first pulsar to be discovered pulsed every 1.337 seconds.

Pulsars are believed to be rapidly rotating **neutron stars**, the aftermaths of **supernova** explosions. A beam of radio waves emitted by the rotating pulsar sweeps past the Earth and is received in the form of pulses, in the same way as a lighthouse is seen to flash. One theory supposes that the beam is emitted along the magnetic axis of the star. With the Crab Pulsar, in the **Crab Nebula**, two pulses are observed for each rotation, which suggests that the magnetic axis is pointed almost directly at us so that we receive a beam from each pole as the star spins. The Crab Pulsar was the first to be seen flashing optically, in 1969. In some pulsars, the pulse amplitude and shape can change with time, and the pulsation can fade out temporarily.

Pulsars are gradually slowing down, but some, such as the Vela Pulsar, occasionally increase their spin rate abruptly; such an event is called a **glitch**. Over 500 pulsars are now known, flashing at rates from about 4 seconds to 1 millisecond (SEE **millisecond**

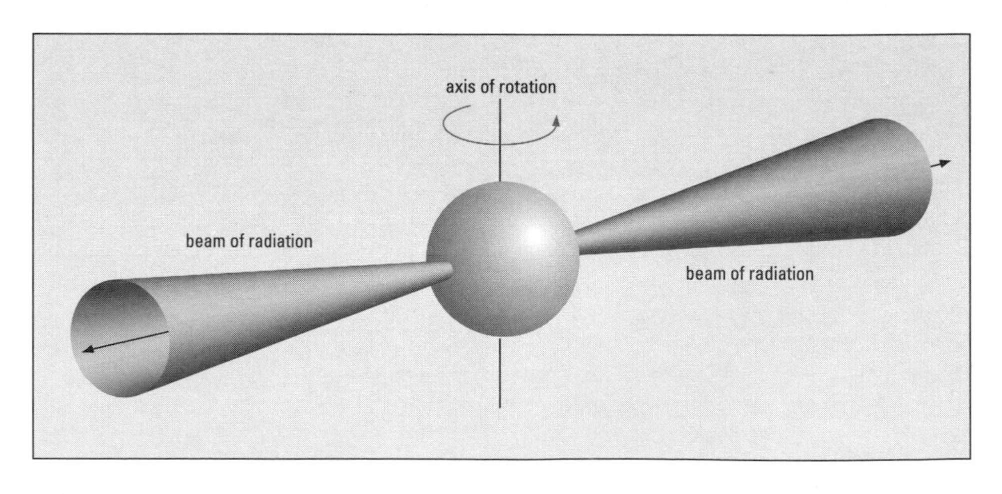

Pulsar: a rapidly rotating neutron star emitting a twin beam of radio waves

pulsar). The most common period is just under 1 second. Pulsars are concentrated towards the disk of our Galaxy, which is where supernovae most commonly occur. Most pulsars are thought to be over a million years old, so that the supernova remnants that once surrounded them would have long since dispersed and faded from view. SEE ALSO **binary pulsar**

pulsating star A type of variable star whose fluctuating brightness arises from a regular pulsation of the star's size. **Cepheid variables** and **RR Lyrae stars** are examples of pulsating stars.

Puppis A southern constellation representing the stern of the ship Argo Navis. Its brightest star is Zeta (Naos), magnitude 2.2. L^2 is a red giant semiregular variable with a range from 2.6 and 6.2, and a period of 140 days or so. Puppis is a rich constellation, with several bright open clusters such as M46, M47, M93, NGC 2451 and NGC 2477.

Purbach, Georg von (1423–61) Austrian astronomer and mathematician. He wrote an influential book on the Solar System, drew up tables of eclipses, and began a translation into Latin of the ***Almagest***, which after his death was completed by **Johann Müller**. His name is also spelt 'Peurbach' and 'Peuerbach'.

Purkinje effect The shift in the maximum colour sensitivity of the human eye, from yellow towards green, that takes place as the level of illumination falls, during **dark adaptation**. It can affect the observer's perception of star colours. (After Czech physiologist Jan Evangelista Purkinje, 1787–1869.)

Pyxis A southern constellation adjoining Vela and Puppis, representing a ship's compass. There is no star above magnitude 3.7, but there is one object of special interest, the **recurrent nova** T Pyxidis.

PZT ABBREVIATION FOR **photographic zenith tube**

Q

Quadrantids A **meteor shower**, one of the most prolific, which occurs in January. Its radiant lies in the constellation Boötes, near the borders with Hercules and Draco. The shower gets its name from a constellation called Quadrans Muralis which once occupied this region, but is no longer officially recognized.

quadrature The position of the Moon or a planet when it makes an angle of 90° with the Sun (SEE the diagram at **elongation** on page 67). The Moon is at quadrature at first and last quarter.

quasar A compact object which is an extremely luminous source of energy, radiated over a wide range of wavelengths from X-rays through optical to radio wavelengths. The name is a shortened form of *quasi-stellar object* (QSO), which is what they were first called because of their starlike appearance on photographs. Quasars were discovered in 1963 when astronomers identified optical counterparts of certain strong radio sources. However, only a small proportion of quasars are actually radio sources.

Every quasar shows a very large **redshift** which, if it results from the expansion of the Universe, indicates that quasars are the oldest and most distant objects in the Universe. In some cases, galaxies that lie between us and a quasar cause **gravitational lens** effects. Quasars seem to inhabit the centres of very remote galaxies. Since they vary in brightness with periods of a few days, they can be no larger than a few hundred astronomical units wide. A surrounding galaxy has been seen around the nearest quasars, almost lost in the glare from the brilliant centre. Quasars thus appear to be a more extreme example of the same kind of activity found in **Seyfert galaxies**.

Quasar activity is thought to be caused by a **black hole** accreting material at the centre of the host galaxy. The most luminous quasars would need to be powered by black holes of 100 million solar masses, swallowing stars at a rate of about one every year. Quasars are bright ultraviolet sources, and this hot radiation seems to come from a swirling **accretion disk**. Also associated with the inner regions of quasars are emission lines of hydrogen, helium, and iron that indicate motions as rapid as 5000 km/s, as would be expected near a massive black hole.

Quasar-like activity may be possible within every galaxy, but at any given time the black hole is accreting matter in only a small proportion of galaxies. Activity in the gas-free elliptical galaxies may produce a **radio galaxy** or **BL Lacertae object**. In galaxies such as our own, where the black hole is small, even at its most active it is swamped by the light of surrounding stars.

Because quasars are the most luminous objects in the Universe, they can be seen to greater distances, and hence further back in time, than anything else. At redshifts near $z = 4$ we see the Universe as it was about one-tenth of its present age. Therefore quasars give us a way of examining the Universe in its youth.

R

RA ABBREVIATION FOR **right ascension**

radar astronomy The use of radar techniques to study bodies in space, by transmitting continuous waves or pulses and observing the reflections, as opposed to radio astronomy, which involves only the reception of radio waves emitted naturally by objects in the Universe. Because of the restrictions of signal strength, radar astronomy is limited to objects in the Solar System.

Radar astronomy began with studies of ionized **meteor** trails in the Earth's atmosphere after World War II. These revealed the existence of daytime meteor showers, and settled a long-standing controversy by demonstrating that sporadic meteors are moving in orbit around the Sun and do not originate from interstellar space, as many astronomers had maintained.

Radar echoes from the Moon were first obtained by John H. De Witt of the US Army Signals Corps on 10 January 1946, followed the next month by Zoltan Bay in Hungary. Radar reflections from Venus were received in 1961 at NASA's Goldstone tracking station and Jodrell Bank. The resulting measurements provided an accurate new value of the solar **parallax** and hence of the scale of the Solar System. In 1965 the giant dish at **Arecibo Observatory** was used to establish the true rotation rates of Venus and Mercury.

As the power of radars grew, techniques were developed for radar mapping of planetary surfaces, which proved particularly important for cloud-covered **Venus**. Radar altimeters have subsequently been carried aboard space probes to map Venus in even greater detail. In 1991, radar mapping of Mercury from the Earth revealed what could be water-ice in craters near the poles. Other objects in the Solar System studied by radar include asteroids, Saturn's rings, planetary satellites and the nuclei of comets.

radial velocity The velocity of a body in the line of sight, either towards (positive) or away from (negative) an observer. The radial velocity is determined from the **Doppler effect** on lines in the object's spectrum. SEE ALSO **redshift**

radiant The point on the celestial sphere from which a **meteor shower** seems to radiate. Most meteor showers are identified by the constellation in which their radiant is situated.

radiation belts (radiation zones) Doughnut-shaped regions of a planetary **magnetosphere** in which charged particles become trapped by the planet's magnetic field. The Earth's radiation belts are the **Van Allen Belts**. All the giant planets have radiation belts; Jupiter's are huge in extent and 10,000 times as intense as the Earth's.

radiation pressure The pressure exerted by **electromagnetic radiation**. A beam of electromagnetic radiation can also be considered as a beam of particles called photons. When a beam of photons strikes a surface, the photons transfer their momentum to the surface and so exert a pressure. It is radiation pressure that keeps a star's outer layers from collapsing inwards under its own gravity. In interplanetary space, small particles are subjected to considerable radiation pressure from the **solar wind** (which to some extent counteracts the **Poynting–Robertson effect**). For example, the tail of a comet always points away from the Sun.

radio astronomy The study of radio emissions, **electromagnetic radiation** with wavelengths from about 1 mm to many metres, reaching the Earth from space. Observations down to about 2 cm can be made at sea level without serious interference from the atmosphere. At shorter wavelengths it is necessary to observe from high mountains to avoid atmospheric absorption, mainly by water vapour. At the long-wavelength end, 10 m to 20 m, radio waves are absorbed and scattered in the ionosphere. Between these extremes, interference from terrestrial communications is the main difficulty. International agreements set aside certain wavebands for the sole use of radio astronomy.

Radio noise from the Milky Way was discovered in 1931, by **Karl Jansky**. The first radio telescope purposely built to study such emission was made in the US by Grote Reber, who made a radio map of the sky. In 1942, James Stanley Hey in England found that the Sun was an emitter of radio waves. The subject grew rapidly after World War II as it was realized that significant information about the Universe could be obtained in this region of the spectrum.

The radio emissions first detected from the Milky Way were produced by the process of **synchrotron radiation** – the motion of high-speed electrons in the Galaxy's magnetic field. A number of localized radio sources were also found. Some of these turned out to be **supernova remnants** within our Galaxy, but optical identifications of sources led to the realization that most were far off in the Universe. These sources are now known as **radio galaxies** and **quasars**. The number of radio sources increases with distance, demonstrating that the Universe has been evolving with time. This, combined with the discovery at radio wavelengths of the **cosmic microwave background**, is strong evidence in favour of the **Big Bang** theory of the origin of the Universe. In 1967 a new class of individual radio sources was discovered in our Galaxy: the fast-flashing **pulsars**, which are ultra-dense neutron stars formed when massive stars die.

Individual spectral lines can also be detected at radio wavelengths, similar to optical spectral lines. In 1951 the emission from hydrogen at 21 cm (SEE **twenty-one centimetre line**) was detected, and has proved to be an especially useful means of studying of the structure of the Galaxy. In 1965 the first **interstellar molecule** was discovered at radio wavelengths: hydroxyl (OH). Spectral lines at radio wavelengths have now been observed from over a hundred interstellar molecules. The study of these molecular lines has stimulated the extension of radio observations into the centimetre and millimetre range.

radio galaxy A galaxy that is an abnormally high emitter of radio waves, around a million times stronger than the weak emissions from galaxies such as our own. Radio galaxies are usually giant ellipticals, such as M87 in the constellation Virgo. The emission from a typical radio galaxy is concentrated in two *lobes* well outside the galaxy, often located symmetrically either side, which appear to have been ejected from the galaxy in explosions. The biggest radio galaxies have lobes around 15 million light years across, comparable in size to a typical cluster of galaxies. The extended radio lobes are almost invariably connected to the main galaxy by jets which seem to originate from the galactic nucleus. Strong, point-like radio sources exist in the nuclei of many of these objects.

It is not understood why some elliptical galaxies emit such enormous amounts of radio energy. However, it does seem clear that the source of the activity lies in the galaxy's nucleus, and in this respect radio galaxies are similar to radio **quasars** and **BL Lacertae objects**. A radio galaxy's enormous energy output comes from **synchrotron radiation**. The central 'engine' produces vast quantities of energetic electrons travelling near the speed of light. When these electrons encounter a magnetic field, they spiral around and, in so doing, lose energy. It is this energy that we detect as radio waves.

radio interferometer A type of radio telescope that operates on the same principle as an **optical interferometer**, to achieve far higher resolution than a single dish. Single-dish telescopes typically have beamwidths of around one arc minute, which means in effect that they can only resolve angular sizes larger than this and measure positions to a few tens of arc seconds. A radio interferometer consists of two or more antennae which are spaced along a baseline and receive radio waves from the same source. The output signals from pairs of antennae are fed to a common radio receiver. The length of the baseline can be changed by varying the distance between the pairs of antennae. A small dish can be very effective as an interferometer in conjunction with a second dish of large collecting area. For *very long baseline interferometry* (VLBI), the radio telescopes can be on different continents, thereby providing a baseline the size of the Earth and achieving resolutions of a thousandth of an arc second or so.

radio telescope An instrument used to collect and record radio waves from space. The basic design is the large single dish or parabolic reflector, similar to an optical reflector, but much larger – up to 100 m (330 ft) in diameter. The smoothness and accuracy of the reflecting surface determines the shortest wavelength that can be detected, since surface irregularities must be no more than a small fraction of this wavelength.

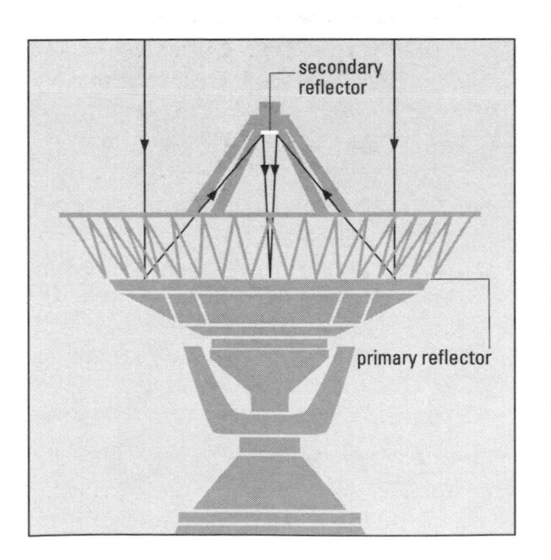

Radio telescope: a parabolic reflector

However, for longer wavelengths, such as around 21 cm, it is perfectly adequate to use open wire mesh for the reflector. Radio waves are reflected by the dish via a secondary reflector to a focus, where they are converted into electrical signals. The signals are amplified and sent to the main control room, where there is further amplification before analysis and recording. A single-dish telescope does not produce a radio 'photograph' of a source: it only collects radio energy in a narrow beam along its axis. To build up a picture of a radio source, the dish must be scanned to and fro, and the signals processed by computer.

A radio telescope needs to have high *sensitivity* (the ability to detect faint signals), and high *resolution* (the ability to see fine detail). The largest individual dishes, such as those at **Effelsberg Radio Observatory**, **Jodrell Bank** and **Parkes Observatory**, are very sensitive. They collect radio energy at a rate of only one hundred-million-millionth of a watt. Resolution depends on wavelength and aperture. Metre wavelengths are a million or so times longer than the wavelengths of light, and resolution is correspondingly poorer. At 1 m wavelength, large radio telescopes have a resolving power of around 1°, about 20 times inferior to that of the naked eye. At a few centimetres their resolution equals that of the eye, which is still much poorer than optical telescopes. Better resolution can be obtained by linking two or more dishes to form a **radio interferometer**, or by combining many of them to form an **aperture synthesis** array, which gives resolutions of a few hundredths of an arc second, 10 to 100 times better than that of the largest optical telescopes on Earth. The sensitivity remains that of the individual dishes added together. An example of such a large array is the **Very Large Array**.

radius vector The imaginary straight line between an orbiting celestial body and its primary, such as the line connecting a planet and the Sun. SEE ALSO **Kepler's laws**

Ramsden eyepiece A basic telescope eyepiece consisting of two simple elements. It suffers from **chromatic aberration**, so the **Kellner eyepiece** is usually preferred. (After English optician Jesse Ramsden, 1735–1800.)

Ranger The first series of US Moon probes. After several failures, Rangers 7, 8 and 9, in 1964 and 1965, transmitted thousands of photographs of the lunar surface as they approached predetermined impact points. The photographs showed details of boulders and craters as small as 1 m (3 ft), a much higher resolution than with Earth-based telescopes.

rays Bright streaks radiating from a crater. They consist of *ejecta* – material ejected by the impact that produced the crater. They are bright because the shock wave accompanying the impact is sufficient to vaporize rock, which re-solidifies in a glassy form. Ray craters are numerous on the Moon. The most prominent is Tycho, whose rays are seen at full Moon to extend across much of the Moon's disk. Ray craters are also found on other bodies in the Solar System, such as Jupiter's moon **Callisto**.

R Coronae Borealis star A member of a class of variable stars that are subject to sudden, unpredictable fadings. They normally spend most of their time at maximum, sometimes pulsating slowly by about half a magnitude in periods of five to seven weeks. The fadings can be anywhere between one and nine magnitudes and occur completely at random; the resultant deep minima can last for a few weeks or months, or for more than a year. There is no way of determining in advance how far the star will fade or how long it will remain at minimum. Decline is usually rapid, and the subsequent rise back to maximum is usually very much slower, with many fluctuations.

The type star, R Coronae Borealis, was first found to be variable in 1795. As shown in the diagram, it is normally at 6th magnitude, where it may remain for as long as ten years before suddenly fading. The minimum may be anywhere between 7th and 15th magnitude and can last for a few weeks or several years. Only about three dozen stars of this sort are known, although none of the others is as bright at maximum as R Coronae Borealis itself. They are thought to be stars of 1 solar mass which have evolved past the red giant stage and have ejected their hydrogen-rich envelopes to expose a carbon-rich, hydrogen-poor core. Their declines are due to obscuration by clouds of carbon particles they puff

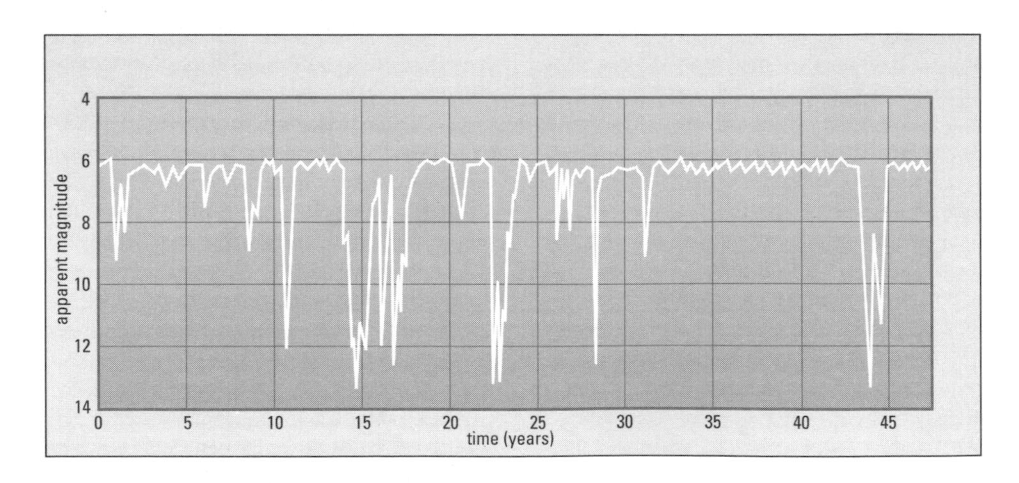

R Coronae Borealis star: light curve, showing irregular and variable minima

off. Each cloud covers only a few per cent of the star's surface, and we see a fading when a cloud is ejected in our line of sight.

recurrent nova A **nova** that is seen to erupt more than once. Recurrent novae are binary stars in which the primary is a late-type (G, K, or M) giant filling its Roche lobe and transferring material to a white dwarf secondary. At intervals of years to tens of years this material, which has accumulated on the surface of the white dwarf, explodes in a thermonuclear reaction, causing the brightness of the star to increase by 7 to 10 magnitudes for about 100 days. The longer the time between outbursts, the brighter the outbursts are. Examples are T Coronae Borealis, which erupted in 1866 and 1946, and T Pyxidis (1890, 1902, 1922, 1944 and 1966). All novae probably recur given time.

reddening A phenomenon exhibited by starlight as it is selectively absorbed and scattered on passing through clouds of interstellar dust and the Earth's atmosphere. Blue light is scattered more than red light, so an unscattered beam of starlight that continues on its way is effectively reddened. The longer the path and the greater the density of the clouds of dust, the greater the amount of reddening. A correction factor for interstellar reddening must be introduced when spectra are used for estimating the distances of remote stars, otherwise they will seem to be more distant than they are. This reddening is not to be confused with **redshift**.

red dwarf A star at the lower end of the main sequence, of spectral type K or M. Red dwarfs have masses of between 0.8 and 0.08 of a solar mass. They are of small diameter, relatively low surface temperature (between 2500 and 5000 K), and low absolute magnitude. Barnard's Star and Proxima Centauri are examples.

red giant A giant star of spectral type K or M, having a surface temperature of less than 4700 K, a diameter 10 to 100 times that of the Sun, and a luminosity 100 to 10,000 times that of the Sun. A red giant is a star in a late stage of evolution, having exhausted the hydrogen fuel in its core. Red giants lie in the upper right hand part of the **Hertzsprung–Russell** diagram. Examples are Arcturus and Aldebaran.

redshift (symbol z) A lengthening of the wavelength of light or other electromagnetic radiation from a source, caused either by the source moving away (the **Doppler effect**) or by the expansion of the Universe. It is defined as the change in the wavelength of a particular spectral line, divided by the unshifted, or rest, wavelength of that line. For example, a redshift of $z = 0.1$ means that all the wavelengths of a source are lengthened by 10%; for $z = 0.2$ they are lengthened by 20%, and so on. The speed (v) at which a source is receding can be obtained from its redshift by the formula $v = z \times c$, where c is the speed of light. For speeds greater than about one-third that of light, **Einstein**'s Special Theory of Relativity gives a less simple relation between z and v, since it is not possible to travel at the speed of light. An object with a redshift of $z = 2$, for example, is not receding from us at twice the speed of light. Instead, a redshift of 2 corresponds to a speed of 80% of the velocity of light, and a redshift of 4 corresponds to 92% of the velocity of light.

Redshifts caused by the expansion of the Universe, called *cosmological redshifts*, have nothing to do with the Doppler effect. The Doppler effect results from motion through space, whereas cosmological redshifts are caused by the expansion of space itself, which literally stretches the wavelengths of light that is travelling towards us. The longer the travel time of light, the more its wavelength is stretched, as embodied in the **Hubble law**. The most highly redshifted photons are those in the **cosmic microwave background** radiation. These photons have been travelling towards us since matter and radiation decoupled from each other some 300,000 years after the **Big Bang**. They exhibit a redshift of z = 1000, meaning that their wavelengths have been stretched a thousandfold. Since the most remote quasars have redshifts of about 5, there is a vast region of space in which no objects have yet been seen.

Gravitational redshift is a phenomenon predicted by Einstein's General Theory of Relativity. Light emitted by a star has to do work to overcome the star's gravitational field. As a result there is a slight loss of energy and a consequent increase in wavelength, so that all the spectral lines are shifted towards the red. The effect was first observed in 1925 in the spectrum of the white dwarf companion of Sirius.

reflecting telescope (reflector) A telescope that forms an image by the reflection of light from a primary concave mirror (usually parabolic) at the back of the tube, via a secondary mirror, to a focus whose position depends on the design of the instrument. The chief optical designs for smaller, amateur reflectors are the **Newtonian** and **Schmidt–Cassegrain**. Larger, professional instruments include the **Cassegrain**, **coudé** and **Schmidt camera** configurations. Many large professional reflectors incorporate two or more different optical configurations, and with some it is possible to site instruments at the **prime focus**. Reflectors have the advantage of being free from **chromatic aberration**.

A reflector's mirrors are usually made of low-expansion glass or a ceramic material and coated with a thin layer of aluminium, which is highly reflective. Reflectors can be made larger than refractors because the primary mirror can be supported at the back, whereas a lens can be supported only around its rim, and is consequently liable to distort under its own weight. Also, new ways of building and controlling mirrors, as in the segmented mirrors of the **Keck Telescopes**, and the technique of **active optics**, have made possible further increases in the size and performance of big reflectors.

The Scottish mathematician James Gregory (1638–75) was the first to describe a reflecting telescope; the first to buid one was **Isaac Newton**, in 1668. In the late 18th century **William Herschel** showed the instrument's potential by making large, accurately figured mirrors. SEE ALSO **Dobsonian telescope**, **Maksutov telescope**, **Ritchey–Chrétien telescope**

reflection nebula A nebula in which starlight is reflected off the dust particles mixed in with the gas. Reflection nebulae generally appear blue in photographs; they are always bluer than the star that illuminates them because blue light is scattered more efficiently than red light. Some reflection nebulae, notably the infrared cirrus discovered by the Infrared Astronomical Satellite (SEE **infrared astronomy**), are lit not by an individual star but by the combined light of the whole Galaxy.

refracting telescope (refractor) A telescope that forms an image by the refraction of light by a **lens** at the front of the tube. This lens, called the *objective*, is a compound lens, of perhaps more than one type of high-quality glass in order to minimize **chromatic aberration**, and it may be coated to reduce reflection and so increase the amount of light transmitted to the eyepiece. The tube is long so as to reduce spherical aberration. Refractors are still the first choice of some amateur observers for subjects such as the planets, but the difficulties of making a large, optically perfect lens and supporting its weight mean that all large professional instruments are now **reflecting telescopes**.

The origins of the refractor are uncertain. The Dutch spectacle-maker Hans Lippershey (*c.* 1570–1619) applied for a patent on one in 1608, although others were making similar instruments around the same time, and telescopes of some sort may have been in use at the end of the 15th century. **Galileo** was among the first to use the refractor in astronomy. Improvements were made by a

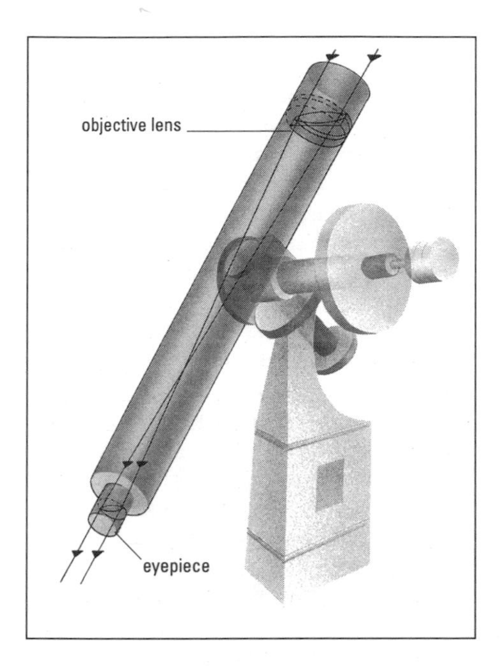

Refracting telescope: light path

succession of opticians, including **John Dolland** and **Joseph von Fraunhofer**, who devised means of reducing aberration, and **Christiaan Huygens** and Jesse Ramsden (SEE **Ramsden eyepiece**), who devised better eyepieces. The largest refractor, at the Yerkes Observatory, has an aperture of 1 m (40 inches) and was built in 1897.

refraction SEE **atmospheric refraction, refracting telescope**

Regiomontanus SEE **Müller, Johann**

regolith The surface dust and debris covering the surface of the Moon and other bodies in the Solar System. It is produced by the disintegration of crustal rocks caused by the impact of meteorites.

regression of the nodes The slow westward motion of the **nodes** of the Moon's orbit, due to the gravitational pull of the Sun. A full circuit takes 18.6 years.

Regulus The star Alpha Leonis, magnitude 1.35, distance 69 light years, luminosity 110 times that of the Sun. It is a main-sequence star of type B7.

relative sunspot number (symbol *r*) An index of sunspot activity which reduces the sunspot counts of different observers to a common, statistical basis. For a particular day, it is given by

$$r = k(f + 10g),$$

where g is the number of groups of sunspots, irrespective of the number of spots each contains, and f is the total number of spots in all the groups; k is a factor based on the estimated efficiency of observer and telescope. The work of recording sunspots in this way was begun by Swiss astronomer Rudolf Wolf (1816–93) at the Zurich Federal Observatory, and the terms *Wolf number* and *Zurich relative sunspot number* were both formerly used.

relativity SEE **Einstein, Albert**

residuals The differences between predicted and observed values, frequently found on analysis of astronomical observations. Residuals in, say, obervations of the position of a planet, if collected over a sufficiently long period of time and then analysed statistically, may reveal some long-term trend. They may also result from inaccurate observations.

resolution (1) SEE **resolving power**

resolution (2) The size of the smallest detail distinguishable by an imaging system at a particular distance, or of the smallest detail visible in an image (of, for example, a photograph taken by a space probe).

resolving power (resolution) The ability of an optical system, such as a telescope or the human eye, to distinguish close but separate objects such as the components of a close binary star, or a single small object. It is measured in angular units. The resolving power of a telescope in seconds of arc is found by dividing 110 by the aperture in millimetres (or 4.56 by the aperture in inches). This measure is called *Dawes' limit*, after the English amateur astronomer William Rutter Dawes (1799–1868). The resolving power of the human eye is about one minute of arc. SEE ALSO **radio telescope**

resonance An effect produced by the gravitational interaction between two bodies orbiting the same primary in orbits whose periods are **commensurable**. Resonance between bodies of similar size can lock bodies into commensurable orbits, as has happened with the **Galilean satellites** of Jupiter. Between bodies of highly dissimilar size, however, it can prevent small bodies from occupying commensurable orbits. This is what keeps clear the **Kirkwood gaps** in the asteroid belt and the various gaps in the rings of **Saturn**.

Reticulum A small but distinctive southern constellation, introduced by **Lacaille**, representing a reticle used for measuring star positions. Its leading star is Alpha, magnitude 3.4, which forms a compact pattern with Beta, Gamma, Delta and Epsilon.

retrograde motion (1) Orbital or rotational motion in the opposite direction to that of the Earth: clockwise as seen from above the Sun's north pole. Some satellites, such as the outermost ones of Jupiter, have a retrograde orbital motion, and so do certain comets, notably Halley's Comet. Venus, Uranus, and Pluto all have retrograde rotation. Orbital or rotational motion is retrograde if the orbital or axial inclination is more than 90°. COMPARE **direct motion** (1)

retrograde motion (2) The temporary east-to-west motion traced out by a superior planet or asteroid on the celestial sphere as the Earth catches up and overtakes it. This reversal is called a *retrograde loop*. The places at which the planet reverses direction are termed *stationary points* (points 4 and 6 in the diagram). COMPARE **direct motion** (2)

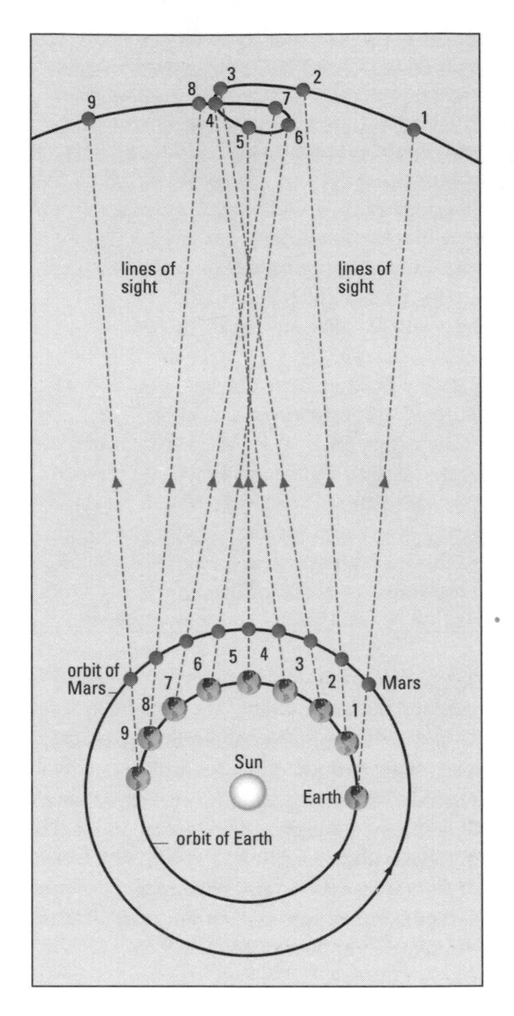

Retrograde motion (2)

revolution The movement of a planet or other celestial object around its orbit, as distinct from rotation of the object on its axis.

Rhea Saturn's second-largest **satellite**, diameter 1530 km (950 miles), discovered by Giovanni Domenico Cassini in 1672. It is a typical heavily cratered icy world, but with few large craters, the largest being the 250 km (150 mile) diameter Izanagi. Some resurfacing seems to have taken place in the past, but there is little sign of more recent geological activity, although wispy streaks poorly imaged by Voyager 1 may be signs of tectonic activity.

rich-field telescope (richest field telescope) (RFT) A low-power telescope equipped with a wide-angle eyepiece (such as a **Nagler eyepiece**) designed to show a wide field of view at a relatively low magnification. Such an instrument is ideal for studying starfields and hunting for novae.

Rigel The star Beta Orionis, the brightest star in the constellation Orion, and marking Orion's left leg. It is of magnitude 0.12 (the seventh-brightest star in the sky), and a supergiant of spectral type B8. Its distance is not accurately known, but is thought to be about 1000 light years. Rigel's luminosity is estimated to be over 100,000 times that of the Sun.

right ascension (RA, symbol α) A coordinate used to define the position of a celestial object. It is the angular distance measured eastwards from the vernal equinox (also known as the First Point of Aries) to the point where the hour circle of an object meets the celestial equator. RA is usually expressed in hours, minutes, and seconds, although it may be expressed as an angle, 1 hour of RA being equivalent to 15°. SEE ALSO **celestial sphere**

Rigil Kentaurus Another name for **Alpha Centauri**.

rille A cleft on the surface of a lunar mare. Rilles that are curved or straight are geological faults. The winding *sinuous rilles* are most likely collapsed lava tubes. Lava was able to flow on the Moon for hundreds of kilometres. The top and sides of a flow can solidify, insulating the molten lava inside which continues to flow. When such a tube has been drained, the roof is liable to collapse, leaving what we see today as a sinuous rille. There are similar features on Uranus's satellite **Ariel**.

ring, planetary A narrow ring or thin disk of large or small particles, or both, orbiting in the equatorial plane of a planet. **Jupiter**, **Saturn**, **Uranus**, and **Neptune** all have systems of rings. Saturn's, by far the most prominent and extensive, were discovered telescopically in the 17th century. Jupiter's were found in 1979 by Voyager 1;

PLANETARY RINGS

Ring or gap	Distance from centre (thousand km)	Width (km)
JUPITER		
Halo	100–122.8	22800
Main	122.8–129.2	6400
Gossamer	129.2–214.2	85000
SATURN		
D	67–74.5	7500
C (Crepe)	74.5–92	17500
Maxwell Gap	88	270
B	92–117.5	25500
Cassini Division	117.5–122.2	4700
A	122.2–136.8	14600
Encke Gap	133.6	325
Keeler Gap	136.5	35
F	140.2	30–500
G	165.8–173.8	8000
E	180–480	300000
URANUS		
1986U2R	37–39.5	2500
6	41.8	1–3
5	42.2	2–3
4	42.6	2–3
Alpha	44.7	7–12
Beta	45.7	7–12
Eta	47.2	0–2
Gamma	47.6	1–4
Delta	48.3	3–9
1986U1R	50	1–2
Epsilon	51.1	20–100
NEPTUNE		
Galle	41.9–43.6	1700
Le Verrier	53.2	15
Plateau	53.2–59.1	5900
Adams	62.9	50

those of Uranus and Neptune were first observed during **occultations**, and were confirmed by Voyager 2 in the 1980s.

Several important factors governing ring systems were established in the 19th century. Saturn's rings were found in 1849 to lie inside the **Roche limit**; any sizeable satellite inside this limit would be broken up by gravitational forces, and any smaller bodies would be prevented from accreting into a larger one. In 1857 the Scottish physicist James Clerk Maxwell (1831–79) showed that on theoretical grounds the rings had to be made up of many separate particles, and this was proved to be so in 1895 by the American James Keeler (1857–1900), who found that Doppler shifts in the spectra of sunlight reflected from the inner and outer parts of the ring system were different.

The ring systems around the four giant planets show similarities and dissimilarities. For example, Saturn's main rings are made up of bright, icy particles, whereas the other three planets have dark rings. Some rings are very narrow and distinct, others are continuous sheets, but still a kilometre or less thick, and in the equatorial plane, while others are large, tenuous, doughnut-shapes haloes. Small **shepherd moons** keep some rings in check, while larger satellites create gaps. Particle sizes range from micrometre-sized in Jupiter's main ring to as much as 10 m (30 ft) in Saturn's B Ring. Continual collisions between ring particles make rings spread inwards and outwards; particles can be moved outwards by a magnetic field (as in Saturn's spokes), and inwards by the **Poynting–Robertson effect**.

ring galaxy A galaxy having the form of an elliptical ring around the nucleus, like the rim of a wheel. Ring galaxies are thought to form as the result of a collision when a small galaxy passes through a larger one.

Ring Nebula The planetary nebula M57 in the constellation Lyra, also known as NGC 6720. It appears as a hollow ellipse; in three dimensions it is shaped not like a spherical shell but like a toroid, or ring doughnut. The nebula is centred on a blue star of 0.2 of a solar mass. It is expanding at 19 km/s (12 mile/s) and is estimated to be about 5500 years old.

Ritchey–Chrétien telescope A modified form of **Cassegrain telescope**, designed originally by George Willis Ritchey (1864–1945) and Henri Chrétien (1879–1956). The primary and secondary mirrors are hyperboloidal, although there are variants with near-hyperboloidal, paraboloidal, or ellipsoidal components. The system is free from **coma** (1) over a wide field, though there is some astigmatism and the field is curved.

Roche limit (symbol R) The minimum distance at which a planetary satellite can remain in orbit without being torn apart by gravitational forces; alternatively, the minimum distance at which smaller bodies in orbit round a planet can combine by **accretion** to form a larger one. A planet's Roche limit is about 2.5 times its radius (r). It depends, though not strongly, on the densities of planet (d_p) and satellite (d_s):

$$R = 2.44\, r^3 \sqrt{(d_p/d_s)}.$$

Actually the Roche limit applies only to 'fluid' satellites. Rocky bodies with a high structural integrity, such as Neptune's innermost satellites, can exist inside the Roche limit. All planetary ring systems lie inside their planet's Roche limit. The concept was originated by the French astronomer and mathematician Edouard Roche (1820–83).

Roche lobe A surface defining the maximum size of a star in a binary system. If one star's surface overflows this size as it evolves, the companion's gravity will pull matter off it.

Römer, Ole Christensen (1644–1710) Danish astronomer. He was the first to measure the speed of light. Working at Paris in 1675, he found that the predictions of the times of eclipses of the **Galilean satellites**

made by **Giovanni Domenico Cassini** did not match the observed times. Intervals between successive eclipses decreased as the orbital motions of the Earth and Jupiter brought them closer together, and increased as they moved further apart. He deduced that the differences were caused by light taking time to travel the intervening distance. His timings gave the speed of light as 225,000 km (140,000 miles) per second, three-quarters of the true value. Römer also invented the first **transit instrument**.

Roque de los Muchachos Observatory A Spanish observatory on La Palma in the Canary Islands, the home of telescopes owned by several European nations, foremost among them the UK's **William Herschel Telescope**. Other major instruments include the 2.5 m (100-inch) Isaac Newton Telescope, moved from England in 1984, and the 2.5 m Nordic Optical Telescope, opened in 1989. A 3.5 m (138-inch) Italian telescope called Galileo is expected to be set up in 1995.

Rosalind One of the small inner **satellites** of Uranus discovered in 1986 during the Voyager 2 mission.

Rosat A German satellite containing an X-ray telescope and a British wide-field camera that made the first survey the sky at extreme ultraviolet wavelengths. It was launched in 1990.

Rosetta A European Space Agency probe planned for launch in the 21st century. It will rendezvous with Comet Wirtanen in 2016, taking samples from the nucleus and returning them to Earth.

Rosette Nebula A diffuse nebula about 5000 light years away in the constellation Monoceros, surrounding the open cluster NGC 2244. The stars of the cluster were born within the past half million years from the Rosette Nebula and now illuminate it. Radiation pressure from the stars has created a cavity about 12 light years across in the central part of the nebula. This cavity will expand through the nebula, which will thus disperse in a few million years.

Rosse, Third Earl of (William Parsons) (1800–67) Irish amateur astronomer. In the grounds of Birr Castle, near Parsonstown in Ireland, and aided only by estate labourers whom he trained, he built a 72-inch (1.83 m) reflector. Completed in 1845, it was the largest telescope in the world for the rest of its lifetime. Lord Rosse studied in particular the hazy cloud-like objects then known collectively as nebulae. In 1845 he discovered the spiral structure of what is now called the **Whirlpool Galaxy** and other spiral nebulae (recognized as galaxies in the 1920s). He was able to resolve certain nebulae into groups of stars. He also discovered the ring structure of planetary nebulae. His son **Laurence Parsons** (1840–1908), the fourth earl, continued the work at Birr Castle; another son, **Charles Algernon Parsons** (1854–1931), was an outstanding engineer who continued the telescope-making business of **Howard Grubb**.

rotation The turning of a celestial body on its axis, as distinct from its orbital revolution.

Royal Greenwich Observatory (RGO) The UK national astronomical observatory, founded at Greenwich, London, in 1675. After World War II the observatory moved to Herstmonceux in Sussex, and in 1990 it moved again, to Cambridge. The RGO operates the UK telescopes at the **Roque de los Muchachos Observatory** on La Palma in the Canary Islands.

RR Lyrae star A pulsating variable star of a class also known as *short-period Cepheids* or *cluster variables* since the first were found in globular clusters. However, RR Lyrae, the type star, was the first to be found outside a globular cluster. RR Lyrae stars are Popula-

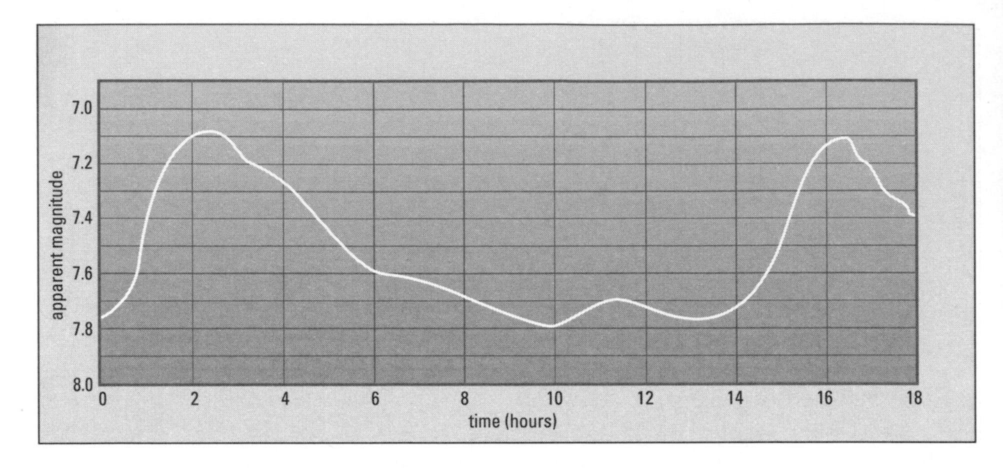

RR Lyrae star: light curve, showing rapid rise and slow decline

tion II giants of spectral type between A and F. Their periods range from 0.2 to 1.2 days and their amplitudes from 0.2 to 2.0 magnitudes, although most have periods of between 9 and 17 hours. Their light curves differ from those of the so-called classical Cepheids; that of RR Lyrae itself is shown in the diagram. Most rise quickly to maximum in no more than a tenth of their total period. Their minima are comparatively long, so that for a few hours their light remains roughly constant. The absolute magnitude of all RR Lyrae stars is about +0.5, so they can be used as 'standard candles' for distance finding. However, they are too faint to be seen beyond the nearest galaxies. SEE ALSO **Cepheid variable**

runaway star A star of spectral type O or B which has an unusually high velocity in space. Such stars are former members of massive binaries, one member of which has exploded as a supernova, flinging off the other like a slingshot. Three runaway stars with a common origin in Orion – 53 Arietis, AE Aurigae, and Mu Columbae – may result from a supernova in what was a quadruple star system three million years ago. It has been estimated that as many as 20% of O stars are runaways.

Russell, Henry Norris (1877–1957) American astronomer. In 1913 he produced a diagram plotting the absolute magnitude of stars against their spectral type, which showed the division between main-sequence stars and giant stars. Unknown to him, **Ejnar Hertzsprung** had done the same some years previously. The diagram, now called the **Hertzsprung–Russell diagram**, is the starting-point for all modern theories of stellar evolution, although Russell first thought, erroneously, that stars evoloved along the main sequence. In 1928 he determined the composition of the Sun's atmosphere from a study of its spectrum.

RV Tauri star A member of a class of pulsating variable stars with alternating primary and secondary minima. RV Tauri stars are luminous supergiants of spectral type between F and M, with periods ranging from 30 to 150 days and amplitudes of up to 4 magnitudes. The diagram shows the light curve of AC Herculis, which has a constant mean magnitude, but with some other stars of this class, such as DF Cygni, the mean varies by up to 2 magnitudes over a period of 600 to 1500 days. RV Tauri stars are undergoing pulsations of more than one period, and appear to be giving off material

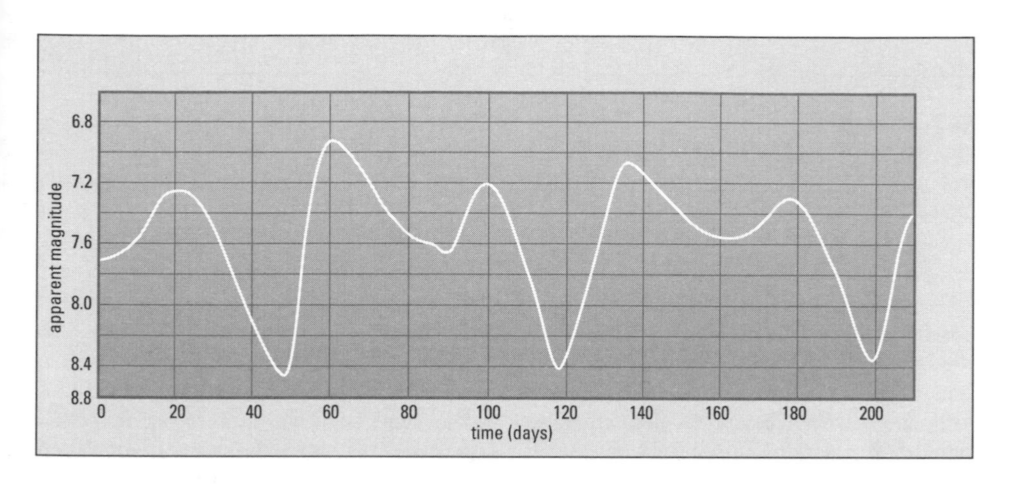

RV Tauri star: light curve, showing alternating deep and shallow minima

in a strong **stellar wind** in what may be the early stages of formation of a **planetary nebula**.

Ryle, Martin (1918–84) English radio astronomer. In 1946 he discovered **Cygnus A**, the first radio galaxy to be found. In 1955 he built the first **radio interferometer**, and in 1960 developed the technique of **aperture synthesis**. For his pioneering work in radio astronomy he received a share of the 1974 Nobel Prize for Physics.

S

Sagan, Carl Edward (1934–) American astronomer. He has studied many aspects of the Solar System, including the physics and chemistry of planetary atmospheres and surfaces, particularly of Venus and Mars. He has investigated the origins of terrestrial life (discovering in 1963 that ATP, an important biochemical, is formed when chemicals thought to be present on the early Earth are mixed together) and the possibility of life existing elsewhere. Sagan has done much to popularize astronomy and science in general.

Sagitta A small constellation representing an arrow, adjoining Aquila. Its brightest star is Gamma, magnitude 3.5. The arrow shape is completed by the 4th-magnitude stars Alpha, Beta, and Delta

Sagittarius The southernmost of the zodiacal constellations, notable for containing the dense star clouds that mark the direction of the centre of our Galaxy. The outline of its main stars is sometimes likened to a teapot. Its brightest star is Epsilon, magnitude 1.8, followed by Sigma (Nunki), 2.0. There are plenty of variable stars, including the Cepheids X and W Sagittarii, both of which range between magnitudes 4 and 5. The Milky Way is at its richest in Sagittarius, and there are no less than 15 Messier objects – globular clusters, open clusters, and gaseous nebulae, including the Omega Nebula (M17), the **Lagoon Nebula** and the **Trifid Nebula**.

Sagittarius A A strong radio source at the centre of our Galaxy. The exact centre of the Galaxy is thought to be marked by a small part of this source, referred to as *Sagittarius A**, less than 20 AU across. Possibly it is an **accretion disk** around a black hole. Estimates of the hole's mass range from as high as 5 million solar masses to as low as 100 solar masses.

Sagittarius dwarf galaxy The closest companion galaxy of the Milky Way, 50,000 light years beyond our Galaxy's nucleus, first detected in 1994. In 100 million years it could be absorbed into the Galaxy by **cannibalism**.

Saha, Meghnad (1894–1956) Indian astrophysicist and nuclear physicist. His theoretical work on spectra, the solar corona, radiation, and ionization resulted in the 1920s in an equation now known as the *Saha equation*. This relates the degree of ionization of atoms (i.e. how many electrons they have lost) in a star's atmosphere to the temperature. It provides an important key to the differences in the spectra of stars of different spectral types, and allows their temperatures to be deduced from the lines present in their spectra. Saha's later work in nuclear physics led to the founding of an Indian institute for the study of the subject.

Salpeter process SEE **triple-alpha process**

Sandage, Allan Rex (1926–) American astronomer. Initially an assistant to **Edwin Hubble**, he has continued to work towards establishing the value of the **Hubble constant** and thus fixing the age of the Universe. In 1960 Sandage made the first identification at optical wavelengths of what would prove to be a **quasar**, and five years later he found the first radio-quiet quasar.

Sakigake A Japanese probe to Halley's Comet. SEE **Suisei**

saros A cycle of lunar and solar eclipses lasting just over 18 years. A saros consists of the same sequence of eclipses repeated at nearly the same time intervals. This is because 223 **lunations** (6585.32 days) are almost equal to 19 **eclipse years** (6585.78 days), after which time the Sun, the Moon, and the Moon's

nodes return to very nearly the same positions relative to one another. This period was known to many ancient peoples, including the Babylonians and the Mayans. Knowledge of it enabled astronomers to predict eclipses, since, if an eclipse of the Sun or the Moon is observed, then after a complete saros there will be a similar eclipse. The difference of 0.46 of a day between the number of lunations and eclipse years in one saros means that each eclipse in the sequence will recur about 165° further west. In predicting a date, allowance must be made for leap days. SEE ALSO **Metonic cycle**

satellite (moon) A celestial body in orbit around a planet. Of the major planets in the Solar System, Mercury and Venus have no known satellite. The Earth has one, Mars two, Jupiter sixteen, Saturn eighteen, Uranus fifteen, Neptune eight, and Pluto one. There are probably more small satellites of the giant planets awaiting discovery. At least one asteroid, **Ida**, is known to have a satellite, and other satellites of asteroids are suspected.

The satellites of the Solar System vary enormously in their size, orbit, surface features, and supposed origin. One, Jupiter's

SATELLITES OF THE MAJOR PLANETS

	Diameter (km)	Distance from centre (thousand km)	Orbital period (days)	Mean opposition magnitude
EARTH				
Moon	3,476	384	27.32	−12.7
MARS				
Phobos	27 × 22 × 19	9.4	0.32	11.3
Deimos	15 × 12 × 11	23.5	1.26	12.4
JUPITER				
Metis	40	128	0.29	17.5
Adrastea	25 × 20 × 15	129	0.30	19.1
Amalthea	270 × 166 × 150	181	0.50	14.1
Thebe	110 × 90	222	0.67	15.7
Io	3,630	422	1.77	5.0
Europa	3,138	671	3.55	5.3
Ganymede	5,262	1,070	7.15	4.6
Callisto	4,800	1,883	16.69	5.7
Leda	16	11,094	238.7	20.2
Himalia	186	11,480	250.6	14.8
Lysithea	36	11,720	259.2	18.4
Elara	76	11,737	259.6	16.8
Ananke	30	21,200	631 *R*	18.9
Carme	40	22,600	692 *R*	18.0
Pasiphae	50	23,500	735 *R*	17.0
Sinope	36	23,700	758 *R*	18.3
SATURN				
Pan	20	134	0.58	—
Atlas	40 × 20	138	0.60	18
Prometheus	140 × 100 × 80	139	0.61	16
Pandora	110 × 90 × 70	142	0.63	16

R indicates retrograde orbit

continued

SATELLITES OF THE MAJOR PLANETS *continued*				
	Diameter (km)	Distance from centre (thousand km)	Orbital period (days)	Mean opposition magnitude
SATURN *(cont.)*				
Epimetheus	140 × 120 × 100	151	0.69	15
Janus	220 × 200 × 160	151	0.69	14
Mimas	392	186	0.94	12.9
Enceladus	500	238	1.37	11.7
Tethys	1,060	295	1.89	10.2
Telesto	34 × 28 × 26	295	1.89	18.5
Calypso	34 × 22 × 22	295	1.89	18.7
Dione	1,120	377	2.74	10.4
Helene	36 × 32 × 30	377	2.74	18
Rhea	1,530	527	4.52	9.7
Titan	5,150	1,221	15.95	8.3
Hyperion	410 × 260 × 220	1,481	21.28	14.2
Iapetus	1,460	3,561	79.33	10.2–11.9
Phoebe	220	12,952	550.5 *R*	16.5
URANUS				
Cordelia	26	50	0.34	24.1
Ophelia	30	54	0.38	23.8
Bianca	42	59	0.43	23.0
Cressida	62	62	0.46	22.2
Desdemona	54	63	0.47	22.5
Juliet	84	64	0.49	21.5
Portia	108	66	0.51	21.0
Rosalind	54	70	0.56	22.5
Belinda	33	75	0.62	22.1
Puck	154	86	0.76	20.2
Miranda	470	129	1.41	16.3
Ariel	1,158	191	2.52	14.2
Umbriel	1,172	266	4.14	14.8
Titania	1,580	436	8.71	13.7
Oberon	1,524	584	13.46	13.9
NEPTUNE				
Naiad	58	48	0.29	24.7
Thalassa	80	50	0.31	23.8
Despina	148	53	0.33	22.6
Galatea	158	62	0.43	22.3
Larissa	208 × 178	74	0.55	22.0
Proteus	436 × 416 × 402	118	1.12	20.3
Triton	2,706	355	5.88 *R*	13.5
Nereid	340	5,513	360.1	18.7
PLUTO				
Charon	1,270	20	6.39	16.8

R indicates retrograde orbit

Ganymede, is bigger than Mercury, and seven are bigger than Pluto, while the smallest are tiny bodies a few kilometres across. Their orbits vary just as much, from Phobos, which circles Mars's equator more than three times a day, to the outer satellites of Jupiter, which take more than two Earth years to complete their inclined, eccentric orbits. In their surface features they are remarkably diverse, ranging from dead, densely cratered worlds to ones which show a fascinating variety of geological processes. Indeed, Jupiter's Io is the most geologically active body known. As to origin, some, such as Jupiter's **Galilean satellites**, may have condensed out of the same part of the solar nebula as their parent planet, while several small ones are believed to be captured asteroids. The **Moon** is currently thought to have formed out of debris flung out when a Mars-sized body collided with the Earth.

Data for all named planetary satellites are given in the table, and each has an entry of its own. SEE ALSO **artificial satellite**

Saturn The sixth major planet from the Sun, and the second-largest of the four giant planets. It was the most remote planet known in the pre-telescope era. Viewed through a telescope it appears as a flattened golden yellow disk encircled by white rings. The disk has surface markings and belts similar to, but fainter than, those on Jupiter. Because of Saturn's axial inclination, the angle at which the ring system is presented to us depends on the planet's position in its orbit, and this has a considerable effect on its brightness. At perihelic opposition with the rings fully open, Saturn is magnitude −0.3; at aphelic opposition with the rings edge-on, it is magnitude 0.8. The planet's apparent size varies from 15 to 21 arc seconds (rings 35 to 48 arc seconds). Saturn has 18 known satellites, more than for any other planet. The main data for Saturn are given in the table.

Saturn's magnificent system of rings is its dominant feature. They were first observed by **Galileo** in 1610, but Christiaan Huygens was the first to identify them as rings, in 1656. (For further historical details and for the dimensions of the rings, SEE **ring, planetary**.) As seen from the Earth, the ring system consists of a few principal components, separated by a number of gaps. These were discovered gradually, as telescopes improved, over a 300-year period from Huygens' time until just before the first fly-by, by Pioneer 11 in 1979. The three main rings, starting from the outside, are Ring A, which is greyish-white, Ring B, which is bright white, and Ring C, which is fainter and blue-grey. Ring D, the innermost, was discovered in 1969; Rings E, F and G, which lie outside Ring A, were discovered by space probes. The most prominent gap is the Cassini Division, between Rings A and B; the Encke Gap in Ring A is also quite prominent. These gaps may be produced by the perturbing effect of Saturn's inner satellites, in the same way that Jupiter's gravitational influence maintains

SATURN: DATA	
Globe	
Diameter (equatorial)	120,000 km
Diameter (polar)	107,100 km
Density	0.69 g/cm^3
Mass	5.69×10^{26} kg
Sidereal period of axial rotation (equatorial)	10h 14m
Escape velocity	35.5 km/s
Albedo	0.70
Inclination of equator to orbit	26° 43′
Temperature at cloud-tops	95 K
Surface gravity (Earth = 1)	1.19
Orbit	
Semimajor axis	9.539 AU = 1427×10^6 km
Eccentricity	0.056
Inclination to ecliptic	2° 29′
Sidereal period of revolution	29.46y
Mean orbital velocity	9.65 km/s
Satellites	18

the **Kirkwood gaps** in the asteroid belt. The rings are made up of particles ranging from dust to objects a few metres in size, all in individual orbits around Saturn. The main rings are only a kilometre or so thick, although the very tenuous Rings G and E may be up to 1000 km (600 miles) thick. Ring A's diameter is about 275,000 km (170,000 miles); this is the extent of the ring system as seen from Earth. Following Pioneer 11, the two Voyager probes flew past Saturn, Voyager 1 in 1980 and Voyager 2 in 1981. They revealed the ring system to be made up of thousands of separate ringlets. The gaps were not empty, but simply regions with a lower ring density. Some rings were eccentric, not circular. Ring F has a braided structure, being made up of intertwined ringlets and controlled by **shepherd moons**. *Spokes*, dark radial features in Ring B, apparently mark the trajectories of ring particles being moved outward along magnetic field lines.

Saturn's disk shows features similar to Jupiter's, but rather less dynamic and more muted in colour because of a high-altitude haze layer. There are dark and light yellow bands parallel to the equator, termed *belts* and *zones*, as on Jupiter (with a similar nomenclature – SEE the table at **Jupiter** – except that there are no 'north north' or 'south south' components). Generally the bands are more straight-edged and less turbulent than Jupiter's; there is no equivalent of the Great Red Spot, although short-lived white spots occasionally appear. The most prominent erupt near the equator, and with a 57-year period: there were major eruptions in 1876, 1933, and 1990. The last of these grew until it was spread around much of the equator, and was referred to as the Great White Spot. There is **differential rotation** on Saturn: features at the equator rotate in about 10¼ hours, those near the poles in nearly 10¾ hours (the rotation period of the interior). Winds at the equator can reach 1800 km/h (1100 mile/h).

Saturn has an internal heat source, which probably drives its weather systems. Its density, 0.7 g/cm^3, is much lower than for any other planet (and its oblateness is higher than for any other), and it also has much more of its mass concentrated at the centre. It is therefore assumed to be composed predominantly of hydrogen, and to have an iron–silicate core about five times the Earth's mass, surrounded by an ice mantle of perhaps twenty Earth masses. Near the mantle the hydrogen would be in metallic form (so compressed it behaves as a conducting metal); above this it would exist as liquid molecular hydrogen. The upper atmosphere contains 97% hydrogen and 3% helium, with traces of other gases including methane, ethane, ammonia and phosphine.

The magnetic field of Saturn is 500 to 1000 times the strength of the Earth's, and unlike the other giant planet's fields the magnetic axis is nearly coincident with the axis of rotation. The magnetic field generates a **magnetosphere** intermediate between those of the Earth and Saturn in its shape, extent, and the intensity of its radiation belts of trapped particles.

Saturn has eighteen known **satellites**, although several more have been inferred from Voyager photographs. They include **Titan**, the only satellite in the Solar System to have a significant atmosphere.

Saturn Nebula The planetary nebula NGC 7009 in Aquarius. Handle-like protuberances give it the appearance of Saturn. The nebula is small, of 8th magnitude, and its appearance probably indicates an internal structure consisting of a series of shells ejected by the central star.

scattering The deflection of light or other types of electromagnetic wave by particles. Where the particles are very much larger than the wavelength, scattering consists of a mixture of reflection and **diffraction**, and the amount of scattering depends very little on wavelength. Where the particles are very much smaller than the wavelength, the amount of scattering (*d*) is inversely proportional to the fourth power of the wavelength (λ) : $d \propto 1/\lambda^4$. Thus blue light is scattered by small particles ten times as much as red light. Scattering of sunlight by atoms and molecules in the atmosphere is what makes the sky blue. It is the main cause of **atmospheric extinction**, making the Sun appear red at sunrise and sunset by preferentially scattering blue light out of the line of sight.

Schiaparelli, Giovanni Virginio (1835–1910) Italian astronomer. He was an assiduous observer of the inner planets, Mars in particular. He prepared a map of the surface of Mars, introducing a nomenclature for the various features, that remained standard until the planet was mapped by space probes. His use of the term *canali* (SEE **canals**) was the cause of much subsequent controversy. He also prepared maps of Venus and Mercury, believing – mistakenly – that they had **synchronous rotation**. Schiaparelli discovered the connection between **meteor streams** and comets, and studied double stars.

Schiefspiegler SEE **tri-Schiefspiegler**

Schmidt, Maarten (1929–) Dutch–American astronomer. He is known for his work on **quasars**, in particular for his discovery in 1963 of the immense **redshift** of the lines in the spectrum of the quasar designated 3C 273. This, and his subsequent finding that the number of quasars increases with distance, provided important support for the Big Bang theory. Schmidt also ascertained the spiral structure of the Galaxy and the distribution of mass within it.

Schmidt camera (Schmidt telescope) A wide-field **catadioptric** telescope used mainly for photography, developed in 1930 by the Estonian instrument-maker and astronomer Bernhard Schmidt (1879–1935).

The large telescopes of the time were reflectors whose primary mirrors were paraboloidal in order to overcome spherical **aberration** (2). This meant, however, that their effective field of view was limited, sometimes to as little as a few minutes of arc, because away from the centre of the field the image became rapidly affected by **coma** (1), especially in instruments having the small **focal ratio** necessary for good images of extended objects. Schmidt's solution was to use a spherical primary mirror but with a correcting plate, placed at the mirror's centre of curvature, figured so as to minimize optical aberrations in the image. The image is formed on a curved focal surface in front of the mirror, necessitating the use of a special plate-holder to curve the photographic plate or film, and is sharp over a diameter of 10° or more.

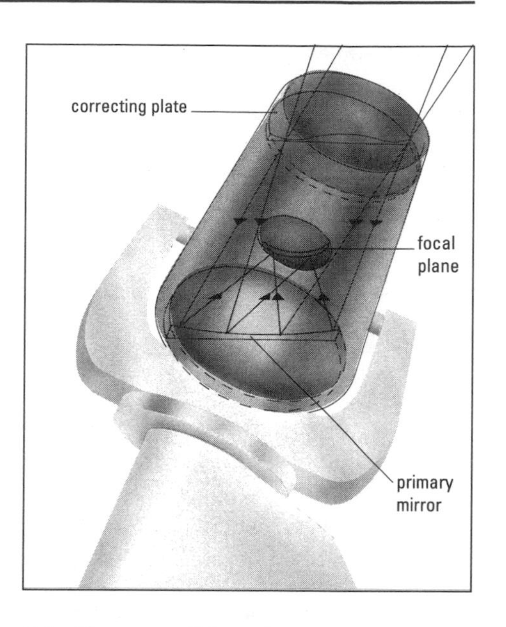

Schmidt camera: light path

The Schmidt camera and its numerous variations enabled whole-sky photographic surveys to be carried out, and has enabled significant advances to be made in photographic astronomy. A typical instrument is the UK Schmidt Telescope at the Anglo-Australian Observatory. Its aperture of 1.2 m (48 inches) is the diameter of the correcting plate; the primary mirror is 1.8 m (72 inches) across. Modifications to the Schmidt camera design that have made popular amateur telescopes have been the **Maksutov telescope** and, more recently, the **Schmidt–Cassegrain telescope**. The *super Schmidt telescope*, developed in the 1950s by adding further correcting plates, had an extremely short focal ratio.

Schmidt–Cassegrain telescope (SCT) A short-focus telescope combining features of the **Schmidt camera** and the **Cassegrain telescope**. The SCT has the Schmidt's spherical primary mirror and specially figured correcting plate, but light is reflected back down the tube by a convex secondary mirror mounted behind the plate and through a hole in the primary to a Cassegrain focus. This makes for a highly compact and portable telescope that has become very popular with amateur astronomers.

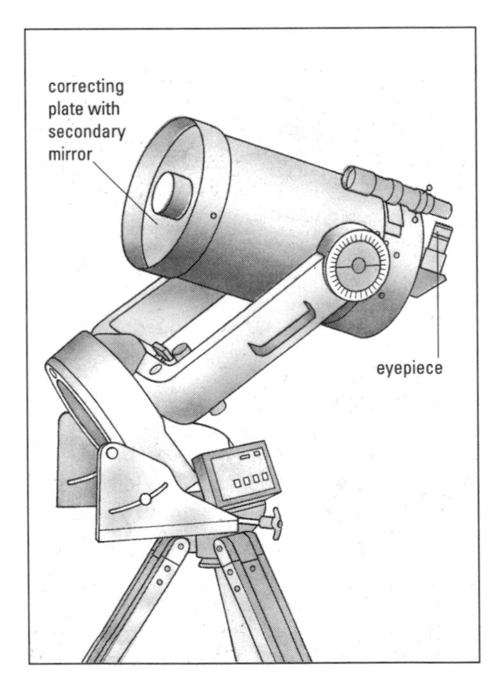

Schmidt–Cassegrain telescope

Schmidt telescope SEE **Schmidt camera**

Schröter effect A difference between the observed phase of Venus near dichotomy and the phase it should theoretically show. At western elongation dichotomy comes a day or two late, while at eastern elongation it is a day or two early. The effect, whose cause is unknown, was first observed in the 1790s by the German astronomer Johann Schröter (1745–1816).

Schwarzschild, Karl (1873–1916) German astronomer. He worked on methods of measuring stellar magnitudes from photographs, and studied stellar atmospheres and motions. In the last year of his life he solved equations in **Einstein**'s General Theory of Relativity that dealt with the gravitational field of a point mass, from which he developed the concepts of what are now called the **Schwarzschild radius** and the **black hole**. His son, **Martin Schwarzschild** (1912–), became a naturalized American and an astrophysicist, working on stellar evolution.

Schwarzschild radius The critical radius at which a very massive body under the influence of its own gravitation becomes a **black hole**. It is the radius of the event horizon of a black hole, from which nothing can escape, not even light. The radius is given by the expression

$$R = 2GM/c^2,$$

where G is the gravitational constant, M is the mass of the body, and c is the speed of light. The Schwarzschild radius for the Sun is about 3 km, and for the Earth it is about 1 cm. (After Karl Schwarzschild.)

scintillation The twinkling of the stars. When observed with the naked eye, stars appear to change in brightness and colour, particularly if they are low down in the sky; through the telescope, the positions of stars are seen to undergo short rapid changes. The effect is caused by the non-uniform density of the Earth's atmosphere, which produces uneven refraction of starlight. Planets do not exhibit this phenomenon (except that Mercury and Mars when in certain phases may do so slightly when close to the horizon) because the scintillations from different points on the surface are not in phase, and the fluctuations are lost in the general illumination. For astronomical purposes the inconvenience of the effects of scintillation is reduced by siting observatories at high altitudes, and it is almost entirely avoided by using telescopes carried by balloons, rockets or artificial satellites.

It has also been found that radio sources scintillate. In this case it is the non-uniformity of the refractive indices of the interstellar medium, the interplanetary medium, and the ionosphere that causes the strength of radio waves to fluctuate. SEE ALSO **seeing**

Scorpius A constellation of the zodiac, in mythology representing the scorpion that killed Orion. Its long line of bright stars has the appearance of a scorpion and its sting, with its heart marked by first-magnitude **Antares**, the constellation's brightest star. The 'sting' is Lambda (Shaula, magnitude 1.6). Both Mu and Zeta are optical doubles. In addition to being immersed in a very rich part of the Milky Way, Scorpius contains some magnificent clusters, notably the globular M4 near Antares, and M6 and M7, open clusters both visible with the naked eye.

Scorpius X-1 The first cosmic X-ray source, discovered in 1962, and the brightest known apart from occasional transients (SEE **X-ray burster**). It is thought to be a low-mass X-ray binary with an orbital period of 19.2 hours, one component being a neutron star; the nature of the other is unknown. X-rays are thought to originate from the neutron star and from a thin **accretion disk** around it.

SCT ABBREVIATION FOR **Schmidt–Cassegrain telescope**

Sculptor A barren southern constellation representing a sculptor's studio, introduced by **Lacaille**. It contains no star above magnitude 4.3, but is rich in faint galaxies, including a dwarf member of our Local Group.

Scutum A small constellation adjoining Aquila, representing a shield. Alpha, magnitude 3.9, is the only star above magnitude 4, but it contains the splendid open cluster M11, nicknamed the Wild Duck, and is crossed by the Milky Way. Delta Scuti is the prototype of a class of variable stars that pulsate in periods of 0.01 to 0.2 days with very small amplitudes, usually only several hundredths of a magnitude.

season One of a number of periods making up a cycle of changes in the surface conditions of a planet or satellite (average temperature and 'weather') as its axial tilt gives different parts of its surface longer or shorter periods of sunlight over the course of one revolution around the Sun. The Earth's seasons are defined by the different positions of the Sun with respect to the ecliptic. In the northern hemisphere, spring is reckoned from the vernal equinox (21 March) to the summer solstice (21 June); summer from the summer solstice to the autumnal equinox (23 September); autumn from the autumnal equinox to the winter solstice (22 December); and winter from the winter solstice to the vernal equinox. Spring in the northern hemisphere corresponds to autumn in the southern hemisphere, summer to winter, autumn to spring, and winter to summer.

Of the other planets, Mars has a similar axial inclination to the Earth's (25° 11′ compared to 23° 27′), and shows seasonal variations such as changes in its **polar caps**. The polar cap on Neptune's moon Triton migrates from pole to pole over the course of Neptune's 165-year orbital period. **Pluto**'s temporary atmosphere, which appears for a 60-year period around perihelion, is a seasonal change, but has to do with the planet's orbital eccentricity rather than its axial tilt.

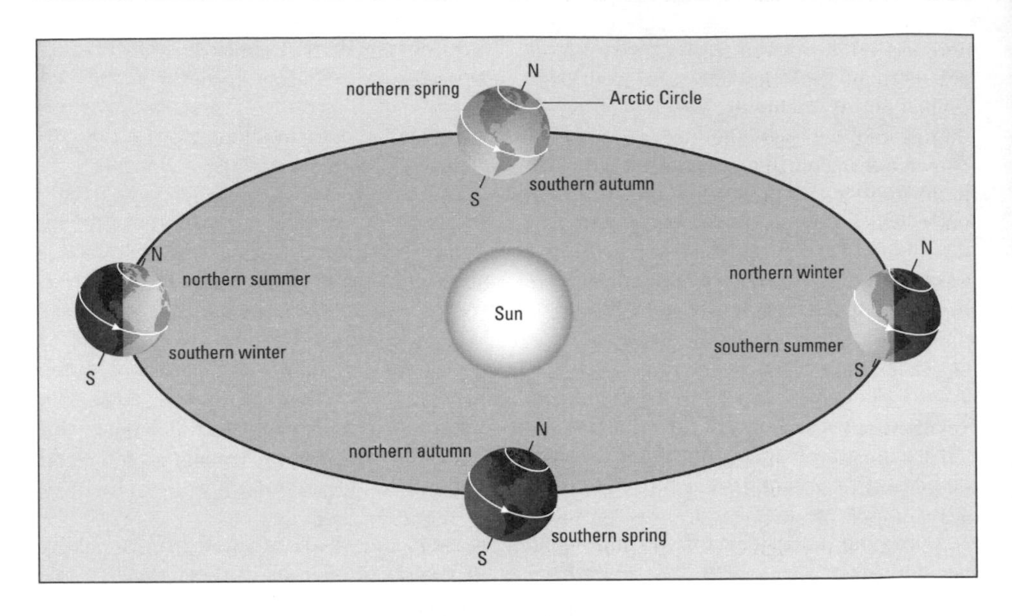

Seasons: the Earth's cycle of changes over the course of a year

Secchi, (Pietro) Angelo (1818–78) Italian astronomer and priest. He was a pioneer of spectroscopy, which he used to study the Sun and the stars. In the 1860s he carried out the first spectroscopic survey of stars, cataloguing the spectra of over 4000 and dividing them into four classes according to their colour and spectral characteristsics (SEE **spectral classification**).

secondary (1) Any of the smaller members of a system of celestial bodies that orbit around the largest, the **primary** (1).

secondary (2) **(secondary mirror)** The small mirror in a **reflecting telescope** that diverts the converging beam of light from the primary mirror towards the eyepiece.

second contact In a total **solar eclipse**, the moment when the leading edge of the Moon's disk covers the last visible part of the Sun's disk, and totality begins. In a **lunar eclipse**, it is the moment when the trailing edge of the Moon enters the umbra of the Earth's shadow. The term is also used for the corresponding stage in eclipses involving other bodies.

second of arc SEE **arc minute, second**

secular acceleration The gradual long-term acceleration of the Moon's apparent motion around the Earth. Examination of ancient lunar eclipse records has revealed that the Moon's period is less now than in former times, and is decreasing by about 10 arc seconds per century. The acceleration is caused by **perturbations** by other planets, and tidal friction (SEE **tides**) between the Earth and the Moon which is slowing the Earth's rotation and transferring angular momentum from the Earth to the Moon.

secular parallax The apparent displacement of a celestial body in the sky over time as a result of the Sun's motion in space.

seeing The quality of the image produced by a telescope, as affected by the steadiness of the air. Seeing depends on the calmness of the atmosphere immediately overhead, and also on air movements at ground level, and even in the telescope tube itself. The evaluation of seeing is largely subjective, though numbers assigned according to the **Antoniadi scale** can help achieve a degree of uniformity in the reporting of conditions under which observations are made.

seleno- Prefix referring to the Moon, as in *selenography*, the description and mapping of the Moon's surface features. (From 'Selēnē', Greek for Moon.)

semimajor axis (symbol *a*) Half the longest diameter of an **ellipse**. The semimajor axis of an elliptical orbit is the average distance of an object from its primary, and is one of the elements used when defining an **orbit**.

semiregular variable A pulsating giant or supergiant star of late spectral type with a period from 20 to 2000 days or more and an amplitude from a few hundredths of a magnitude to several magnitudes. Some semiregular variables show a definite periodicity, interrupted at times by irregularities. Others differ little from long-period variables except in having smaller amplitudes.

separation The angular distance between two members of a visual binary or multiple star system. It is measured in arc seconds.

Serpens A constellation in two parts, Caput (the head) and Cauda (the body), separated by the serpent-bearer Ophiuchus; Serpens represents a snake coiled around Ophiuchus. Alpha (Unukalhai), its brightest star, is magnitude 2.7. Theta (Alya) is a wide, easy pair with components of magnitudes 4.6 and 5.0. Delta is also an easy telescopic double, magnitudes 4.2 and 5.2. The Mira-type variable R Serpentis can reach magnitude 5.2 (minimum 14.4, period 356 days). Thsere are two bright Messier objects: M5, a globular cluster, and M16, an open cluster embedded in the **Eagle Nebula**.

Seven Sisters Popular name for the **Pleiades**, an open star cluster in Taurus.

Sextans An obscure constellation between Leo and Hydra representing a sextant. Its brightest star is only magnitude 4.5.

Seyfert galaxy A class of galaxies which have extremely bright, compact nuclei and whose spectra show strong emission lines. Most are spiral galaxies. About 1% of all galaxies are Seyferts; M77 (NGC 1068) is one of the brightest. Seyferts emit strongly at ultraviolet and infrared wavelengths, and exhibit a degree of short-term variability. Some are strong X-ray sources but few are particularly strong radio emitters. The gas clouds in Seyfert galaxies move at several thousand kilometres per second, possibly due to the gravitational influence of a massive object such as a black hole at the nucleus. They are named after American astrophysicist Carl Keenan Seyfert (1911–60), who first studied them.

shadow bands Grey ripples seen moving across light-coloured surfaces in the minutes either side of a total eclipse of the Sun. Close to totality the bands are more ordered and show higher contrast. They are caused by turbulence in the atmosphere affecting light from the very thin crescent of the Sun, just as it affects starlight to produce **scintillation**.

Shapley, Harlow (1885–1972) American astronomer. In 1914 he explained how the variability of **Cepheid variables** is caused by their pulsations. In the years that followed he calibrated the period–luminosity law discovered by **Henrietta Leavitt**, using Cepheid variables in globular clusters, which enabled him to make the first accurate estimate of the size of the Galaxy and the position in it of the Solar System. He also made estimates of the sizes of eclipsing binaries, and worked on photometry and spectroscopy.

shell star A hot B-type star with an equatorial disk thrown out by rapid rotation.

Strong shell spectra, which include emission lines from the surrounding disk, are seen in B0e to B3e giants and dwarfs. Examples are Gamma Cassiopeiae, BU Tauri (Pleione), and 48 Librae (FX Librae).

shepherd moon A minor moon that through its gravitational influence keeps in check the particles in a planetary ring. Shepherd moons often act in pairs. Examples are Prometheus and Pandora, orbiting either side of Saturn's F Ring, and Cordelia and Ophelia, orbiting either side of Uranus's Epsilon Ring.

Shoemaker–Levy 9, Comet A **comet** discovered in 1993 jointly by Eugene and Carolyn Shoemaker, David Levy, and Philippe Bendjoya. It was in a temporary orbit around Jupiter, and a close approach to the planet in 1992 had disrupted its nucleus into over 20 separate fragments. Over a six-day period in July 1994 these fragments collided with Jupiter, creating fireballs and leaving temporary dark scars at the impact sites. This was a unique event in the history of observational astronomy, and was observed worldwide by amateur and professional astronomers, and by the Galileo space probe.

shooting star A popular name for a **meteor**.

short-period comet A **comet** whose apparitions are sufficiently frequent to permit correlation of orbital data; formally defined as one whose period is less than 200 years.

sidereal Of or pertaining to the stars; measured or determined with reference to the stars.

sidereal day The interval between two successive passages of a given star, or the vernal equinox, across the observer's meridian. It is equal to 23h 56m 4.091s of mean solar time, and divided into 24 *sidereal hours*.

sidereal month One revolution of the Moon around the Earth relative to a fixed star. It is equal to 27.32166 mean solar days.

sidereal period The orbital period of a planet or other celestial body with respect to a background star. It is the true orbital period. COMPARE **synodic period**

sidereal time Local time reckoned according to the rotation of the Earth with respect to the stars. The time is 0h when the vernal equinox crosses the observer's meridian. The sidereal day is 23h 56m 4s of mean solar time – nearly 4 minutes shorter than the mean solar day. Sidereal time is equal to the **right ascension** of an object on the observer's meridian.

sidereal year The time required for the Earth to complete one revolution around the Sun relative to the fixed stars. It is equal to 365.25636 mean solar days.

siderite An older name for an **iron meteorite**.

siderolite An older name for a **stony-iron meteorite**.

siderostat A mirror mounted equatorially and driven so as to counteract the apparent movement of a star or other celestial object and direct its light into a fixed telescope. A more sophisticated version is the **coelostat**.

Siding Spring Observatory An astronomical observatory in the Warrumbungle mountains of New South Wales, founded in 1962 by the Australian National University, which operates several telescopes there including a 2.3 m (90-inch) reflector opened in 1986. The site is shared by the **Anglo-Australian Observatory**.

singularity A point at which the known laws of physics break down. For example, the singularity at the centre of a **black hole** is a point at which matter is compressed into an infinitely small volume and so has an infinitely large density.

Sinope One of Jupiter's four small outermost **satellites**, discovered in 1914 by Seth Nicholson. It is in a retrograde orbit – SEE **retrograde motion** (1) – and may well be a captured asteroid.

sinuous rille SEE **rille**

Sirius The star Alpha Canis Majoris. It is the brightest star in the sky, magnitude −1.47, distance 8.7 light years, luminosity 23 times that of the Sun. Sirius is a main-sequence star of spectral type A0, and is the sixth-closest star system to us. It has a binary companion, Sirius B, whose existence was deduced by **Friedrich Bessel** in 1844 from its gravitational perturbations on Sirius, which affected the proper motion of Sirius. The companion was discovered optically by Alvan G. Clark in 1862. It is a **white dwarf** of 1 solar mass, but with a diameter only 0.022 that of the Sun. Its orbital period is 50 years. Sirius B was the first star to be recognized as a white dwarf.

Sixty-one Cygni (61 Cygni) The first star to have its parallax measured (by **Friedrich Bessel** in 1838). It lies 11.1 light years away and is a binary consisting of two K-type main-sequence stars, magnitudes 5.2 and 6.0, with a period of 650 years.

Slipher, Vesto Melvin (1875–1969) American astronomer. On the death of **Percival Lowell** he became director of the Lowell Observatory. He was a specialist in astronomical spectroscopy, especially of galaxies. In 1912 he became the first to obtain the spectrum of another galaxy, the Andromeda Galaxy, and in 1920 he discovered the **redshifts** in galactic spectra. In 1925, following **Edwin Hubble**'s finding that the redshifts of galaxies increase with distance, he measured the radial velocities of over 40 galaxies. Slipher also showed that light from certain nebulae was reflected from stars embedded in them, thus discovering reflection nebulae.

Small Magellanic Cloud SEE **Magellanic Clouds**

SNC meteorites A group of achondritic meteorites (SEE **achondrites**) believed to have originated on Mars. Their ages of 1.3 billion years or less indicate that they originated on a sizeable planet that was relatively recently geologically active, and gases trapped in them are consistent with measurements of the Martian atmosphere made by the Viking spacecraft. However, it is still unclear how they could have emerged as intact fragments from an impact of sufficient force to accelerate them to Mars's escape velocity. The name (pronounced 'snick') comes from the three subtypes: *shergottites*, *nakhlites* and *chassignites*.

SNR ABBREVIATION FOR **supernova remnant**

SOHO ABBREVIATION FOR **Solar and Heliospheric Observatory**

Solar and Heliospheric Observatory (SOHO) A European Space Agency probe to study the corona of the Sun, oscillations of the Sun's surface, and the solar wind. SOHO is planned for launch in July 1995 and will be placed at the L_1 **Lagrangian point** 1.5 million km from Earth in the direction of the Sun. An associated mission to study the effects of the solar wind on the Earth, called **Cluster**, will be launched at the end of 1995.

solar apex, antapex SEE **apex**

solar constant A measure of the amount of solar energy received by a body a certain distance from the Sun. For the Earth, the solar constant is defined as the solar power received per unit area, at the top of the

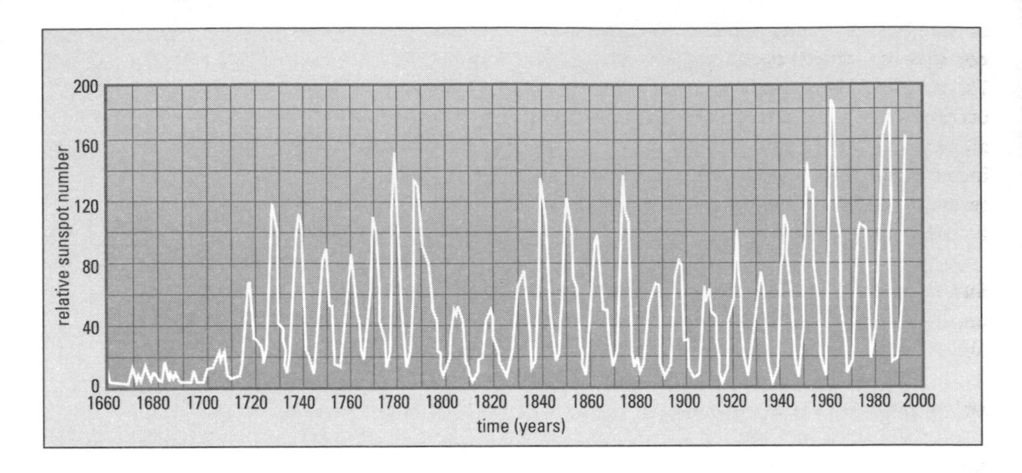

Solar cycle: the 11-year periodicity in the Sun's activity, with longer-term fluctuations apparent

atmosphere, at the average Earth–Sun distance of 1 AU; its value is about 1.35 W/m^2. The 'constant' varies from day to day with sunspot activity, and also in the longer term with the **solar cycle**.

solar cycle The periodic fluctuation in the number of **sunspots** and the level of other kinds of solar activity; the cycle lasts about 11 years. Over the course of a cycle, sunspots vary both in number and latitude in a way which is neatly illustrated in the **butterfly diagram**. The variation in latitude is described by **Spörer's law**. The diagram here shows the cycle of sunspot activity going back to the 17th century, as reckoned by the **relative sunspot number**. At *solar maximum*, when sunspot numbers are greatest, astronomers refer to the *active Sun*; *solar minimum* is also called the *quiet Sun*. There is some variation in the level of activity at solar maximum, and in the period up to 1715 – during the **Maunder miniumum** – activity was low for about 70 years.

solar day SEE **day**

solar eclipse An **eclipse** of the Sun by the Moon. Since the Moon's orbital plane is inclined to the plane of the ecliptic, a solar eclipse can occur only when the Moon is at conjunction (i.e. at new Moon) and at the same time is at or near one of its **nodes**; the Sun, Moon, and Earth are then very nearly in a straight line. In addition, the Sun's angular distance from one of the Moon's nodes at conjunction will determine whether an eclipse can or cannot occur (SEE **ecliptic limits**).

The apparent diameters of the Sun and the Moon as seen in the sky are almost the same, but they do vary slightly, particularly the Moon's, whose maximum and minimum distances from the Earth differ by about 10%. A *total solar eclipse* is seen at places where the umbra of the Moon's shadow-cone falls on and moves over the Earth's surface; at the same time the eclipse will appear *partial* to observers on either side of the central track of totality. Shortly before and after totality the phenomena of **shadow bands** and **Baily's beads** are observable. During the brief period of totality, the **corona** and any prominent **prominences** become visible. When the Moon is near apogee, the tip of its shadow-cone does not reach to the Earth and there is an *annular eclipse*, in which a rim of light is seen around the darkened disk of the Moon.

The overall duration of a solar eclipse from **first contact** to **fourth contact** can be as much as 4 hours; totality, from **second contact** to **third contact**, lasts at most 7½ minutes. There are from two to five solar eclipses each year; if there are five, they will all be partial. The total number of solar and lunar eclipses in a year varies from two to seven; if there are only two, they will both be solar.

solar mass The mass of the Sun, used as a unit of mass for other stars and celestial objects.

solar parallax SEE **parallax**

Solar System The Sun and all the celestial bodies which revolve around it: the nine **planets**, together with their **satellites** and ring systems, the thousands of **asteroids** and **comets**, meteoroids, and other interplanetary material. The boundaries of the Solar System lie beyond the orbit of Pluto, which orbits the Sun at an average distance of 30 AU, to include the **Kuiper Belt**, which may extend to 100 AU, and the inferred **Oort Cloud** of comets, some of which may have aphelion distances of 100,000 AU (1½ light years).

All the planets have direct orbits, i.e. they revolve around the Sun in the same direction as the Sun itself rotates; the orbital motion of most of the satellites is direct too. The Sun is by far the most massive component of the Solar System, its mass being 330,000 times that of the Earth. The nine planets have a total mass equal to 448 times that of the Earth, of which Jupiter accounts for 70%. Interplanetary space contains cosmic dust and extremely tenuous ionized gas (particles of the **solar wind**).

The Solar System came into being nearly 5000 million years ago, probably as the end-product of a contracting cloud of interstellar gas and dust (the *solar nebula*). The exact mechanisms by which the Sun and its retinue were formed remains a major problem for cosmogonists. Similar systems centred on other stars (referred to as 'solar systems', without the capital S's) are expected to exist, as by-products of star formation. **Beta Pictoris** is believed to be at the centre of a solar system in the process of formation.

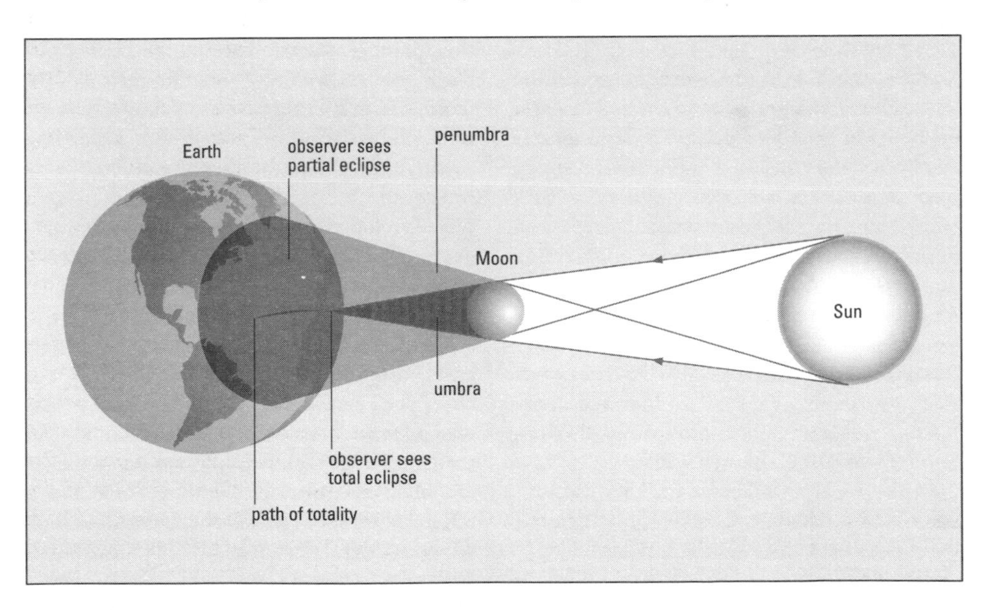

Solar eclipse (not to scale): may be total or partial

solar wind The steady flow of charged particles (mainly protons and electrons) from the solar corona into interplanetary space, controlled by the Sun's magnetic field. Solar wind particles are too energetic to be held back by the Sun's gravity. Some of them get trapped in planetary magnetic fields. At the Earth, some are trapped in the outer **Van Allen Belt**; others reach the Earth's upper atmosphere in the region of the magnetic poles and cause aurorae. The solar wind carries away about 10^{-13} of the Sun's mass per year. Its intensity increases during periods of solar activity. SEE ALSO **stellar wind**

solstice The extreme northern or southern position of the Sun in its yearly path among the stars. At the solstices the Sun reaches its greatest declination, 23½°N or 23½°S. The *summer solstice* occurs around 21 June and the *winter solstice* is on or about 22 December.

solstitial colure The **great circle** passing through the north and south celestial poles and the winter and summer **solstices**.

Sombrero Galaxy A spiral galaxy in the constellation Virgo, also known as M104 or NGC 4594. The galaxy is seen nearly edge-on, and has the appearance of a sombrero hat. It is crossed by a thick dark lane of interstellar dust, and there is an unusually large central bulge containing numerous globular clusters. It lies about 35 million light years away.

South African Astronomical Observatory An observatory at Sutherland in the Karoo semi-desert north-east of Cape Town, founded in 1972. Its main instrument is a 1.88 m (74-inch) reflector which was originally at the Radcliffe Observatory, Pretoria; there is also a 1 m (40-inch) reflector.

Southern Cross The popular name for the southern-hemisphere constellation **Crux**.

space probe An unmanned space vehicle sent to investigate the Moon, planets, or interplanetary space.

spacetime A unified dimensional framework in which the three dimensions of space (length, breadth, and height) and the dimension of time are linked together. An event in spacetime is specified by the three coordinates of space and the time coordinate. The concept arose from **Einstein**'s Special and General Theories of Relativity. In the General Theory, gravity is a distortion of spacetime by matter.

space velocity The true velocity of a star in space with respect to the Sun. It is the hypotenuse of the right-angled triangle formed by its **tangential velocity** (obtained from observation of its proper motion) and its **radial velocity** (obtained by measurement of the Doppler shift in its spectrum).

speckle interferometer SEE **optical interferometer**

spectral classification The categorization of stars based on the characteristics of their spectra. In the 1860s **Angelo Secchi** made the first spectral classification of stars by dividing them into four groups according to colour and spectral lines. As spectroscopy improved, a more comprehensive system was developed at Harvard College Observatory. This was embodied in the ***Henry Draper Catalogue*** of stellar spectra published in 1918–24. The sequence of spectral types was arranged according to the prominence or absence of certain lines in the spectra. There were seven spectral types, designated O, B, A, F, G, K and M (an ordering which results from revising an earlier alphabetical sequence). This sequence is a temperature sequence, from the hottest stars of types O and B, which appear blue-white, to type M, the coolest stars, which appear orange-red. As a result of this ordering, O, B and A stars are called **early-type stars**;

SPECTRAL CLASSIFICATION: MAIN SPECTRAL TYPES		
Type	Surface temperature (K)	Main features
O	40,000–25,000	Blue-white stars. Lines of ionized helium. Weak hydrogen lines.
B	25,000–11,000	Blue-white stars. Neutral helium lines, with hydrogen lines strengthening from B6 to B9.
A	11,000–7500	White stars. Very strong hydrogen lines at A0, decreasing towards A9. Lines of ionized calcium increase in strength from A0 to A9.
F	7500–6000	Yellow-white stars. Ionized calcium continuing to increase in strength, and hydrogen weakening. Lines of other elements begin to strengthen.
G	6000–5000	Yellow stars. Strong lines of calcium, hydrogen weaker. Lines of iron become prominent.
K	5000–3500	Orange stars. Strong metallic lines. Molecular bands of CH and CN become prominent.
M	3500–3000	Orange-red stars. Strong absorption bands of titanium oxide and large numbers of metallic lines.

SPECTRAL CLASSIFICATION: LUMINOSITY CLASSES	
Ia	Supergiants of high luminosity
Ib	Supergiants of lower luminosity
II	Bright giants
III	Normal giants
IV	Subgiants
V	Main-sequence stars (dwarfs)
VI	Subdwarfs

K and M stars are called **late-type stars**. Three new types were added when it was found that some cool stars had strong absorption bands not usually seen in other stars of the same colour. These were classes R and N with strong bands of molecular carbon, and class S with bands of zirconium oxide.

Stellar spectra can be classified into even finer divisions within these seven types, so decimal subdivisions were introduced. G5, for example, indicates a star midway in type between G0 and K0. Further refinements include the use of additional letters as suffixes to the spectral type, giving more information about the star, for example the existence of emission lines (e), metallic lines (m), broad lines due to rotation (n and nn), or a peculiar spectrum (p).

However, the Harvard system could not deal with stars of different luminosities at a given temperature (i.e. dwarfs, giants, and supergiants). In 1943 William Morgan, Philip Keenan and Edith Kellman of Yerkes Observatory redefined the spectral types and added a classification scheme for the luminosity (absolute magnitude) of stars. This is now known as the **Morgan-Keenan system** or *MK system*, and is used universally. For example, a star classified in the MK system as O9.5 IV–V has a spectral type (and therefore a temperature) midway between that of an O9 and a B0 star, and a luminosity between that of a dwarf and a subgiant. The spectral types and luminosity classes of the MK system are listed in the tables.

The MK system is applicable to stars of normal chemical composition, which is to say about 95% of all stars. The various types of peculiar star are given their own special classification schemes. The Harvard R and N types are now combined into one **carbon star** class, the designations for which include a temperature type and a carbon band strength, as for example in C2,4. A similar classification is used for the **S stars**. White dwarfs are usually classified on a Harvard-type scheme, with D preceding the type

spectral type A series of divisions, indicating surface temperature and hence colour, into which stars are classified according to the nature of their spectrum. SEE **spectral classification**; **A star, B star, carbon star, F star, G star, K star, M star, O star, S star**

spectrogram A photographic record of the spectrum of an object, produced by a **spectrograph**.

spectrograph A **spectroscope**, fitted with a camera or an electronic detector such as a CCD, used to obtain a permanent record of the spectrum of a celestial object. Astronomical spectrographs are generally designed for use with a specific telescope. On telescopes with a **coudé** system or **Nasmyth focus** a large spectrograph can be mounted in a permanent position. Spectrographs generally operate in a band of wavelengths from the near-infrared to the near-ultraviolet. There are various designs for different purposes. *High-dispersion* instruments such as the **spectroheliograph** spread the spectral lines widely so that a narrow band of wavelengths can be studied in detail.

spectroheliograph A type of **spectrograph** used to photograph the Sun in the light of one particular wavelength only. A second slit placed in front of the photographic plate or detector images a narrow strip of the Sun's disk at a chosen wavelength. By moving the entrance slit and the second slit in tandem, the whole disk is scanned and an image of it, a *spectroheliogram,* is built up. The spectroheliograph was invented independently by **George Ellery Hale** and Henri Deslandres in the 1890s. The term is also used for a number of modern instruments of different design but working on the same principle.

spectrohelioscope A **spectroheliograph** adapted for visual use. The two slits are oscillated rapidly so that by persistent vison the observer sees a steady, monochromatic image of the Sun's disk.

spectrometer A **spectroscope** equipped to measure accurately the positions and intensities of spectral lines, by means of a device such as a **Fabry–Pérot interferometer**. Spectrometers are also loosely referred to as **spectrographs**. Since spectra are now routinely recorded on CCDs, from which information in digital form can be analysed by computer, the distinction between modern spectrometers and spectrographs lies in their application rather than their design.

spectrophotometer An instrument which scans the lines in a spectrum recorded by a spectrograph and measures their intensities. The instrument outputs a graph of intensity against wavelength called a *line profile*.

spectroscope An instrument for producing spectra for the analysis of electromagnetic radiation (light or other wavelengths). *Spectroscopy* is the use of such an instrument to probe the chemical composition and physical conditions (such as temperature and motion) of an object. From its origins in the pioneering work by **William Huggins**, **Angelo Secchi**, and others in the mid-19th century, spectroscopy developed into the most important investigative technique available to astronomers.

In the arrangement shown in the diagram, a focused beam of light or other radiation is

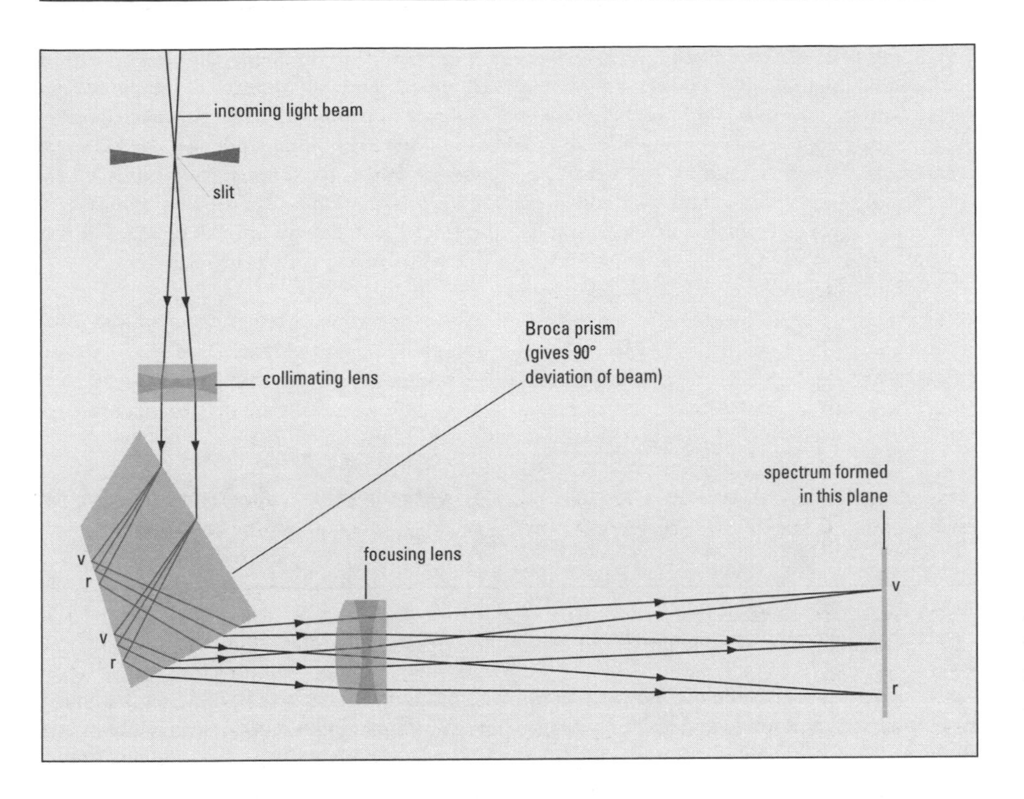

Spectroscope: one variant of the traditional instrument using a prism and lenses

passed through a narrow slit, to prevent images of different wavelengths from overlapping and ensure a sharply defined spectrum, and then through a lens to collimate it (make it parallel). The parallel beam is then split into its component wavelengths by a prism (often not the simple triangular form), and focused by another lens to produce the spectrum. In modern instruments a **diffraction grating** is usually used instead of a prism, and concave mirrors replace the collimating and focusing lenses.

Astronomical spectroscopes are known as **spectrographs** or **spectrometers**. Strictly, a spectrograph is a spectroscope equipped with a camera for recording a permanent record of a spectrum, whereas a spectrometer incorporates devices for accurately measuring the wavelengths and intensities of the spectral lines.

spectroscopic binary A **binary star** system whose two components are too close for them to be resolved visually as separate objects, but whose binary nature can be deduced from the periodic **Doppler shift** of the absorption lines in the combined spectrum. As the diagram (SEE page 198) shows, when the two stars have reached positions in their orbits around each other where one is moving away from us and the other towards us, the spectral lines of the one are slightly redshifted, and those of the other slightly blueshifted, from their mean positions.

spectroscopic parallax A method of determining the distance of a star from its spectral type (SEE **spectral classification**) and apparent magnitude. Analysis of a star's spectrum reveals the spectral type. If the absolute magnitude of a star of that type is

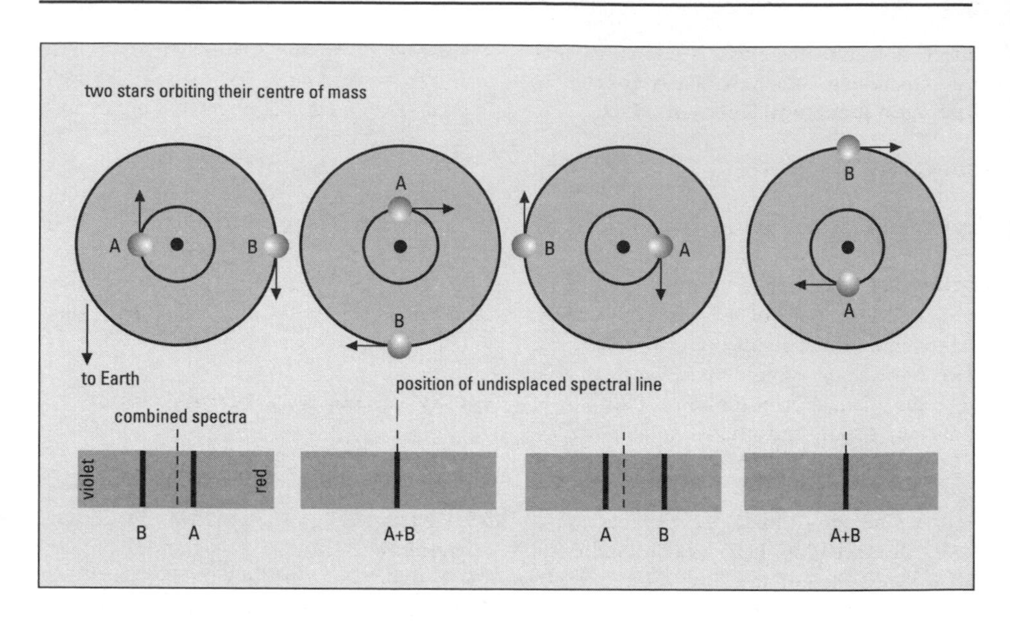

Spectroscopic binary: the periodic variations in spectral lines reveal the binary nature

known, the star's distance can be calculated by comparing its absolute and apparent magnitudes. This is the most common method of determining stellar distances. However, the scale must be calibrated against nearby stars whose distances can be measured directly by **trigonometric parallax**.

spectrum (plural **spectra**) The distribution of intensity of **electromagnetic radiation** with wavelength. When a beam of light from a source is deflected by a prism or by another dispersing agent such as a diffraction grating, the radiation is fanned out into a spectrum because different wavelengths are deviated by different degrees.

Spectra are of various types. A *continuous spectrum* is an unbroken distribution of radiation over a broad range of wavelengths. White light, for example, is split into a continuous 'rainbow' band of colours from red to violet. A *line spectrum* contains lines corresponding only to certain wavelengths or frequencies which are characteristic of the chemical elements present in the source. An *emission line spectrum* contains lines emitted at particular wavelengths, and is produced by substances in an incandescent state at low pressure. An *absorption line spectrum* contains dark lines at particular wavelengths superimposed on a continuous spectrum. It is produced when radiation from a hot source emitting a continuous spectrum passes through a layer of cooler gaseous material. The spectra of most stars are continuous spectra crossed by absorption lines, and sometimes by emission lines. Emission line spectra are typical of luminous nebulae. *Band spectra* contain absorption bands, wider than the spectral lines of atoms, and these are the spectral signatures of molecules.

The intensity of an emission line in a spectrum is a measure of the energy of the radiation at that wavelength. The width of the line indicates the wavelength band over which it is distributed. The intensity of the emission is greatest at the centre of the line and decreases towards the edges; the reverse is true for an absorption line. These variations in the appearance of a line are known as the *line contours*. The pressure of the gas which is producing the spectral line can affect its width; this effect is known as *pressure broadening*. SEE ALSO **flash spectrum**, **hydrogen spectrum**, **Zeeman effect**

speculum SEE **mirror**

spherical aberrration SEE **aberration**

spheroid The surface formed by rotating an ellipse about one of its axes. If the minor axis is chosen, the result is an *oblate spheroid.* This is the shape of fluid bodies such as stars and gas planets like Jupiter rotating at a sufficient speed, and also of solid but non-rigid bodies like the Earth.

Spica The star Alpha Virginis, magnitude 0.98, distance 220 light years, luminosity 1500 times that of the Sun. It is a main-sequence star of type B1 and a spectroscopic binary.

spicule A narrow jet, like a tiny prominence, seen protruding from the Sun's chromosphere when it is viewed edge-on. Spicules are short-lived (about 5–10 minutes) and congregate at the edges of **supergranulation** cells. They are possibly caused by chromospheric gases shooting into the lower corona.

spiral galaxy A **galaxy** consisting of a nucleus of stars from which spiral arms emerge, winding around the nucleus and forming a flattened, disk-shaped region. The arms contain gas, dust, and young stars, while the nucleus contains old stars.

sporadic meteor A meteor which is not part of a recognized **meteor shower**. Its path cannot be traced back to a known **radiant**. The Earth is contantly sweeping up meteoroids, and 3–8 sporadic meteors should be visible each night under ideal conditions. SEE ALSO **radar astronomy**

Spörer's law The appearance of sunspots at lower latitudes over the course of the 11-year **solar cycle**. The average position drifts from mid-latitudes (about 30–40°) towards the equator as the cycle progresses from one minimum to the next. The phenomenon was first studied in detail by German astronomer Gustav Friedrich Wilhelm Spörer (1822–95), after whom it was later named. Despite it being called a 'law', there is no mathematical formulation of the statement. SEE ALSO **butterfly diagram**

spring equinox SEE **vernal equinox**

spring tide SEE **tides**

SS Cygni star An alternative name for a **U Geminorum star**.

SS433 An unusual binary star, number 433 in a catalogue of stars with bright hydrogen emission lines compiled by American astronomers Bruce Stephenson and Nicholas Sanduleak. SS433 lies at the centre of a supernova remnant in Aquila called W50. What makes SS433 puzzling is the variable **Doppler shift** in its emission lines. SS433 is thought to consist of a hot, massive star orbited every 13.1 days by a neutron star,

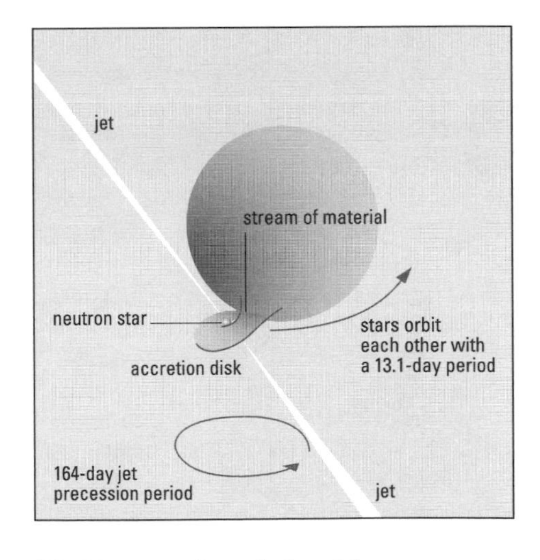

SS433: current theoretical model

which was presumably formed by the supernova that created W50. A stellar wind from the hot star produces one set of emission lines. Matter is transferred from the massive star to the neutron star via an **accretion disk**, but some of this material is ejected at high speed along two jets. The jets sweep around the sky every 164 days, producing the variable Doppler shift in SS433's spectral features.

S star A giant star whose surface temperature lies in the range for M stars, but whose spectrum contains absorption bands of zirconium oxide as well as the titanium oxide characteristic of the M stars. The overabundance of zirconium, and also of carbon and many other heavy elements, is a result of convective mixing which brings to the surface products from nuclear reactions in the interior. As a result, many S stars show spectral lines of the radioactive element technetium. M stars with weak zirconium oxide bands are assigned the intermediate spectral type MS. S stars often show hydrogen emission lines in their spectra, and many are long-period variables.

standard epoch A fixed point in time, or **epoch**, to which star positions are referred in order to remove the effects of precession and proper motion, together with a reference system in which the positions are measured. The standard epoch currently used for coordinates in catalogues and star charts is designated 2000.0. Its fixed points are noon on 1 January 2000, and the mean equator and equinox (i.e. ignoring minor variations in the positions of the celestial equator and the equinoxes).

STARS: THE BRIGHTEST (FIRST MAGNITUDE OR BRIGHTER)

Star	Name	Apparent magnitude	Spectral type	Distance (light years)
α Canis Majoris	Sirius	−1.47	A0	8.7
α Carinae	Canopus	−0.72	A9	74
α Centauri	Rigil Kentaurus	−0.27**	G2+K1	4.3
α Boötis	Arcturus	−0.04	K2	36
α Lyrae	Vega	0.03	A0	25
α Aurigae	Capella	0.08	G6+G2	41
β Orionis	Rigel	0.12	B8	1400
α Canis Minoris	Procyon	0.38	F5	11
α Eridani	Achernar	0.46	B3	70
α Orionis	Betelgeuse	0.50*	M2	1400
β Centauri	Hadar	0.61*	B	320
α Crucis	Acrux	0.76**	B0.5+B1	510
α Aquilae	Altair	0.77	A7	17
α Tauri	Aldebaran	0.85*	K5	68
α Scorpii	Antares	0.96*	M1.5	520
α Virginis	Spica	0.98*	B1	220
β Geminorum	Pollux	1.14	K0	36
α Piscis Austrini	Fomalhaut	1.16	A3	22
β Crucis	Mimosa	1.25*	B0.5	460
α Cygni	Deneb	1.25	A2	1500
α Leonis	Regulus	1.35	B7	69
ε Canis Majoris	Adhara	1.50	B2	570

** variable* *** combined magnitude of double star*

STARS: THE NEAREST (CLOSER THAN 11.5 LIGHT YEARS)				
Star	Apparent magnitude	Spectral type	Absolute magnitude	Distance (light years)
Proxima Centauri	11.2*	M5.5	15.7	4.25
α Centauri A	−0.01	G2	4.4	4.3
B	1.33	K1	5.7	4.3
Barnard's Star	9.5	M5	13.2	6.0
Wolf 359	13.5*	M6.5	16.6	7.8
Lalande 21185	7.5	M2	10.5	8.3
Sirius A	−1.47	A0	1.4	8.7
B	8.4	DA2	11.3	8.7
UV Ceti A	12.6*	M5.5	15.4	8.7
B	13.0*	M5.5	15.8	8.7
Ross 154	10.5	M3.6	13.2	9.4
Ross 248	12.3	M5.5	14.8	10.3
ε Eridani	3.7	K2	6.2	10.7
Ross 128	11.1	M4	13.5	10.9
61 Cygni A	5.2*	K3.5	7.6	11.1
B	6.0*	K4.7	8.4	11.1
ε Indi	4.7	K3	7.0	11.2
Groombridge 34 A	8.1	M1.3	10.4	11.2
B	11.1	M3.8	13.4	11.2
L789-6	12.2	M7	14.5	11.2
Procyon A	0.38	F5	2.6	11.4
B	10.7	DF	13.0	11.4

* *variable*

star A self-luminous ball of gas whose radiant energy is produced by nuclear reactions, mainly the conversion of hydrogen into helium. The temperatures and luminosities of stars are prescribed by their masses (according to the **mass–luminosity relation**). The most massive stars are about 100 solar masses (i.e. a hundred times heavier than the Sun). Above this mass the star is not stable and is likely to break up. Large stars are very luminous and hot, and therefore appear blue. Medium-sized stars like the Sun are yellow, while small stars are a dull red. The smallest stars contain less than one-twentieth of a solar mass; below this the temperature does not become high enough for nuclear reactions to take place, and the object becomes instead a **brown dwarf**. The colours, and hence the temperatures, of stars give rise to the scheme of **spectral classification**. Most stars are born not singly but in pairs or groups (SEE **binary star**, **star cluster**). SEE ALSO **stellar evolution**, **stellar populations**

starburst galaxy A **galaxy** that appears much brighter at infrared wavelengths than would be expected from its visual brightness. Starburst galaxies were first detected in 1983 by the Infrared Astronomical Satellite, IRAS, which found some galaxies that emitted up to 50 times more energy as infrared radiation than they do as visible light; by contrast, the Andromeda Galaxy, a normal spiral, emits only about 4% of its total energy in the infrared. The extra infrared emission from these galaxies is probably caused by an unusually high rate of star formation – hence their name. Some starburst galaxies are near other galaxies, and it may be that the gravitational influence of a neighbour is responsible for triggering the starburst. However, many starburst galaxies have no close companions, so this cannot be the only explanation.

star cluster An assembly of stars moving as a whole through space. Star clusters are classified as globular or open. A *globular cluster* is one in which an extremely large number of stars (10^5 to 10^7) are packed closely into a roughly spherical space typically 50 to 150 light years across. The stars are strongly concentrated towards the centre. Globular clusters contain very old (**Population II**) stars and generally lie in the halo of our Galaxy. The compactness and large total masses of globular clusters make them dynamically very stable systems that remain largely unchanged over many thousands of millions of years. There is, however, a slow but steady loss of stars, mostly of low mass, that are swung out of the cluster through encounters with other stars; this is called *cluster evaporation.*

Open clusters are much less rich than globulars, and the stars are not as concentrated towards the centre. Open clusters contain young (**Population I**) stars and occur in or near the central plane of the Galaxy. They are susceptible to disruption, not only through evaporation but also through tidal interaction with the Galaxy as a whole or encounters with interstellar clouds. Only the richest survive for more than 10^9 years, while the smallest and least tightly bound last no more than a few million years. As all stars in an open cluster are moving together, it is possible to find the distances of nearby clusters by the **moving cluster** method.

star diagonal SEE **diagonal** (1)

Star of Bethlehem The Biblical 'star seen in the east' which guided the Wise Men, mentioned only by St Matthew. It has been ascribed to various astronomical phenomena, including comets, novae, meteors, and the conjunction of two or three planets. The search for an astronomical explanation is complicated by uncertainty over the exact date of Christ's birth. The three separate conjunctions of Jupiter and Saturn in Pisces in 7 BC are widely favoured. However, the single most remarkable astronomical event about that time was the apparent merging in Leo of Venus and Jupiter, the two brightest planets, in June of 2 BC, which would have created a brilliant temporary 'star'.

star streaming The overall effect of the proper motions of stars (corrected for the Sun's motion) not being random, giving rise to two preferred directions of motion in exactly opposite directions. One vertex is in the constellation Orion, the other in Scutum. Star streaming results from the motion of stars in their orbits around the Galaxy. It was discovered by the Dutch astronomer Jacobus Kapteyn (1851–1922) in 1904.

star trailing In **astrophotography**, the appearance of stellar images as curved lines, or trails, on photographs taken by a stationary camera – one that is not driven to compensate for the rotation of the Earth. The length of the trails depends on the exposure time. A camera aimed at the celestial pole will record star trails as concentric arcs centred on the pole.

stationary point The point in the apparent path of a planet when it appears motionless – SEE **retrograde motion** (2).

steady-state theory A cosmological theory put forward by **Hermann Bondi** and **Thomas Gold** in 1948, and further developed by **Fred Hoyle** and others. According to this theory the Universe has always existed; it had no beginning and will continue for ever. Although the Universe is expanding, it maintains its average density – its steady state – through the continuous creation of new matter. Most cosmologists now reject the theory because it cannot naturally explain the **cosmic microwave background** or the observation that the appearance of the Universe has changed with time.

Stebbins, Joel (1878–1966) American astronomer. He pioneered the use of photoelectric photometry in astronomy. In 1910 he attached a primitive photocell to a telescope and discovered the secondary minimum in the light curve of the eclipsing binary star Algol. In later years, as the quality of photocells rapidly improved, he applied the technique to study the solar corona, globular clusters, interstellar dust, and other galaxies.

stellar association SEE **OB association, T association**

stellar evolution The development of a star over its lifetime, which can range from thousands of years to thousands of millions of years, depending on its mass. The theory of stellar evolution is based on mathematical models of stellar interiors. Its account of the evolution of stars can be checked by observations of stars at each of the predicted stages.

Stars form when a cloud of gas and dust collapses under its own gravity. As the cloud collapses its temperature rises because the energy of infalling atoms is released as heat when they collide. At this stage it is termed a *protostar*. A protostar will continue to contract under gravity until its centre is hot enough (about 15 million K) for nuclear reactions to commence, with the release of energy (SEE **carbon–nitrogen cycle, proton–proton reaction**). The outward flow of radiation from deep within halts the collapse (SEE **radiation pressure**). As this is happening, the star's outer layers are still falling in. **T Tauri stars** are low-mass stars in this final stage of collapse.

Once nuclear fusion has commenced in the core of a star, the star adopts a stable structure with its gravity balanced by the radiation from its centre. This is known as the main-sequence phase of a star's life (SEE **Hertzsprung–Russell diagram**). It is the longest phase in the life of a star. The Sun has a main-sequence lifetime of some 10,000 million years, of which half has expired. Less massive stars are redder and fainter than the Sun, and stay on the main sequence for longer. More massive stars are bluer and brighter and have a shorter lifetime. As a star ages on the main sequence, nuclear fusion changes the composition of its core from mainly hydrogen to mainly helium.

Eventually, the hydrogen supply in the core of the star runs out. With the central energy source removed, the core collapses under gravity and heats itself further until hydrogen fusion is able to take place in a spherical shell surrounding the core. As this change occurs the outer layers of the star expand considerably and the star becomes a **red giant** or, in the most massive stars, a **supergiant**. While it is a red giant, the star's core temperature will reach 100 million K, hot enough for the fusion of helium to carbon to begin (the **triple-alpha process**). In low-mass stars the onset of helium fusion is sudden (the *helium flash*), but in high-mass stars it gets under way more gradually. After this stage they can fluctuate in size, becoming **long-period variable** stars like Mira.

When all of the helium fuel in the core has been converted into carbon, the core of the star will again collapse and heat up. In a low-mass star the central temperature will not rise far enough to initiate carbon fusion. The red giant will lose its outer layers as a **planetary nebula**, after which only a **white dwarf** remains. In a high-mass star the contraction of the carbon core will lead to further episodes of nuclear fusion involving heavier elements, during which the star will remain a supergiant. Once fusion has converted all of the core into elements with atomic weights close to that of iron, no further fusion is possible. At this point the core will collapse explosively, and the star will throw off its outer layers in a **supernova** explosion. The explosion takes place not at the centre but on the exterior of a dense core. The explosion throws off several solar masses of gas at speeds of many thousands of kilometres per second, and also pushes inwards on the core with immense force. The effect of this implosion is to crush the already extraordinarily dense core still further, forming a **neutron star** or a **black hole**, both effectively end-points of stellar evolution.

stellar interferometer SEE **optical interferometer**

stellar nomenclature Some stars have proper names given them by Greek, Roman and Arab astronomers, such as Sirius, Capella and Aldebaran. In addition to any proper name, the stars within each constellation are identified by a letter or number followed by the name of the constellation in the genitive (possessive) form (SEE the table of **constellations** on pages 46–7). The brightest stars are indicated by Greek letters, as in α (Alpha) Lyrae or ε (Epsilon) Eridani. These letters are known as *Bayer letters* because they were assigned by **Johann Bayer** in his star atlas *Uranometria* of 1603. Stars that do not have Bayer letters are prefixed by a number, as in 61 Cygni. These are known as *Flamsteed numbers* because they relate to the stars charted in John Flamsteed's *Historia coelestis Britannica* of 1725. Fainter stars are referred to by their designation in any of a number of other catalogues. **Variable stars** have their own system of nomenclature.

stellar populations A classification of stars according to their age and location in the Galaxy. There are two principal populations. *Population I* consists of relatively young stars located in the plane of the Galaxy, in its spiral arms. *Population II* consists of relatively old stars dispersed throughout the entire Galaxy but prominently visible in its centre and halo. There is also a gradation of stars intermediate between these two extremes.

The distribution of populations in our Galaxy is explained in terms of the collapse of the Galaxy during its formation. The oldest population of stars was made during the infall, which is why Population II is distributed throughout the galactic halo and central regions of the Galaxy. Stars in these regions, including those in globular clusters, continue to have elliptical orbits around the Galaxy. After the Galaxy had developed its flat rotating disk and spiral arms, stars of Population I formed, moving in circular orbits. Between these two extremes lie intermediate populations showing progressively flatter and flatter distributions, representing successive stages in the collapse of the Galaxy.

Chemical elements produced in stars are recycled back into the interstellar material by stellar winds and supernova explosions, so Population I stars have greater concentrations of heavier elements than Population II. The Sun's content of heavy elements puts it among the younger Population I stars. The chemical history of the Galaxy suggests that there has existed a Population III, created when the Galaxy first formed, which have now disappeared. These stars would have manufactured the smattering of heavy elements present in Population II before dying to form compact objects like neutron stars.

stellar wind A stream of charged particles, mostly protons and electrons, from the surface of a star. The strength of the wind depends on the type of star, and its velocity can range from a few hundred to several thousand kilometres per second. Young stars evolving towards the main sequence have powerful stellar winds, up to a thousand times stronger than the Sun's **solar wind**. Old stars evolving into red giants also have strong stellar winds.

Stephan's Quintet A group of peculiar galaxies in the constellation Pegasus discovered in 1877 by the French astronomer Edouard Stephan (1837–1923). Four of the objects (NGC 7317, 7318A, 7318B and 7319) have nearly the same redshift (z = 0.02), but a fifth galaxy, NGC 7320, has a much lower redshift (z = 0.0003). Presumably it merely happens to lie in the same line of sight. Some astronomers now refer to the group, omitting NGC 7320, as *Stephan's Quartet*.

stereo comparator SEE **comparator**

Steward Observatory The observatory of the University of Arizona, Tucson. Its main telescope is a 2.3 m (90-inch) reflector, sited on Kitt Peak.

stony-iron meteorite (siderolite) A meteorite consisting of approximately equal proportions of silicates (stony material) and metals, mostly nickel–iron. There are two main subtypes. *Pallasites*, which consist of olivine (a magnesium–iron silicate) mixed with nickel–iron, may have originated near the core/mantle interface of a planetary body. *Mesosiderites* are a much coarser combination of chunks of various silicates and nickel–iron, and could have been produced by impacts on a planetary surface.

stony meteorite (aerolite) A **meteorite** that consists mostly of silicates (stony material), with only a small amount of chemically uncombined metals (typically 5% nickel–iron). There are two subtypes: **chondrites** and **achondrites**.

Stratton, Frederick John Marrion (1881–1960) English astrophysicist. He specialized in solar physics, in particular the chromosphere, but also made important contributions to the study of variable stars and novae, including the nova in Gemini in 1912.

strewnfield SEE **tektites**

Strömgren, Bengt George Daniel (1908–87) Swedish-born Danish astronomer. In 1940 he suggested the idea of what is now called a Strömgren sphere (SEE **H II region**) to explain how some nebulae shine. He also worked on photoelectric photometry, and the internal structure of stars.

Struve Russian–German family that produced four generations of astronomers. They included:

Struve, (Friedrich Georg) Wilhelm von (1793–1864), emigrated from Germany to Russia as a young man. In 1824 he began to observe double stars with a telescope made by **Joseph von Fraunhofer**, and in 1837 published the first good catalogue of double stars, containing nearly 2500 doubles he had discovered himself. In 1840 he made the third good measurement of the parallax of a star (Vega).

Struve, Otto (Wilhelm) (1819–1905), son of Wilhelm. He too studied double stars, and determined an accurate value for the rate of **precession**, taking into account the motion of the Sun with respect to nearby stars.

Struve, (Karl) Hermann (1854–1920), elder son of Otto. He studied the Solar System, refining the orbits of the satellites of Mars, Saturn, and Neptune.

Struve, (Gustav Wilhelm) Ludwig (1858–1920), younger son of Otto. He investigated the proper motion of the Solar System.

Struve, Georg (1886–1933), son of Hermann. He studied the Solar System, in particular the satellites of Saturn and Uranus.

Struve, Otto (1897–1963), son of Ludwig, emigrated and became a naturalized American. A leading observer and astrophysicist of his time, he applied spectroscopy to the study of binary stars, stellar rotation, and discovered interstellar hydrogen and calcium regions.

subdwarf A star that is less luminous by 1 to 2 magnitudes than main-sequence stars of the same **spectral type**. Subdwarfs are mainly of types F, G, and K, and lie below the main sequence on the **Hertzsprung–Russell diagram**. Most are **Population II stars**. They are placed in luminosity class VI, although an alternative designation is to prefix their spectral type with the letters 'sd'.

subgiant A star of smaller radius and lower luminosity than a normal giant star of the same **spectral type**. They are mainly of types G and K, and lie between the main sequence and the giants on the **Hertzsprung–Russell diagram**. Subgiants are placed in luminosity class IV.

subsolar point The point on the surface of the Earth or other body in the Solar System at which, at any given moment, the Sun is directly overhead. Similar points are defined for other pairs of bodies.

Suisei One of Japan's first two space probes, launched in 1985 towards Halley's Comet; the other was **Sakigake**. Suisei, which carried an ultraviolet imager and plasma detector, passed 150,000 km (under 100,000 miles) from the comet in March 1986; Sakigake studied the plasma and magnetic field from a closest approach, in the same month, of 7 million km (over 4 million miles). Both probes were later redirected for further encounters with other comets in the late 1990s.

summer solstice SEE **solstice**

Summer Triangle The prominent triangle formed by the first-magnitude stars Altair (in Aquila), Vega (in Lyra) and Deneb (in Cygnus). It is overhead on summer nights in northern temperate latitudes.

Sun The star at the centre of the Solar System, around which all other Solar System bodies revolve in their orbits. The apparent daily motion of the Sun across the sky and its annual motion along the **ecliptic** are caused by the Earth rotating on its axis and revolving in its orbit. The Sun does, however, have its own motion in the Galaxy of about 20 km/s (12½ mile/s) towards the point in the sky known as the **apex**. The Sun's light is occasionally blocked by the Moon (which, like the Sun, has an apparent diameter of about half a degree) in a **solar eclipse**. The amount of sunlight normally reaching the Earth is quantified as the **solar constant**. Data for the Sun are given in the table.

The Sun is a dwarf **G star**, on the **main sequence** of the Hertzsprung–Russell diagram, and is thus a typical, average star. It consists of about 70% hydrogen (by weight) and 28% helium, with the remainder mostly oxygen and carbon. Its temperature, pressure, and density increase towards the centre, where the values are about 15–20 × 10^6 K, 10^{11} bar, and 155 g/cm^3, respectively. Like all stars, the Sun's energy is generated by nuclear fusion reactions taking place under the extreme conditions in the core, chiefly the **proton–proton reaction**, which converts hydrogen into helium. The mass converted into energy is 4.3 million tonnes per second, but even at this enormous rate the loss amounts to only 0.07% of the total mass in 10^{10} years.

The Sun's core is about 400,000 km (250,000 miles) across. Energy released

SUN: DATA	
Distance from the Earth	
mean (the astronomical unit)	149.6 × 10^6 km
maximum (at aphelion)	152.1 × 10^6 km
minimum (at perihelion)	147.1 × 10^6 km
Diameter	1.392 × 10^6 km
Density (mean)	1.409 g/cm^3
Mass	1.989 × 10^{30} kg
Volume	1.412 × 10^{18} km^3
Period of axial rotation	
at equator	24d 6h
at poles	about 35d
Inclination of axis of rotation to pole of ecliptic	7° 15′
Surface gravity (Earth = 1)	28
Spectral type	G2V
Luminosity	3.86 × 10^{26} W
Magnitude (mean visual)	26.86 (apparent) +4.71 (absolute)
Rotational velocity (mean)	1.9 km/s
Escape velocity	618 km/s
Temperature of surface	5700 K
Temperature of core	14 million K

from the core passes up through the *radiative zone*, which is about 300,000 km (nearly 200,000 miles) thick, by a process of successive absorptions and re-emissions. It then passes through the 200,000 km (125,000-mile) thick *convective zone*, transported by rising and falling cells of gas, to the surface, the **photosphere** (meaning 'sphere of light'), from where it is radiated into space.

Most of the Sun's visible activity takes place in the 500 km (300-mile) thick photosphere. The pattern of convective cells that transport energy from below is visible as the **granulation** (SEE ALSO **supergranulation**). **Sunspots** are darker, cooler regions of the photosphere where the local magnetic field is enhanced. Their passage across the Sun's disk reveals the Sun's **differential rotation**, which might indicate that its core is spinning more quickly than its outer regions. Associated with sunspots are **flares** – sudden, violent releases of energy and material that lead to **aurorae** in the Earth's atmosphere. Other phenomena occurring in the photosphere are **faculae** and **plages**. There is a periodicity in level of solar activity known as the **solar cycle.**

Above the photosphere lies the **chromosphere** (meaning 'sphere of colour', so-called because of its rosy tint when seen during a total solar eclipse), which consists of hot gases and extends for thousands of kilometres. It is generally in a state of turbulence, the temperature rising extremely rapidly with height. This is the realm of **spicules** and **prominences**, huge eruptions of material from the Sun's limb. Extending outward from the chromosphere for millions of kilometres is the extremely tenuous **corona**. A continuous emission from the corona of particles and radiation is known as the **solar wind**. The solar wind and the Sun's magnetic field dominate a region of space called the **heliosphere**, which extends to the boundaries of the Solar System.

The **continuous spectrum** of the Sun, with its dark **Fraunhofer lines**, results from reactions in the photosphere and has yielded much information. The **Zeeman effect** is a splitting of lines of the solar spectrum which indicates the presence of a magnetic field.

The Sun must never be viewed directly through any optical instrument. There are two ways for amateurs to observe the Sun's disk. A *full-aperture filter* has a special metallic coating which cuts out harmful radiation before the Sun's light enters the telescope. Alternatively the Sun's image can be projected on to a white screen. Temporary or permanent blindness can result from direct viewing or from the use of unsuitable filters. Professional instruments for solar work include the **coronagraph** and the **spectroheliograph**.

There have been many probes (such as **Ulysses**) and satellites (such as the **Solar Maximum Mission** and **Yohkoh**) launched to study the Sun.

sundial A simple timekeeping device in which a shadow cast by the Sun falls on a dial graduated in hours. The shadow is cast by a short pillar called a *style* or *gnomon* standing out from the dial, which may be mounted vertically or horizontally. A sundial shows **apparent solar time**.

sundog SEE **parhelion**

sungrazer A **comet** which at perihelion passes through the Sun's corona. Many of the bright long-period comets of the 19th century were sungrazers. Often, such comets do not survive their close passage of the Sun.

sunrise The moment when the Sun's upper limb first appears above the horizon. It is defined as the moment when the Sun's **zenith distance** is 90° 50′, and decreasing. This value is arrived at by making allowance for the Sun's semi-diameter (16′) and for atmospheric refraction (34′).

sunset The moment when the Sun's upper limb disappears below the horizon. Similarly to **sunrise**, it is defined as the moment when the Sun's zenith distance is 90° 50′, and increasing.

sunspot A region in the Sun's **photosphere** which is cooler than its surroundings and therefore appears darker. Sunspots consists of a dark central region, the *umbra*, and a grey outer region, the *penumbra*. They vary in size from around 1000 to 50,000 km (600 to 30,000 miles), and occasionally up to about 200,000 km (125,000 miles), at which size they are visible to the naked eye. Their duration varies from a few hours to a few weeks, or months for the very biggest. Sunspots are seen to move across the face of the Sun as it rotates. The number of spots visible depends on the stage of the 11-year **solar cycle**.

Sunspots occur where there is a local strengthening of the Sun's magnetic field, which cools the area to around 4000 K. The stronger field is believed to suppress the convection of hotter gases from lower levels. Spots are usually found in pairs of opposite magnetic polarity, between latitudes 30° and 40° north or south of the equator, although larger groups are not uncommon (SEE ALSO **following, preceding**).

The spectrum of a sunspot differs from that of the photosphere as a result of its lower temperature. The motion of gases in the penumbra was detected by a Doppler shift of spectral lines (the **Evershed effect**). The **Zeeman effect** in the spectra of some sunspots demonstrates the presence of strong magnetic fields, the polarity of which reverses for each solar cycle.

supercluster SEE **cluster of galaxies**

supergiant An extremely luminous star of large diameter and low density. Supergiants can be of any spectral type, from O to M. Rigel (type B) and Betelgeuse (type M) are examples. The luminosities of supergiants are several magnitudes greater than those of giant stars of the same spectral type, and so they lie at the top of the **Hertzsprung–Russell diagram**. Red (M-type) supergiants like Betelgeuse have the largest diameters, around 1000 times that of the Sun. Supergiants are assigned to luminosity class I. The brightest are often given the separate class Ia, and the others placed in class Ib.

supergiant elliptical A large **elliptical galaxy.**

supergranulation A pattern of convective cells distributed fairly uniformly over the Sun's **photosphere**, and much larger than ordinary photospheric granules (SEE **granulation**). Material has been detected flowing from the centre to the edge of the cells, where most of the magnetic flux coming from the photosphere is concentrated, and it is believed that it is the magnetic field at the edges of the cells that leads to the formation of **spicules**.

superior conjunction SEE **conjunction**

superior planet Any of the planets Mars, Jupiter, Saturn, Uranus, Neptune and Pluto, which orbit outside the Earth's orbit.

superluminal Describing the motion of an object which appears to be moving faster than the speed of light. Parts of some radio sources appear to be moving apart at superluminal velocities. This has been explained as a trick of the geometry of the source as we see it, for such velocities are impossible according to **Einstein**'s Special Theory of relativity.

supermassive black hole SEE **black hole**

supernova (plural **supernovae**) A stellar explosion in which virtually an entire star is disrupted. For a week or so, a supernova may outshine all the other stars in its galaxy. The luminosity (absolute magnitude up to about −19) is some 23 magnitudes (1000 million times) brighter than the Sun, and the energy released in the explosion is the same as is released over the star's entire previous life. A supernova is about 1000 times brighter than a **nova**. The diagram shows the light curve of Supernova 1989B, in the galaxy M66. Supernovae are designated by the year in which they are observed, plus a letter if there is more than one in that year.

Analysis of ancient records identifies several supernovae in our own Galaxy, before the invention of the telescope. One was observed in 1054 by Chinese astronomers. The **Crab Nebula** and its associated pulsar are the remains of this event. In 1572 Tycho Brahe and others saw a new star as bright as Venus in the constellation Cassiopeia (SEE **Tycho's Star**). The last naked-eye supernova in our Galaxy was seen in 1604 in Ophiuchus, and was observed for over a year (SEE **Kepler's Star**). There is probably about one supernova every 30 years in a galaxy like our own, but most of them are concealed by dust.

Supernovae are of two main types, classified by their light curves and spectra. *Type II* supernovae have hydrogen lines in their spectra, whereas *Type I* supernovae do not. Type I supernovae have been found in all kinds of galaxies, but Type II have never been found in ellipticals. Type II supernovae are thought to be caused by the explosion of stars greater than about 8 solar masses at the end of their life. When the nuclear fuel inside the star gives out, the core collapses to form a **neutron star** or **black hole**. The collapse of the star's interior releases energy which is picked up by the outer layers, and these layers are ejected into space at about 5000 km/s (3000 mile/s). Type I supernovae are believed to be formed by the explosion of the white dwarf component of a binary star. Hydrogen from the companion leaks on to the white dwarf and drives it over the critical mass of 1.4 solar masses, causing it to explode.

After a couple of years the supernova has expanded so much that it becomes thin and transparent. For hundreds or thousands of years the ejected material remains visible as a **supernova remnant**.

Supernova 1987A A bright supernova which flared up in the Large Magellanic Cloud in February 1987, reaching naked-eye visibility – the first supernova to do so since Kepler's Star of 1604.

supernova remnant (SNR) A gaseous emission nebula, the expanding shell of matter thrown off into space during the outburst of a **supernova**. These remnants are often strong radio and X-ray sources. The ejected layers of the supernova collide with the surrounding interstellar gas and heat up to

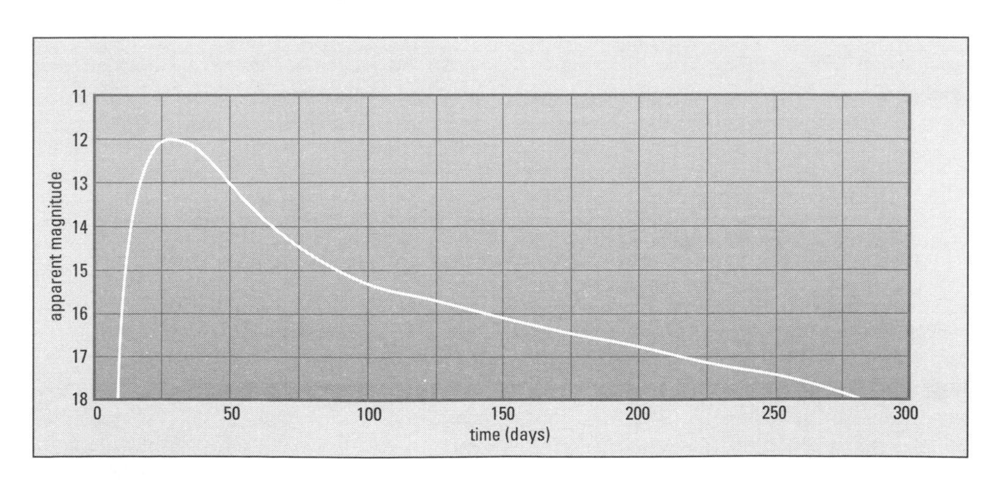

Supernova: light curve, showing a characteristically steep rise to maximum, and a shallower fall

perhaps a million degrees K, emitting X-rays. Electrons spiral in the magnetic fields in the compressed gas and emit **synchrotron radiation** in the form of radio waves. The most intense radio source in the sky, **Cassiopeia A**, is a supernova remnant. Supernova remnants usually appear as hollow shells. However, some (including the Crab Nebula) are filled balls of radio and X-ray emission. Such a remnant is termed a *plerion*, meaning 'filled'. This implies that they contain electrons produced by an active pulsar, formed in the supernova explosion.

super Schmidt telescope SEE **Schmidt camera**

Surveyor A series of seven US spacecraft, five of which made successful soft landings on the Moon between May 1966 and January 1968 in preparation for the Apollo landings. They took photographs and made chemical analyses of the surface.

Swift–Tuttle, Comet A **comet** discovered in 1862 by American astronomers Lewis Swift and Horace Tuttle, subsequently identified with Comet Kogler of 1737, and observed at its return in 1992. It is the second-brightest short-period comet, Halley's Comet being the brightest. The large number of jets seen emerging from the nucleus in 1992 suggest that the comet's changing period (at present 130 years) may be caused by **non-gravitational forces**. It is the parent comet of the **Perseid** meteor shower; this relationship was discovered by Giovanni Schiaparelli, and was the first comet–meteor connection to be established.

symbiotic variable A **binary star** consisting of two stars of widely differing surface temperatures, such as a cool red giant and a hot dwarf. Gas from the cool star falls on to, and heats, the smaller star. Symbiotic variables are also known as *Z Andromedae stars*, after their prototype, which has a range from magnitude 8.3 to 12.4.

synchronous orbit An orbit in which a satellite's period of revolution is the same as the primary's period of axial rotation. From the planet's surface the satellite appears to hover over one point, neither rising nor setting. An example is **Chiron**'s orbit around Pluto.

synchronous rotation The axial rotation of a celestial body in the same period as its period of revolution. Consequently, it always presents the same face towards the body about which it revolves, as in the case of the Moon orbiting the Earth. Tidal friction (SEE **tides**) has locked the Moon into this condition, which is also known as *captured rotation.* Other examples are the regular satellites of Jupiter and Saturn.

synchrotron radiation The electromagnetic radiation emitted by charged particles (usually electrons) that are accelerated by a strong magnetic field to speeds which are a significant fraction of the speed of light. The higher the energy of the particles, the shorter the wavelength of the radiation they emit. Synchrotron radiation is so called because it was first observed in particle accelerators called sychrotrons. Cosmic sources include **radio galaxies** and supernova remnants such as the **Crab Nebula**.

synodic month The period between two identical phases of the Moon; it is the same as the duration of one **lunation**. Its length is 29.53059 mean solar days.

synodic period The period of apparent revolution of one body about another as observed from the Earth, for example from one opposition or conjunction to the next. COMPARE **sidereal period**

synthesized aperture SEE **aperture synthesis**

Syrtis Major A dark, triangular feature near the Martian equator, and the planet's most conspicuous feature in telescopic views from the Earth. It was first recorded by **Christiaan Huygens** in 1659. It is a sloping, cratered area, officially named Syrtis Major Planum.

Systems I and II Regions of **Jupiter**'s upper atmosphere with slightly differing speeds of rotation.

SY Ursae Majoris star SEE **U Geminorum star**

syzygy The approximate alignment of three celestial bodies; in particular, the alignment of the Earth, the Sun, and the Moon or another planet. Thus syzygy occurs at full Moon and new Moon, and at planetary oppositions and conjunctions.

T

TAI ABBREVIATION FOR **International Atomic Time** (from its name in French).

tangential velocity The component of a star's velocity at right angles to the line of sight. It is also known as *transverse velocity*. It can be found from measurements of the annual **proper motion** of the star and its distance in parsecs. If the **radial velocity** is also known, then the **space velocity** of the star can be calculated.

Tarantula Nebula A nebula faintly visible to the naked eye on the southeastern edge of the Large Magellanic Cloud; also known as NGC 2070 or 30 Doradus. It is 1000 light years across and 500,000 times the Sun's mass, bigger and brighter than any nebula in our Galaxy or any other nearby galaxy. It has a complex filamentary structure, and has a cluster of stars, designated R136, at its centre.

T association A region of recent and active star formation consisting of low-mass stars known as **T Tauri stars**. The members of T associations offer a series of snapshots of stars like our Sun at various stages through its infancy. They are still surrounded by the dusty, obscuring material of the cloud from which they are forming, and consequently they are often brightest in the infrared. COMPARE **OB association**

Taurus A constellation of the zodiac, representing the bull into which Zeus changed himself to carry off Princess Europa. It contains two of the most famous open clusters in the sky, the **Pleiades** and the **Hyades**, and also the **Crab Nebula**. Its brightest star is **Aldebaran**. Beta (Elnath), magnitude 1.7, and Zeta, magnitude 3.0, mark the tips of the bull's horns. Lambda is an eclipsing binary (range 3.4 to 3.9, period 3.95 days). There is also the prototype of the **T Tauri stars**.

tectonics Name given to the various processes by which the surface of a planetary body is deformed as a result of heating from within. Tectonic processes operate on bodies that have undergone **differentiation** and so possess a surface crust with a molten layer below. Large-scale movements of the molten layer affect the crust, producing faults and folds, and lifting mountains. Several satellites (e.g. **Ariel** and **Europa**) have surface features which appear to indicate tectonic activity. *Plate tectonics* operates, as far as we know uniquely, on the Earth, whose crust consists of a number of *plates* which move with respect to one another at a few centimetres per year, supported on the mantle below. Where these plates meet there are earthquakes, volcanoes, and other geological activity, producing features such as mid-oceanic ridges and mountain chains.

tektites Small, glassy objects, typically centimetre-sized, found scattered across certain specific areas (*strewnfields*) of the Earth's surface. They consist mainly of silica, with small quantities of metallic oxides, and are from 600,000 to 65 million years old. Their chemical and physical characteristics and their ages are specific to the areas in which they are found, and have nothing in common with the geological characteristics of their surroundings. All tektites have clearly solidified rapidly from a temperature of around 1500–2000 K. Their origin is uncertain, though their flight markings, which are characteristic of solidification while flying through the air at high speed, suggest that they originated as terrestrial rock was vaporized by the impact of large meteorites. Tektites are named after the location of the strewnfield: for example, *Australites* are from Australia, and *Indochinites* from South-east Asia.

telescope An instrument for collecting and magnifying light or other electromagnetic radiation from a distant object. There are two basic types of optical telescope: the **refracting telescope**, in which the light-collecting element is a lens, and the **reflecting telescope**, in which it is a mirror; **catadioptric** systems are hybrids of the two. The image formed by the primary (main) lens or mirror is magnified by an **eyepiece**. The eyepiece may be replaced by one of several devices: for example, a camera, photographic plate, or **CCD** for **astrophotography** (SEE ALSO **Schmidt camera**); a **photometer** for measuring brightness; or a **spectrometer** for examining spectra. A telescope is mounted (SEE **telescope, mounting of**) in such a way that its weight is supported and it is easily aimed at celestial objects.

The **focal ratio** of a telescope is equal to the focal length of the primary lens or mirror divided by its diameter. Refractors have longer focal ratios than reflectors, which, aperture for aperture, are thus shorter. The most compact designs are short-focus instruments like the **Schmidt–Cassegrain telescope**. The image produced by an astronomical telescope is inverted, as opposed to that formed in a terrestrial telescope, which is erect.

The refractor installed in the Yerkes Observatory in 1897, which has an objective 1 m (40 inches) in diameter, is still the largest of its kind. Large telescopes built since then are all reflectors, because large mirrors are easier to make and support than large lenses. The world's largest reflector is the first of the **Keck Telescopes**, which has a composite mirror 10 m (396 inches) across. With the development of spacecraft, it has become possible to send instruments such as the **Hubble Space Telescope** into space above the Earth's atmosphere, and so make observations unaffected by atmospheric absorption. SEE ALSO **radio telescope**

telescope, mounting of A means of supporting the weight of a telescope and enabling it to be easily aimed at different points in the sky. Of necessity, all large professional instruments are on permanent mounts; most amateur telescopes and their mounts are portable. Permanent or portable, rigidity is a prime requirement.

So that a telescope may be pointed at different parts of the sky, the mounting incorporates two axes, perpendicular to each other, about which the telescope can be rotated. In the **altazimuth** mount the axes are parallel and perpendicular to the horizon. The disadvantage of this mounting is that the telescope has to be moved about the two axes simultaneously in order to keep an object in view. This is overcome in **equatorial telescopes**, in which one axis (the polar axis) is parallel to the Earth's axis of rotation, and so points to the celestial pole. The telescope then only needs to be moved about the other axis, in the direction opposite to that of the Earth's rotation, to keep an object in the field of view.

Since the introduction of computer-controlled drive mechanisms, the altazimuth mount has become the choice for all large professional telescopes. A simple altazimuth platform is incorporated in the **Dobsonian telescope**, a large and portable amateur reflecting telescope.

Telescopium A faint southern constellation representing a telescope, introduced by **Lacaille**. Alpha, its brightest star, is magnitude 3.5, but it contains little of interest.

Telesto A small **satellite** of Saturn, discovered by Bradford Smith and others in 1980 between the two Voyager encounters. Telesto is irregular in shape, measuring 34 × 28 × 26 km (21 × 17 × 16 miles). It is a **co-orbital satellite** with Tethys and **Calypso**.

telluric line An **absorption line** in the spectrum of a celestial object produced by molecules such as oxygen and water in the Earth's atmosphere.

terminator The boundary between the illuminated and unilluminated hemispheres of a planet or satellite as viewed from Earth. Mercury, Venus, Mars, and the Moon show phases and therefore their terminators are visible. The mountainous lunar surface gives the Moon a visibly irregular terminator.

terrestrial planets The planets Mercury, Venus, Earth and Mars, so called because they have rather similar characteristics in respect of size, density, and few or no satellites.

Tethys A medium-sized **satellite** of Saturn, diameter 1060 km (660 miles), discovered by Giovanni Cassini in 1684. The whole surface is densely cratered, but evidence of past geological activity is provided by some less heavily cratered areas which must have undergone resurfacing. The two dominant features are Odysseus, a 440 km (270-mile) diameter crater, and Ithaca Chasma, a 100 km (60-mile) wide trough running three-quarters of the way around the satellite; their formation may be linked. Tethys shares its orbit with two co-orbital satellites, **Calypso** and **Telesto**.

Thalassa One of the small inner **satellites** of Neptune discovered in 1989 during the Voyager 2 mission.

Thales of Miletus (*c.* 625 – *c.* 550 BC) Greek philosopher. He provided the first cosmogony – a theory of the Universe's origin – in which everything developed from a primordial mass of water. He knew of the **saros**, and is credited with having used it to predict the eclipse of the Sun which occurred probably on 28 May 585 BC during a battle between the Lydians and the Medes.

Thebe One of the small inner **satellites** of Jupiter discovered in 1979 during the Voyager missions. It is irregular, measuring about 110 × 90 km (70 ×55 miles).

third contact In a total **solar eclipse**, the moment when the trailing edge of the Moon's disk begins to uncover the Sun's disk, and totality ends. In a **lunar eclipse**, it is the moment when the Moon begins to leave the umbra of the Earth's shadow. The term is also used for the corresponding stage in eclipses involving other bodies.

three-body problem A fundamental problem in celestial mechanics: to determine the motions of three bodies under the influence only of their mutual gravitational attractions. There is no exact general solution, only solutions for special cases, but highly accurate approximations can be achieved with modern computers. The three-body problem was examined first by **Isaac Newton**, and subsequently tackled by many astronomers and mathematicians, including **Leonhard Euler**, **Joseph Louis Lagrange** (SEE ALSO **Lagrangian points**), Carl Jacobi, and **Henri Poincaré**. Work on the problem has been stimulated mostly by the need to understand the orbit of the Moon, under the gravitational influence of the Earth and the Sun, and more recently by the need to calculate the orbits of artificial satellites. SEE ALSO **many-body problem**

tides Distortions induced in a celestial body by the gravitational attraction of one or more others. The gravitational attraction a body experiences is greatest on the side nearest the attracting body, and least on the side furthest away, causing it to elongate slightly in the direction of the attracting body, acquiring a *tidal bulge* on each side. If the deformed body is orbiting the attracting body, different parts of its surface periodically experience tidal distortion as it rotates. These tides most obviously affect a gaseous atmosphere or a fluid ocean, but the solid crust of a planet like the Earth, which is supported by a fluid mantle, is able to 'flex' and so also experiences tidal distortion, but to a much lesser degree. The enormous gravitational forces exerted by Jupiter on its satellite **Io** produce a large amount of flexure and interior heating, and this *tidal heating* is

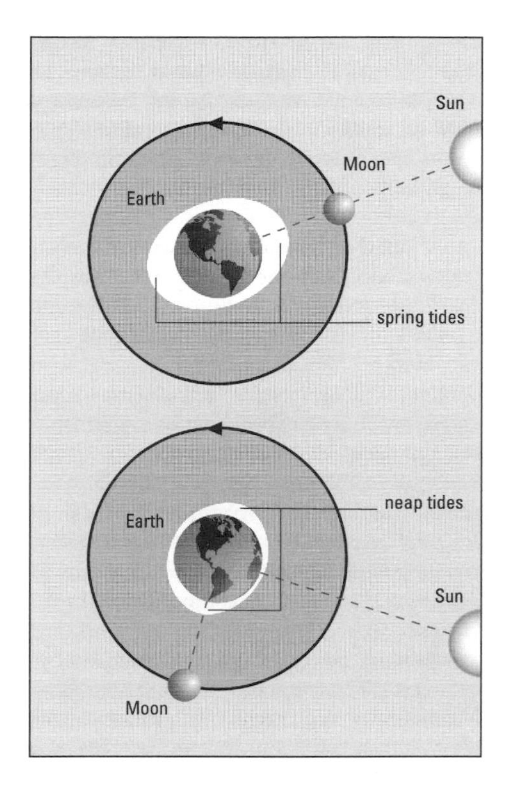

Tides (not to scale)

responsible for Io being highly volcanically active. Smaller and completely rigid bodies can under certain circumstances be broken up by tidal forces.

The familiar ocean tides on the Earth are raised by the gravitational attraction of the Moon and the Sun, the Moon's influence being about three times the Sun's. There are two high tides and two low tides each day. When the Sun and the Moon are exerting a pull in the same direction (as at new Moon or full Moon) their effects are additive and high tides are higher (*spring tides*). When the pull of the Sun is at right angles to that of the Moon (as at first quarter or last quarter), high tides are lower (*neap tides*). *Tidal friction* caused by the tidal ebb and flow of water over the ocean floor acts as a brake on the Earth's rotation, as a result of which the length of the day is increasing by about one millisecond per century. The angular momentum which is thus being lost is transferred to the Moon (SEE **secular acceleration**).

time A scale for measuring the duration of events by reference to some regularly recurring phenomenon. From remote antiquity the basis of time has been the passage of the Earth once around its orbit, giving the unit of one **year**, and the rotation of the Earth on its axis, giving the unit of one **day**. However, the rotation of the Earth is not constant, and so it cannot be used as an accurate and invariable standard of time. Modern timescales are now based on a continuous count of time units defined by atomic resonators (**International Atomic Time**). The basis of civil timekeeping is **Greenwich Mean Time** (known technically as **Universal Time**). In addition, astronomers reckon time by the stars: **sidereal time**. SEE ALSO **apparent solar time, equation of time, local time, mean solar time, time zone**

time zone One of 24 divisions of the Earth's surface, each 15° of longitude wide, within which the time of day is reckoned to be the same. At a conference held in Washington in 1884 the meridian of Greenwich was adopted as the zero of longitude, and zones of longitude were established. Standard time in each successive zone westwards is one hour behind that in the preceding zone. Large territories such as the USA and Russia span several time zones. The adoption of time zones results in a discrepancy at longitude 180° (the International Date Line), which is resolved by omitting one day from the calendar when crossing from west to east, or repeating one day if the crossing is from east to west.

Titan The largest **satellite** of Saturn, and the second largest in the Solar System, discovered by Christiaan Huygens in 1655. Titan is unique among planetary satellites in having a substantial atmosphere. At 1.88 g/cm^3 Titan is the densest of Saturn's large satellites, and is composed of rock and water-ice in roughly equal proportions. The space probe Voyager 1 found no gaps in an opaque, reddish cloud layer 200 km (125 miles) above the surface. The atmosphere consists mostly of nitrogen, with some methane and other hydrocarbon compounds, and exerts a surface pressure 1.5 times that on the Earth. The surface temperature is 95 K, at which methane can exist as solid, liquid or gas, so methane may play the role that water does on Earth, forming, clouds, rain, lakes, and even snow. The surface may therefore have been shaped by the same types of erosion that have operated on Earth. More will be known early in the 21st century when the probe called Huygens is due to be landed on Titan by the **Cassini** spacecraft.

Titania The largest of the five main **satellites** of Uranus, discovered by William Herschel in 1787. It is 1580 km (980 miles) in diameter, and has a density of 1.68 g/cm^3, indicating that it consists of a mixture of rock and water-ice. Titania is Oberon's twin in size, but its surface resembles that of **Ariel**, with similar features and a similar geological history. The most prominent trough is named Messina Chasma, and is 1500 km (950 miles) long.

Tombaugh, Clyde William (1906–) American astronomer. In 1930, nearly a year into a search based on predictions by **Percival Lowell**, he discovered the planet **Pluto**. He continued to search for other planets for more than ten years, discovering in the process star clusters, clusters of galaxies, a comet, and hundreds of asteroids. After World War II he developed telescopic cameras for tracking rockets after launch.

topocentric Pertaining to observations made from a point on the surface of the Earth, e.g. topocentric coordinates as opposed to geocentric coordinates (which are measured from the centre of the Earth).

transient lunar phenomenon (TLP) A short-lived, localized change in the appearance of a lunar surface feature, such as a glow or an obscuration. TLPs were reported by lunar observers for many years but treated with scepticism, but they are now generally regarded as real events. In 1958 a carbon spectrum of activity in the crater Alphonsus was obtained, and instruments set up by the Apollo astronauts established a connection between TLPs and moonquakes. Since they are most common when the Moon is at perigee, TLPs are most likely releases of gas and dust induced by tidal forces (SEE **tides**).

transit (1) The passage of a celestial body across the observer's **meridian**. Such transits, observed with a **transit instrument**, were formerly important in the measurement of time and for navigational purposes.

transit (2) The passage of a body directly between the Earth and the Sun. The planets Mercury and Venus do so on occasion, when they appear as a black spot crossing the solar disk (SEE ALSO **black drop**). Accurate measurements of such transits helped to establish the scale of the Solar System.

transit (3) The passage of a planetary satellite across the planet's disk. Jupiter's four main satellites frequently transit its disk, and Saturn's inner satellites are occasionally seen in transit. Observations of transits of Jupiter's satellites by Ole **Römer** established the enormous but finite speed of light.

transit (4) The passage of a surface or atmospheric feature of a body across its **central meridian** as it rotates, the timing of which gives a means of measuring the body's rotation period.

transit instrument A telescope mounted on a horizontal east–west axis and movable only in the vertical plane of the meridian. Transit instruments were once used to time the **transits** (1) of bodies across the local meridian and measure their altitudes. Such measurements were used for precision timekeeping, but have been superseded by devices such as the **photographic zenith tube**.

transverse velocity SEE **tangential velocity**

Trapezium Popular name for the multiple star Theta Orionis in the Orion Nebula. Its four main components, magnitudes 5.1, 6.7, 6.7 and 8.0, are arranged in the shape of a trapezium. The stars are very hot, and light up the nebula.

Triangulum A small but easily found constellation between Andromeda and Aries. Its brightest star, Beta (magnitude 3.0), forms a well-marked triangle with Alpha (3.4) and Gamma (4.0). The main feature of interest is the spiral galaxy M33, 2.7 million light years away in our Local Group, visible with binoculars under clear, dark skies.

Triangulum Australe A southern constellation representing a triangle, marked by the stars Alpha (magnitude 1.9), Beta and Gamma (both 2.9). The open cluster NGC 6025 is visible with binoculars.

Trifid Nebula An emission nebula in the constellation Sagittarius, apparently divided into three main sectors by dark lanes of dust. It is also known as M20 or NGC 6514. At its centre is an 8th-magnitude double star. The Trifid lies 5000 light years away.

trigonometric parallax A means of determining the distance of a star using the principle of triangulation. As the Earth revolves around the Sun, nearby stars show a small change in position with respect to more distant stars. This change in position can be measured from photographs taken six months apart, i.e. when the Earth is on opposite sides of the Sun. As the diagram shows, the angular displacement is a direct measure of the star's parallax (π). Since the baseline (the distance from the Sun to the Earth) is known, the distance d of the star can be determined by trigonometry. The first star to have its distance measured by this method was 61 Cygni, by **Friedrich Bessel** in 1838. Parallax is measured in arc seconds. The inverse of the parallax is the distance in parsecs. Alpha Centauri is the star with the largest known parallax, 0.752″, which corresponds to a distance of 1.3 parsecs. Typical errors in measuring parallax photographically are about 0.01″, but by

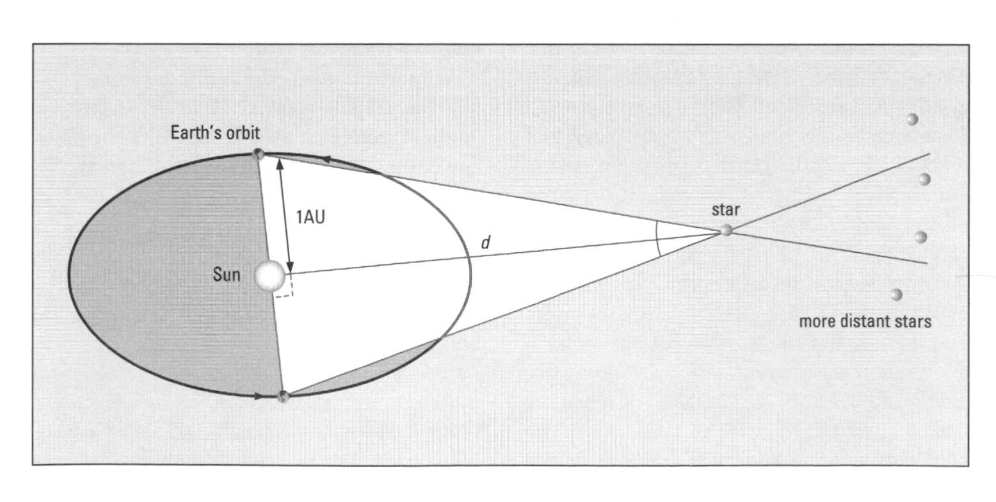

Trigonometric parallax: measuring the distance of nearby stars

using many plates this error can be reduced to about 0.004″. The parallax of a star at 25 parsecs is 0.04″, so the uncertainty is around 10%. The most accurate parallax determinations have been made by the **Hipparcos** satellite.

triple-alpha process A nuclear reaction in which three alpha particles (helium nuclei) are transformed into carbon with the release of energy. It takes place after all the hydrogen of a star's core has been exhausted. The core contracts and its temperature rises until it exceeds 100 million K, when the triple-alpha reaction begins. In each reaction two helium nuclei combine to from a beryllium nucleus which, in turn, captures a further helium nucleus to form a carbon nucleus. Additional helium captures can produce oxygen, neon, and a number of heavier elements. The reaction is believed to be the dominant energy-producing process in red giants. It is also known as the *Salpeter process* after the US physicist Edwin Salpeter (1924–), who described it.

triplet A compound lens consisting of an assembly of three component lenses, which may be air-spaced or cemented.

Triton The largest **satellite** of Neptune, discovered in 1846 by William Lassell. It is in an inclined, retrograde orbit and has almost certainly been captured by Neptune. Its diameter is 2705 km (1680 miles), and its density is 2.07 g/cm^3, consistent with a composition largely of three parts rock to one of water-ice. Triton is the coldest and most distant world so far visited: Voyager 2's instruments measured a surface temperature of 35 K (−238°C). Its surface is complex and varied, and includes *cantaloupe terrain* – a dimpled landscape crossed by broad fissures and resembling the skin of a melon. There is much evidence of very recent volcanic activity in a variety of features formed by the upwelling of fluid material. In the south polar region, which is covered with a thin **polar cap** of nitrogen ice, dark eruptive plumes send fine dark material rising to altitudes of 8 km (5 miles) before being carried downwind for 100 km (60 miles) or more in the tenuous nitrogen atmosphere.

Trojan asteroid A member of one of two groups of **asteroids** sharing Jupiter's orbit. One group lies ahead of Jupiter, and the other behind, oscillating about the L_4 and L_5 **Lagrangian points** of Jupiter's orbit around the Sun. The Trojans ahead of Jupiter are known as the *Achilles group*, after the first one to be found, **Achilles** in 1906; those behind are the *Patroclus group*, named after the second Trojan to be discovered, also in 1906. Two hundred Trojans are known, and no doubt there are many more undiscovered ones.

tropical month One revolution of the Moon around the Earth relative to the vernal **equinox**. It equals 27.32158 days.

tropical year The time taken for the Earth to orbit the Sun relative to the vernal equinox. Its length is 365.24219 mean solar days. Because of the precession of the vernal equinox, which moves along the ecliptic in a direction opposite to the Sun's motion, the tropical year is about 20 minutes shorter than the sidereal year. It is sometimes also known as the *solar year*.

Tropic of Cancer The parallel of latitude on Earth, 23½° north, that marks the most northerly declination reached by the Sun at the summer **solstice**, on or about 21 June.

Tropic of Capricorn The parallel of latitude on Earth, 23½° south, that marks the most southerly declination reached by the Sun at the winter **solstice**, on or about 22 December.

true anomaly SEE **anomaly**

Trumpler classification The classification of open star clusters devised by Swiss–

American astronomer Robert Julius Trumpler (1886–1956) on the basis of three characteristics: the number of stars in the cluster, the concentration of stars towards the centre of the cluster, and the range of brightness within each cluster.

T Tauri star A very young star still settling on to the main sequence and belonging to a class of irregular variables, named after the prototype, T Tauri. T Tauri stars are found in nebulae or young clusters, and are characterized by high-velocity infall or outflow of gas as they adjust to the onset of nuclear reactions. Emission lines in their spectra indicate an extended atmosphere of gas. T Tauri stars are usually less massive than the Sun. Heavier stars either pass through the T Tauri stage while they are still obscured in their clouds, or have a different appearance at the same stage of life. SEE ALSO **Hertzsprung–Russell diagram, T association**

Tucana A far southern constellation representing a toucan. Its overall faintness is redeemed by the presence of the Small Magellanic Cloud and the superb globular cluster 47 Tucanae (NGC 104). At magnitude 4 and half a degree across, 47 Tucanae is inferior only to Omega Centauri. It lies 15,000 light years away. NGC 362 is another globular, visible in binoculars.

Tully–Fisher relation A relationship that links the absolute brightness of a galaxy to its speed of rotation as revealed by the width of its 21 cm hydrogen line. It is one of the most important methods for estimating distances to galaxies. It was formulated by R. Brent Tully and J. Robert Fisher in 1977.

Tunguska Event A huge aerial explosion on 30 June 1908 just north of the Stony Tunguska River in Siberia, which devastated the surrounding forest. The explosion, which was heard more than 800 km (500 miles) away and was recorded seismographically all around the world, was preceded by a fireball as bright as the Sun. Trees up to 40 km (25 miles) away were felled, and in the central 1000 square km (400 square miles) trees and animals were incinerated.

The absence of impact craters or meteoritic fragments puzzled earlier investigators, but the best current theory is that a **near-Earth asteroid** roughly 50 m (150 ft) in diameter entered the atmosphere obliquely, and shattered and vaporized at an altitude of about 8 km (5 miles), blanketing the area with dust. Dust particles preserved in resin from trees in the area have the composition of known stony meteorites, and the damage pattern is consistent with the impactor having an asteroidal density. An incoming comet, which would be less dense, would vaporize too high up to cause much damage on the ground; a former theory had part of Encke's Comet as the impactor.

tuning fork diagram A diagram (SEE page 220) of **galaxy** types, originated by Edwin Hubble. The 'handle' of the fork consists of elliptical galaxies (E), numbered according to their degree of elongation, while the two prongs of the fork are made up of ordinary (S) and barred (SB) spirals, designated a, b or c according to how tightly wound the spiral arms are. The classification was originally interpreted as a sequence of evolution, with ellipticals developing into spirals, but this is now known to be incorrect.

twenty-one centimetre line The emission line of neutral hydrogen in interstellar clouds. It lies in the radio spectrum at a wavelength of about 21 cm; the frequency is 1420 MHz. Its existence was predicted by **Hendrik van de Hulst** in 1944 and discovered by him and others in 1951. SEE ALSO **HI region**

twilight The intermediate period during which the illumination of the sky gradually increases before sunrise, and decreases after sunset. The phenomenon is caused by the

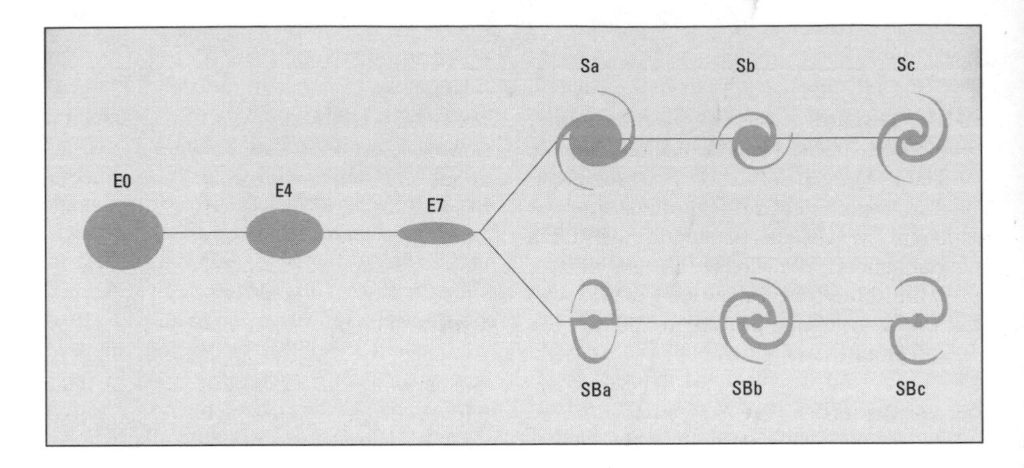

Tuning fork diagram: illustrating the basic classification of galaxies

scattering of sunlight by molecules of air and particles of dust in the Earth's atmosphere. The duration of twilight depends on the steepness of the Sun's apparent path with respect to the horizon, so that twilight lasts longer at higher latitudes. The following forms of twilight are distinguished. *Civil twilight* ends, or begins, when the centre of the Sun's disk is 6° below the horizon, and is regarded as a period during which normal daytime activities cease to be possible. *Nautical twilight* ends, or begins, when the centre of the Sun's disk is 12° below the horizon, and the marine horizon is no longer visible. *Astronomical twilight* begins in the morning and ends in the evening when the centre of the Sun's disk is 18° below the horizon, and is the time when the faintest stars can be seen with the naked eye.

Tycho A large lunar crater, 85 km (53 miles) in diameter, in the Moon's southern uplands. It is the centre of the most conspicuous system of lunar **rays**, stretching across much of the Moon's nearside, which suggests that Tycho may be a relatively young crater.

Tycho Brahe SEE **Brahe, Tycho**

Tychonian system A planetary theory put forward by **Tycho Brahe**.

Tycho's Star A supernova in Cassiopeia in 1572 that was observed and described by **Tycho Brahe**. It was brighter than Venus at maximum and was visible during daytime. Its remnant has been detected as a radio and X-ray source. Its light curve identifies it as a Type I supernova.

U

UBV system A system of three-colour photometry devised by Harold Johnson and William Morgan of Yerkes Observatory. Stellar magnitudes are determined at three different wavelength bands through three colour filters: ultraviolet (U) peaking at 360 nm; blue (B) peaking at 440 nm; and yellow (V for visual) peaking at 550 nm. SEE ALSO **colour index**

U Geminorum star A member of a class of dwarf novae which show sudden outbursts of between 2 and 6 magnitudes, followed by a slower return to minimum where they remain until the next outburst. Periods range from 10 days to several years. They are also known as *SS Cygni stars*. The best known examples are SS Cygni, whose light curve is shown in the diagram, and U Geminorum. All are close binaries, comprising a subgiant or dwarf star of type K or M which has filled its Roche lobe, and a white dwarf surrounded by an **accretion disk** of infalling matter. Almost all members show rapid irregular flickering of about half a magnitude at minimum. As they are all binaries, some show eclipses at minimum, such as U Geminorum itself. The *SY Ursae Majoris* subclass has normal maxima and 'supermaxima' that last about five times as long and are twice as bright. It has been suggested that they evolve from **W Ursae Majoris stars**. They are closely related to **Z Camelopardalis stars**.

Uhuru The first X-ray astronomy satellite, launched by the US in December 1970. Many X-ray sources bear numbers prefixed with the letter U, being their designation in the Uhuru catalogue.

ultraviolet astronomy The study of **electromagnetic radiation** from space with wavelengths between those of the visible spectrum and X-rays, i.e. from about 400 nm down to about 10 nm. Apart from the longest wavelengths (called the *near ultraviolet*), ultraviolet radiation does not penetrate the atmosphere, so observations have to be made from rockets and satellites. The first ultraviolet satellites to study the Sun and stars were the Orbiting Solar Observatory and Orbiting Astronomical Observatory series in the 1960s. A major observatory satellite, the **International Ultraviolet Explorer** (IUE),

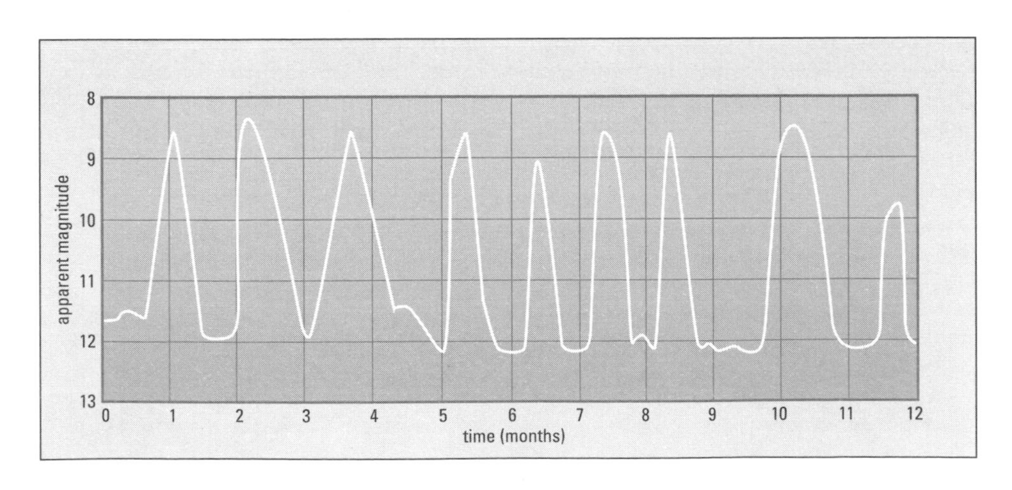

U Geminorum star: light curve, showing irregular maxima and 'flickering' minima

was launched in 1978. Coverage was extended to the shortest ultraviolet wavelengths by **Rosat** in 1990 and the **Extreme Ultraviolet Explorer** (EUVE) in 1992.

Unlike other wavelengths such as radio, infrared, and X-rays, ultraviolet astronomy has discovered few new sources. Instead, the main application is in spectroscopy. Many of the abundant atoms and ions in celestial sources have their strongest spectral lines in the ultraviolet. Observations at ultraviolet wavelengths have been significant for studies of the composition and motions of interstellar gas. Hot stars of spectral types O and B emit strongly in the ultraviolet, and they excite surrounding nebulae by their ultraviolet emissions. In the Solar System, ultraviolet spectroscopy is used to study planetary atmospheres and the gas in comets. Further off in the Universe, ultraviolet astronomy is important in the study of luminous galaxies and quasars, which contain hot stars and large quantities of gas.

Ulugh Beg (1394–1449) Mongol ruler and astronomer. He established an observatory at Samarkand, and compiled the first star catalogue to surpass those of Hipparchus and **Ptolemy** in precision.

Ulysses A European Space Agency probe launched in October 1990 to study the polar regions of the Sun. It was thrown out of the ecliptic by a fly-by of Jupiter in February 1992, and looped over the south pole of the Sun in summer 1994, flying over the north pole of the Sun a year later.

umbra (1) The inner part of the shadow cast by a celestial body illuminated by an extended source such as the Sun. An observer in the umbral region of a shadow sees all of the source obscured, as, for example, in a total **solar eclipse**. SEE ALSO **lunar eclipse**. COMPARE **penumbra** (1)

umbra (2) The dark central area of a sunspot. COMPARE **penumbra** (2)

Umbriel One of the five main **satellites** of Uranus, discovered by William Lassell in 1851. Its diameter is 1172 km (728 miles), and its density of 1.51 g/cm^3 is consistent with a composition of mostly water-ice and rock. Umbriel is a geologically inactive, cratered satellite with a predominantly dark surface. The only bright spot is the floor of the 140 km (85-mile) diameter crater Wunda.

United States Naval Observatory The US government observatory with headquarters at Washington, DC. Its main telescope there is a 0.66 m (26-inch) refractor. The Observatory has an observing station at Flagstaff, Arizona, with a 1.55 m (61-inch) reflector for astrometric work.

Universal Time (UT) Standard time used for scientific purposes throughout the world, popularly known as **Greenwich Mean Time** (GMT). There are several versions of Universal Time. *UT0* is the observed rotation of the Earth relative to the stars. *UT1* is UT0 corrected for a slight wandering of the Earth's geographical poles, which affects the longitude of the place of observation. *Coordinated Universal Time* (UTC; the abbreviation is of the term in French) is a regular system of time based on the caesium atomic clock, and is used for radio time signals. As the Earth's rotation is gradually slowing, the time shown by an atomic clock gradually diverges from UT1. To keep atomic time in step with UT1 to the nearest second, time signals are retarded when necessary by one **leap second** at midnight on 30 June and 31 December. UTC is what is generally known as GMT.

Universe All of space and everything contained in it. SEE **cosmogony**, **cosmology**

upper culmination SEE **culmination**

Uranus The seventh major planet from the Sun, and the second-smallest of the four giant planets. With a mean magnitude of 5.5, Uranus is visible to the naked eye under good conditions. Its apparent diameter varies from 3.3 to 4.1 arc seconds, and through a telescope it appears as a small, featureless, greenish blue disk. In size, mass, atmosphere and colour, Uranus resembles Neptune. Like all the giant planets, it possesses a ring system and a retinue of satellites. The main data for Uranus are given in the table below.

Uranus was the first planet to be discovered since ancient times. Although it had been observed on several occasions (and had once been catalogued as a star), its non-stellar nature was first recognized by **William Herschel**, who observed its disk on 13 March 1781 during a telescopic survey of faint naked-eye stars. He first took it to be a comet, but as observations accumulated it became clear that the object was in a circular, planetary orbit and was indeed a planet.

Like Pluto, Uranus's axis of rotation is steeply inclined, and lies close to the ecliptic plane. Its poles therefore spend 42 years in sunlight, followed by 42 years in darkness. Highly exaggerated seasonal variations are therefore experienced by both the planet and its satellites. The cause of the tilt is not known; a possibility is a collision during the final stages of its **accretion**.

URANUS: DATA	
Globe	
Diameter (equatorial)	51,118 km
Diameter (polar)	49,947 km
Density	1.29 g/cm^3
Mass (Earth = 1)	14.53
Volume (Earth = 1)	62.18
Sidereal period of axial rotation (equatorial)	17h 55m (retrograde)
Escape velocity	21.3 km/s
Albedo	0.51
Inclination of equator to orbit	97° 52′
Temperature at cloud-tops	55 K
Surface gravity (Earth = 1)	0.79
Orbit	
Semimajor axis	19.191 AU = 2871 × 10^6 km
Eccentricity	0.046
Inclination to ecliptic	0° 46′
Sidereal period of revolution	84.01y
Mean orbital velocity	6.81 km/s
Satellites	15

The fly-by of the Voyager 2 probe in 1986 provided most of our current knowledge of the planet. The upper atmosphere is about 83% molecular hydrogen, 15% helium, and the other 2% mostly methane, whose strong absorption of red light gives Uranus its predominant hue. Strong enhancement of Voyager images was needed to show anything other than the featureless disk visible from Earth. Faint banding and a haze over the Sun-facing pole, possibly of hydrocarbons produced by the effect of solar ultraviolet radiation on methane, were discernible, along with traces of methane clouds.

Near-infrared images obtained from the ground and from the Hubble Space Telescope have since revealed dark spots and bright clouds similar to those visible when Voyager flew past **Neptune**. It may be that Uranus's 'weather' was unusually calm at the time of the Voyager visit. However, the planet has little internal heat, and certainly insufficient to drive a dynamic meteorology like Neptune's. **Differential rotation** operates on Uranus, atmospheric features taking 16 to 17 hours to rotate, compared with the internal rotation period of over 17 hours. Wind speeds in the atmosphere vary from about 150 to 600 km/h (90 to 370 mile/h).

Uranus's interior structure remains conjectural. There may be an iron–silicate core of anywhere between 1 and 10 Earth masses, surrounded by a deep 'mantle' of water, ammonia and methane. The mantle may be liquid, or 'mushy' – a mixture of solids, liquids and gases in proportions that vary with depth. It is in this region that the planet's magnetic field, which is of comparable strength to Saturn's, originates. The magnetic axis is tilted by 59° to the axis of rotation, and is displaced by nearly 8000 km

(5000 miles) from the planet's centre. (In this respect Uranus is also similar to Neptune.) The magnetic field gives rise to an appreciable **magnetosphere**.

The five largest **satellites** were known before the Voyager encounter, which led to the discovery of ten more. All 15 are regular satellites orbiting in or close to Uranus's equatorial plane. They are all darkish bodies composed of ice and rock.

The main components of Uranus's ring system were discovered from the Kuiper Airborne Observatory in 1977 when the planet occulted a star (SEE **occultation**). Others were imaged by Voyager. The brightest and outermost is the Epsilon Ring; the innermost is a very diffuse sheet of material. Between the two are nine narrow, darkish rings. SEE ALSO **ring, planetary**

Ursa Major A famous northern constellation, the Great Bear, whose main pattern is known as the Plough or Big Dipper. This feature consists of seven stars: Alpha (Dubhe), magnitude 1.8; Beta (Merak), 2.4; Gamma (Phekda), 2.4; Delta (Megrez), 3.3; Epsilon (Alioth), 1.8; Zeta (**Mizar**, a famous double), 2.3; and Eta (Alkaid), 1.9. Dubhe and Merak are the 'pointers' to the pole star. Five of the Plough stars make up a **moving cluster**, the exceptions being Dubhe and Alkaid. There are many galaxies in Ursa Major, including several Messier objects (M81, M82, M101, M108 and M109) as well as the **Owl Nebula**.

Ursa Minor The constellation that contains the north celestial pole. Its brightest star is Alpha (**Polaris**), the pole star, magnitude 2.0. The constellation's seven main stars make a pattern that gives the impression of a faint and distorted Plough, or Little Dipper.

UT ABBREVIATION FOR **Universal Time**

UV Ceti star Another name for a **flare star**. UV Ceti itself is a binary consisting of 13th-magnitude red dwarf components, which exhibits large flares every few hours. It is among the closest stars to us, lying 8.7 light years away.

V

Valhalla A large impact basin on Jupiter's satellite **Callisto**, surrounded by a series of concentric rings representing 'ripples' from the impact.

Valles Marineris SEE **Mariner Valley**

Van Allen Belts Two zones of plasma (high-energy charged particles) in the Earth's **magnetosphere**. The concentric, tyre-shaped belts trap charged particles which then spiral around magnetic field lines, back and forth between the two magnetic poles. The diagram, which is roughly to scale, shows the form of the belts. The outer belt contains mainly electrons from the **solar wind**, and the inner belt mainly protons from the same source. Within the inner belt is a radiation belt consisting of particles produced by interactions between the solar wind and heavier **cosmic ray** particles. The activity of the Van Allen Belts varies with the level of solar activity (SEE **solar cycle**). Because the Earth's magnetic axis is offset from its rotational axis, the belts are not a uniform distance from the Earth's surface. The inner belt comes very close to the surface over the South Atlantic Ocean; this is called the *South Atlantic Anomaly*. The two main belts were discovered by American physicist James Alfred Van Allen (1914–) and his collaborators during investigations using Geiger counters on board the first American Explorer satellites in 1958. The cosmic-ray zone was found by Russian Cosmos satellites in 1991.

variable star A star whose brightness varies with time. *Intrinsic variables* are stars that vary because of some inherent feature, such as pulsations in size or events in the atmosphere. In *extrinsic variables*, external factors such as eclipses or obscuring dust affect the amount of light reaching us from

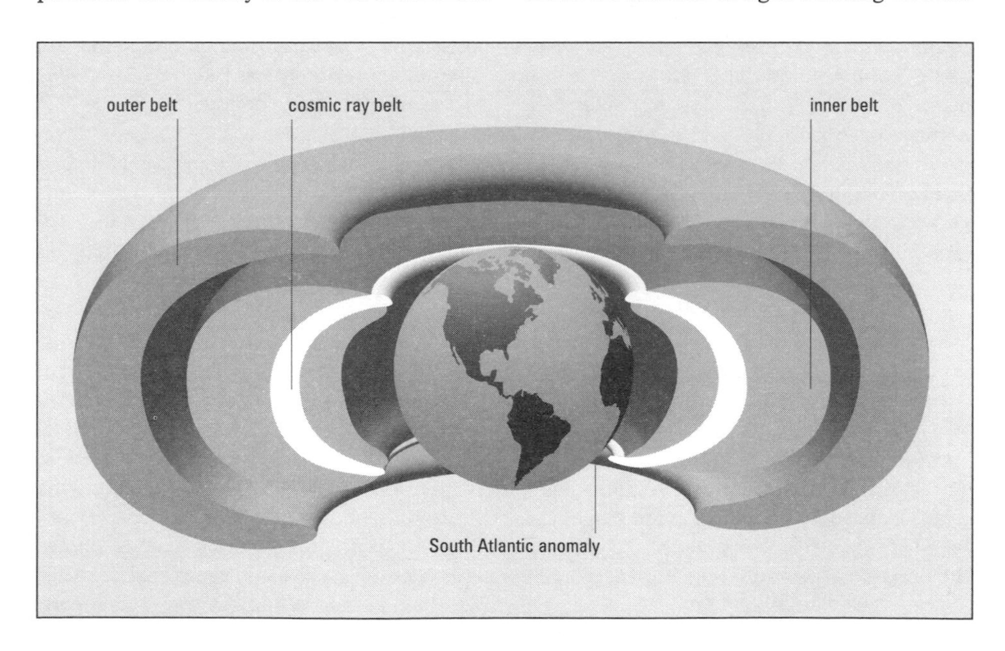

Van Allen Belts: the radiation zones which surround the Earth

the star. There are three main classes of variable. **Eclipsing binaries** are extrinsic variables in which members of a binary system periodically pass in front of each other as seen from Earth. **Pulsating stars** are intrinsic variables which expand and contract in size, either regularly or irregularly. **Cepheid variables** and **RR Lyrae stars** are examples of pulsating stars with regular cycles of variation. Less regular variation is found among the **long-period variables** (or Mira stars), **semiregular variables**, and **irregular variables**. Eruptive (or cataclysmic) variables are a mixed class, including **supernovae, novae, dwarf novae, flare stars, R Coronae Borealis stars**, and **shell stars**. Each main group also has its own sub-classification.

Variable stars have their own peculiar naming system. The letter R is assigned to the first variable to be discovered in a particular constellation (unless it already has an existing designation, such as a **Bayer letter** or Flamsteed number). Further discoveries are denoted by S to Z. Then come RR to RZ, then SS to SZ, and so on down to ZZ. After that come AA to AZ, BB to BZ, and so on down to QZ, but the letter J is never used. At this point the 334th variable has been reached. Subsequent variables are numbered V335, V336, and so on.

variation A perturbation in the Moon's motion caused by the Sun's changing gravitational pull on the Moon as the Moon orbits the Earth.

Vega The star Alpha Lyrae, fifth-brightest in the sky, magnitude 0.03, distance 25 light years, luminosity 50 times that of the Sun. It is the standard star of spectral type A0 and luminosity class V in the **UBV system**. The Infrared Astronomy Satellite, IRAS, found a disk of gas and dust extending out to 85 AU from it, possibly a planetary system in the process of formation.

Vega probes Two space probes launched by the Soviet Union in December 1984 which dropped balloons into the atmosphere of Venus in June 1985 and encountered Halley's Comet in March 1986. They flew past the comet's nucleus at distances of about 8000 km (5000 miles), photographing it and analysing the dust and gas given off.

Veil Nebula The brightest part of the **Cygnus Loop**, a large supernova remnant in Cygnus. The portion known as the Veil Nebula is designated NGC 6992.

Vela A constellation representing the sails of the dismembered constellation Argo Navis. Its leading star is Gamma, magnitude 1.8 but slightly variable; this is the brightest **Wolf–Rayet star**, with a luminosity 40,000 times that of the Sun. Vela is a rich region; Delta (magnitude 2.0) and Kappa (2.5) make up the **False Cross** with Iota and Epsilon Carinae. Other features of the constellation are the **Gum Nebula** and the **Vela Pulsar**.

Vela Pulsar A pulsar with a period of 89 milliseconds, discovered in 1968 in the constellation Vela; known as PSR 0833-45. The optical counterpart, discovered in 1977, flashes with the same period.

Venera A series of space probes launched by the Soviet Union to the planet Venus between 1961 and 1983. Venera 7 in 1970 was the first craft to land successfully on Venus, although it transmitted for only 23 minutes before being destroyed by the extreme temperatures and pressures at the surface. Venera 8 in 1972 proved more resilient and transmitted for 50 minutes. Veneras 9 and 10 in 1975 took the first photographs of the surface. Veneras 11 and 12 landed in 1978 and sent back information about the atmosphere, but no pictures. Veneras 13 and 14 landed in 1982, taking colour photographs of the surface and sampling the rocks. Veneras 15 and 16, both radar mappers, went into orbit around the planet in October 1983 and operated until the following July.

Venus The second major planet from the Sun, and the second-largest of the four inner planets. Visible around dawn or dusk as the so-called *morning star* or *evening star*, it is the most conspicuous celestial object, after the Sun and Moon. At its brightest (magnitude −4.7) it is even visible to the naked eye in the daytime when the Sun is high in a clear sky. At greatest elongation Venus is 45° to 47° from the Sun, and has an apparent diameter of 25 arc seconds. Like the Moon and Mercury, it is seen to go through **phases**. **Transits** of Venus are rare, and occur in pairs 8 years apart, with over a century between each pair. The next pair are on 7 June 2004 and 5 June 2012. There is no known satellite. The main data for Venus are given in the table (right).

A telescope shows the planet's dazzling yellowish white cloud cover, with faint markings. In ultraviolet images these markings show clearly as a Y-shaped feature, enabling the clouds' rotation period of just under 4 days (nearly 60 times as fast as the rotation period of the planet itself) to be measured. The **ashen light**, the **cusp caps**, and the **Schröter effect** may be watched for by amateur observers; their causes have been long debated, but they are most likely real atmospheric phenomena rather than optical effects, as has sometimes been claimed.

Venus's very high surface temperature was indicated by measurements at radio wavelengths in 1958, but space probes were needed to reveal more about the surface. The most successful were the long Soviet **Venera** series, the two US **Pioneer Venus** probes, and, most recently, **Magellan**. After early failures, Venera craft were built to withstand the extreme surface conditions – a temperature of 750 K and a pressure of over 90 bars – for long enough to transmit measurements and pictures. Magellan has produced highly detailed radar mappings of the surface.

A gently undulating plain covers two-thirds of Venus. Highlands account for a further quarter, and depressions and chasms the remainder. The two principal highland regions are Ishtar Terra in the northern hemisphere, and Aphrodite Terra, largely in the southern. On Ishtar Terra is the Maxwell Montes mountain range, rising to 11 km (7 miles) above the average level, and the highest point on Venus. The plains are cratered; the largest crater is Mead, 280 km (170 miles) in diameter, but there are no craters smaller than a few kilometres across as smaller objects would have been destroyed by their passage through the dense atmosphere. However, most of the surface features are volcanic in origin. There are large shield volcanoes like Rhea Mons in the Beta Regio highlands, and calderas like Colette, extensive lava flows from which have covered the plateau Lakshmi Planum. The volcanic peak Maat Mons, at 8.5 km (5.3 miles) high the second-highest point on Venus, is surrounded by lava flows which have been estimated at no more than 10 years old. If this is so, then Venus is almost certainly still volcanically active. There are many other types of near-circular

VENUS: DATA

Globe	
Diameter	12,104 km
Density	5.25 g/cm^3
Mass (Earth = 1)	0.8149
Volume (Earth =1)	0.8568
Sidereal period of axial rotation	243d 0h 30m (retrograde)
Escape velocity	10.4 km/s
Albedo	0.76
Inclination of equator to orbit	177° 20′
Surface temperature	750 K
Surface gravity (Earth =1)	0.90

Orbit	
Semimajor axis	0.723 AU = 108.2 × 10^6 km
Eccentricity	0.007
Inclination to eliptic	3° 24′
Sidereal period of revolution	224.701d
Mean orbital velocity	35.02 km/s
Satellites	0

volcanic feature, which have been given names such as *arachnoids*, *coronae*, and *ovoids*. With a handful of exceptions, the planet's surface features are named after famous women, some from legend, from all the world's cultures.

The atmosphere consists of 96% carbon dioxide and 3½% nitrogen, with traces of helium, argon, neon, and krypton. Its great density has produced the very high surface temperature by an extreme **greenhouse effect**. At various altitudes there are haze layers of sulphuric acid, sulphur dioxide, water vapour and sulphur; and ultraviolet light from the Sun causes some carbon dioxide to break down into carbon monoxide and oxygen. Pioneer Venus orbiter measurements showed that the sulphur dioxide level in the cloud-tops dropped tenfold between 1979 and 1986, which may reflect changes in the level of volcanic activity. A chemically active atmosphere is probably the only agent for 'weathering' of surface features. Wind speeds in the atmosphere, which are 350 km/h (220 mile/h) at the cloud-tops, fall to 10 km/h (6 mile/h) at the surface.

The internal structure is largely unknown. Venera measurements indicate a low-density crust. There may be a silicate mantle and iron–nickel core, but that is just guesswork based on analogy with the similar-sized Earth. Certainly there is no appreciable magnetic field, although the solar wind does interact strongly with Venus's **ionosphere** to produce a well-defined bow shock – a **magnetosphere** of sorts.

vernal equinox The point at which the Sun crosses the celestial equator from south to north. It is also known as the *spring equinox* and the *First Point of Aries*. It is the zero point of the celestial coordinate called right ascension. The vernal equinox moves westwards by about one-seventh of a second of arc daily because of the effect of **precession**.

vertical circle A **great circle** on the **celestial sphere** that passes through the zenith and the nadir, and is thus at right angles to the observer's horizon. The altitudes of celestial objects are measured along great circles.

Very Large Array (VLA) The most complex radio telescope on a single site in the world. It consists of twenty-seven dishes of 25 m (82 ft) diameter, movable along the arms of a giant Y on the plains near Socorro, New Mexico. Each arm can be up to 21 km (13 miles) long, using the technique of **aperture synthesis** to produce a virtual dish some 34 km (21 miles) wide.

Very Long Baseline Array (VLBA) An array of ten 25 m (82 ft) radio telescopes spread across the continental USA and Hawaii, giving a baseline for **aperture synthesis** 8000 km (5000 miles) long. Resolutions of 0.001 arc second are possible. The instrument began operation in 1993. SEE ALSO **radio interferometer**

Vesta **Asteroid** no. 4, diameter 501 km (311 miles), discovered in 1807 by Heinrich Olbers. At magnitude 6.4 it is the brightest asteroid and the only one ever visible (under ideal conditions) to the naked eye. Vesta's spectrum suggests that its surface, unique among the larger asteroids, was once molten (SEE **achondrite**).

vignetting A defect in an optical instrument in which light is unevenly distributed across the field of view, being typically darker near the edges than at the centre. The most common cause is an obstacle in the light path.

Viking Two American space probes to Mars, launched in 1975. Each was a combined orbiter and lander. Viking 1 went into orbit around Mars in June 1976. The lander spacecraft was separated from the orbiter and landed on 20 July 1976. Viking 2 went into orbit in August 1975 and its lander reached the surface on 3 September 1976.

Both landers photographed their surroundings, reported the weather at their locations, and studied the planet's soil. Experiments conducted by the landers found no indications of life at the landing sites. The orbiters photographed the surface of the planet and its two small moons, Deimos and Phobos.

Virgo The largest constellation of the zodiac, lying on the celestial equator, and representing the goddess of justice. Its brightest star is **Spica**. Gamma is a splendid binary with equal components of magnitude 3.5, period 169 years; they are closest in the year 2005. Virgo is one of the richest areas for faint galaxies since it includes the **Virgo Cluster**. Closer to us than this cluster is the **Sombrero Galaxy**.

Virgo Cluster A rich concentration of galaxies in the direction of the constellation Virgo, although some lie across the border in Coma Berenices. The centre of the cluster is about 45 million light years away, and is marked by the giant elliptical galaxy M87, also known as the radio source Virgo A, which is ejecting a jet of gas. In all there are 16 Messier objects in the cluster; the total number of galaxies in the cluster is about 3000.

visible spectrum The range of wavelengths in the electromagnetic spectrum perceptible to the human eye. It varies slightly from one person to another, but extends typically from 385 nm (violet) to 700 nm (red). SEE ALSO **electromagnetic radiation**

visual binary A binary system that can be observed as a double star with a telescope, as distinct from a spectroscopic binary. SEE **binary**

visual magnitude (symbol m_V) The **apparent magnitude** of a celestial object in the colour region to which the human eye is most sensitive, i.e. about 560 nm.

VLA ABBREVIATION FOR **Very Large Array**

VLBA ABBREVIATION FOR **Very Long Baseline Array**

Vogel, Hermann Carl (1841–1907) German astronomer. He was a skilled user of astronomical instruments, and specialized in spectrometry and spectral analysis. From stellar spectra he measured radial velocities, and in 1889 he discovered **spectroscopic binaries**, deriving the masses and orbits of the double stars Algol and Spica.

Volans A small, faint southern constellation adjacent to Carina, representing a flying fish. Its brightest stars are of 4th magnitude.

volatile A substance that solidifies at a low temperature. Substances such as water and carbon dioxide, familiar as gases on Earth, are examples. Bodies in the Solar System that formed nearer the Sun have a lower volatile content than those that formed further away. Volatiles in comets, which formed in the outer reaches of the Solar System, evaporate as the gas tail when the comet approaches the Sun.

volcanism The eruption of molten material at the surface of a planetary body. Volcanism in which molten silicate rock erupts, cools, and becomes new surface material has taken place on Venus and Mars as well as the Earth, and similar material produced the lunar maria (SEE **mare**) by volcanic flooding. In the outer Solar System materials such as water- and ammonia-ices can behave as lava, and have produced, for example, the volcanic flood-plains on **Ariel**. Sulphur volcanism operates on **Io**, the most volcanically active world known. On **Triton**, dark particles are carried aloft by nitrogen geysers termed *plumes*.

Voyager Two American space probes to the outer planets. Both were launched in 1977. They passed Jupiter in March and July

1979, photographed the planet and satellites, and discovered Jupiter's ring system. Voyager 1 reached Saturn in November 1980, followed by Voyager 2 in August 1981, photographing the rings in detail, as well as the moons. Voyager 1's path was bent out of the ecliptic by its approach to Titan, but Voyager 2 proceeded to Uranus, reaching it in January 1986, and Neptune, in August 1989. Both Voyager probes are now on their way out of the Solar System, their last task to locate the heliopause, the boundary of the **heliosphere**.

Vulcan A supposed planet within the orbit of **Mercury**. It was originally invoked to explain, by its gravitational effect, the residual in the advance of Mercury's perihelion, since accounted for by Einstein's General Theory of Relativity. In 1859 a French physician, Edmond Lescarbault, observed a small body in transit across the Sun's disk and made subsequent careful observations of it. **Urbain Le Verrier** examined Lescarbault's observations closely, and was convinced of his integrity and the planet's existence, but no further trace of it has ever been found.

Vulpecula A dim constellation next to Cygnus, representing a fox. Its brightest stars are of 4th magnitude. It contains the **Dumbbell Nebula**, reputedly the easiest planetary nebula to see in the sky.

Weizsäcker, Carl Friedrich von (1912–) German theoretical physicist and astrophysicist. In 1938 he and **Hans Bethe** independently proposed a detailed theory for the production of energy in the Sun and other stars (SEE **carbon–nitrogen cycle**) in which hydrogen is converted into helium by nuclear fusion. In 1944 he revived the *nebular hypothesis* of the origin of the Solar System originally proposed by Immanuel Kant and **Pierre Simon de Laplace**.

West, Comet One of the brightest comets of recent years, discovered by Danish astronomer Richard West in 1975. The following year it became a prominent naked-eye object with a fan-shaped tail. The comet was extremely active, and the nucleus broke into at least four fragments as it passed within 30 million km (20 million miles) of the Sun.

Westerbork Radio Observatory A radio astronomy observatory located near Groningen in the Netherlands, operated by the Netherlands Foundation for Research in Astronomy. Its major instrument is an **aperture synthesis** telescope consisting of fourteen dishes 25 m (82 ft) in diameter, along an east–west baseline 3 km (2 miles) long.

Whipple, Fred Lawrence (1906–) American astronomer. He is best known for his 'dirty snowball' theory of comets, proposed in 1949 and shown to be correct in 1986 when space probes were sent to Halley's Comet. Whipple discovered several comets and worked on cometary orbits; he also studied planetary nebulae, stellar evolution, and the Earth's upper atmosphere.

Whirlpool Galaxy A well-defined spiral galaxy of type Sc which appears face-on to us and has very prominent arms; also known as M51 and NGC 5194. It lies in Canes Venatici, not far from the end of the tail of Ursa Major, and was the first galaxy in which spiral structure was noted, by Lord **Rosse** in 1845. A small companion galaxy, NGC 5195, appears to be connected to it by an extension of one of the spiral arms. The Whirlpool Galaxy is about 15 million light years away.

white dwarf A type of star about the size of the Earth, but with a mass about that of the Sun. As a result, its density is enormously greater than that of any terrestrial material (0.1 to 100 tonnes/cm^3). This is because the normal atomic structure is broken down completely, with electrons and nuclei packed tightly together (the state known as **degenerate matter**). A white dwarf cannot have a mass of more than about 1.4 solar masses (the **Chandrasekhar limit**). For larger masses, gravity would always overwhelm the pressure of the electrons and the star would collapse under its own weight, forming a **neutron star** or **black hole**. White dwarfs are of low luminosity and gradually cool down to become cold, dark objects. They represent the final stage in the evolution of low-mass stars after they have lost their outer layers. The first white dwarf to be discovered was the companion of **Sirius**.

Widmanstätten pattern The distinctive pattern on a sectioned and polished surface of an **iron meteorite** that has been etched with acid. It is named after Count Aloys Joseph von Widmanstätten (1754–1849), who observed it in 1808. The pattern is only seen in meteoritic material, and consists of intersecting plates of the iron–nickel minerals kamacite and taenite.

William Herschel Telescope A reflecting telescope with a 4.2 m (165-inch) mirror at the **Roque de los Muchachos Observatory** in the Canary Islands. It was opened in 1988 and is owned and operated by the **Royal Greenwich Observatory**.

Wilson, Robert Woodrow (1936–) American physicist. With **Arno Penzias**, he detected in 1964 the **cosmic microwave background** radiation that provides the strongest evidence for the **Big Bang**. They shared the 1978 Nobel Prize for Physics for their discovery.

Wilson effect The foreshortening of a **sunspot** when it is near the Sun's limb, accompanied by a widening of the penumbra on the side nearest the limb, and a narrowing on the side farthest from the limb. The phenomenon was discovered by the Scottish astronomer Alexander Wilson (1714–86), who took it as an effect of perspective, indicating that sunspots are saucer-shaped depressions. However, not all spots show the Wilson effect. The cause is now thought to be that sunspots are more transparent than the surrounding photosphere, and the umbra more transparent than the penumbra.

winter solstice SEE **solstice**

Wolf, Maximilian Franz Joseph Cornelius (Max) (1863–1932) German astronomer. He was a pioneer of photographic methods in astronomy. In 1891 he made the first photographic discovery of an asteroid (Brucia, no. 323); in all he made 232 asteroid discoveries. He detected several new nebulae, including the **North America Nebula**, the periodic comet (1884 III) which bears his name, the nebulosity around the Pleiades, numerous dark nebulae, and the first cluster of galaxies (in Coma Berenices) to be identified as such, as well as the **Virgo Cluster**.

Wolf number SEE **relative sunspot number**

Wolf–Rayet star (WR star, W star) A star whose spectrum contains bright emission lines rather than dark absorption ones. WR stars are divided into two kinds. In the WN type, emission lines from nitrogen dominate the spectrum. In the WC type, emission lines of carbon and oxygen predominate. Both types have strong lines of helium and a few have moderate or weak lines of hydrogen as well. All the emission lines are broad. WR stars are very hot, with surface temperatures between 25,000 and 50,000 K, luminosities between 100,000 and 1 million times the Sun's, and masses from 10 to 50 solar masses. They have strong **stellar winds** that carry away 3 solar masses per million years. Many central stars of planetary nebulae are WR stars. The type of WR star that emerges after the original hydrogen-rich envelopes have been lost depends on how far the star's evolution had progressed. Wolf–Rayet stars are named after their French discoverers, Charles Joseph Etienne Wolf (1827–1918) and Georges Antoine Pons Rayet (1839–1906).

wrinkle ridge A winding ridge on the surface of a lunar mare, having sloping sides rising to a typical height of 200 m (650 ft), and several hundred kilometres long. They are often associated with **rilles**, and may have originated when the lava that formed the maria colled and contracted.

W star Another name for a **Wolf–Rayet star**.

W Ursae Majoris star A member of a class of eclipsing binary stars, almost in contact as they orbit one another. They differ from **Beta Lyrae stars** in that the two components are smaller, less luminous stars of nearly identical brightness.

W Virginis star A member of a group of pulsating variable stars superficially similar to Cepheids. They are giant stars, typically in the spectral range G0 to M0 and with absolute magnitudes from −1 to −4, on the instability strip in the **Hertzsprung–Russell diagram**. Their masses can be as low as 0.5 solar masses, suggesting that they have evolved from low-mass main-sequence stars. Light curves of W Virginis stars can be distinguished from those of classical Cepheids by their less regular shape and a double-peaked maximum. Their periods range from about 1 to 100 days, and their **period–luminosity law** is distinctly different from that of normal Cepheids; however, they can still be used as distance estimators for galactic and extragalactic objects. Some belong to Population I and some to Population II. Occasionally W Virginis stars show small period changes but in one case, RU Camelopardalis, the variations suddenly stopped in 1964 for about three years. It is not understood how pulsations can stop, or start, in such a short time.

XMM ABBREVIATION FOR **X-ray Multi-Mirror Mission**

X-ray astronomy The study of **electromagnetic radiation** of short wavelength (about 0.01 to 10 nm), shorter than ultraviolet radiation but longer than gamma-rays. The Earth's atmosphere is opaque to X-rays, so this radiation must be studied by means of instruments aboard rockets or satellites. For astronomical purposes X-rays are usually described according to the energy associated with the X-ray photons; this energy is expressed in **electron-volts**.

X-ray astronomy began in 1949 with the discovery, by a team at the United States Naval Research Laboratory using captured German rockets, that the Sun emits X-rays. By the time that the first X-ray satellites were launched, the general properties of the Sun's X-ray emission had been established. Other celestial X-ray sources were discovered by rocket flights in the 1960s. The first all-sky survey at X-ray wavelengths was made by the US satellite Uhuru, launched in 1970. Uhuru, and later the UK satellite Ariel V, discovered the existence of **X-ray binaries**, in which one component is a neutron star or a black hole, as well as X-ray emission from supernova remnants, active galaxies, quasars, and hot gas pervading clusters of galaxies. An **X-ray telescope** aboard the Einstein satellite, launched in 1978, produced the first true images of X-ray sources, rather than simply counting X-rays as previous instruments had done. Objects in nearby galaxies such as M31 and the Magellanic Clouds were resolved and studied for the first time. It was also discovered that almost all types of ordinary star emit X-rays. Subsequent X-ray satellites have included Exosat, Rosat and **Yohkoh**. Future major X-ray satellites will include the US Advanced X-ray Astrophysics Facility (AXAF) and the European X-ray Multi-Mirror Mission (XMM).

X-ray binary A binary system consisting of a normal star and a collapsed star – a **neutron star**, **black hole**, or, in less intense sources, a **white dwarf**. The two components are very close together. In some cases, where the normal star is a giant or supergiant, a stellar wind blows directly on to the compact companion, causing the X-ray emission. In other cases, expansion of the normal star in the course of its evolution results in a flow of gas towards the collapsed star. This matter forms an **accretion disk** of hot material which emits X-rays. SEE ALSO **X-ray burster**

X-ray burster A source of intense flashes of X-rays. These X-ray bursts have rise times of about 1 second, fall times of about 60 seconds, and total luminosities equivalent to one week's energy output of the Sun. The intervals between bursts are irregular, ranging from hours to days, while many sources undergo burst-inactive phases that can last for weeks or even months. Burst sources are associated with old Population II objects and emit no pulsations. They are thought to be binaries containing an old neutron star with no magnetic field, which explains the absence of pulsations. The neutron star is paired with an old low-mass star, and individual bursts come from thermonuclear explosions after material accreted from the companion star has exceeded a critical mass on the neutron star's surface. *X-ray transients* are nova-like outbursts of X-rays believed to come from similar binaries in which the mass transfer is very uneven.

X-ray Multi-Mirror Mission (XMM) A European Space Agency satellite for X-ray astronomy, planned for launch in 1999.

X-ray telescope An instrument for imaging X-ray sources. All X-ray telescopes are satellite-borne because X-rays do not penetrate the Earth's atmosphere. Unlike less energetic electromagnetic radiation, X-rays cannot be focused by reflection from a conventional concave mirror (as in a **reflecting telescope**) because, for all but the shallowest angles of incidence, they penetrate the surface. In *grazing incidence* X-ray telescopes the focusing element is a pair of coaxial surfaces, one paraboloidal and the other hyperboloidal, off which incoming X-rays are reflected at a very low 'grazing' angle towards a focus. The detecting element is often a **CCD** adapted for X-ray wavelengths. An alternative instrument is the *microchannel plate detector*. The incident radiation falls on a plate which is made up of many fine tubes, rather like a short, wide fibre-optic bundle. The plate is charged, so that the radiation generates electrons which are accelerated down the tubes and together form an image that can be read off.

X-ray transient SEE **X-ray burster**

Yagi antenna A basic form of antenna used in simple **radio telescopes**. It consists of several parallel elements mounted on a straight member, and often forms the basis of cheap arrays used in **aperture synthesis**; it is also a familiar form of TV aerial. It was developed by the Japanese engineer Hidetsugu Yagi (1886–1976).

year The time taken by the Earth to complete one revolution around the Sun. Various years, defined according to the choice of reference point, are given in the table. The *civil year* (the **calendar** year) averages 365.2425 **mean solar days**. SEE ALSO the entries for each type of year listed in the table below.

TYPES OF YEAR	
anomalistic year	365.25964 mean solar days
eclipse year	346.62003 mean solar days
sidereal year	365.25636 mean solar days
tropical year	365.24219 mean solar days

Yerkes Observatory The observatory of the University of Chicago, at Williams Bay, Wisconsin. It was founded by **George Ellery Hale**. Its main instrument is a 1 m (40-inch) refractor, opened in 1897 and still the largest in the world.

Yohkoh A satellite launched in 1991 to study the Sun, particularly solar flares, at X-ray and gamma-ray wavelengths. The mission was a Japanese–British–American collaboration; Yohkoh is Japanese for 'sunbeam'. The satellite returned many X-ray images of the Sun, the first obtained from orbit since Skylab missions in 1973. Its instruments revealed much about the rapid changes that occur in the corona, and how flares originate and evolve. In 1993 it recorded an 'X-ray transit' of Mercury, as the planet passed across the solar corona, blocking X-ray emission from it.

Young, Charles Augustus (1834–1908) American solar astronomer. He made some of the earliest spectroscopic studies of the corona, and discovered the **flash spectrum** of the chromosphere during a total solar eclipse in 1870. He also used spectral line measurements to determine the Sun's rotation period.

Z

Z Andromedae star SEE **symbiotic variable**

Z Camelopardalis star A member of a small subgroup of **dwarf novae** similar to **U Geminorum stars**, except that they experience occasional 'standstills', remaining more or less constant at some intermediate brightness. The occurrence of the standstills, and their duration, ranging from a few days to many months, are unpredictable.

Zeeman effect The splitting of a spectral line into several components by a strong magnetic field. In circumstances where these components cannot be resolved, the effect is apparent as a widening of the original line. The Zeeman effect occurs in the spectra of sunspots and stars. It demonstrates the existence of magnetic fields in celestial bodies and, since the field strength depends on the degree of splitting, it allows the fields to be measured. The phenomenon was predicted by Hendrik Lorentz, and discovered in 1896 by Pieter Zeeman (1865–1943). The two Dutch physicists shared the 1902 Nobel Prize for Physics.

Zelenchukskaya Observatory An observatory on Mount Pastukhov in the Caucasus Mountains of southern Russia, site of the 6 m (236-inch) Large Azimuthal Telescope, which was opened in 1975. Also at Zelenchukskaya is the RATAN-600 radio telescope, which consists of a ring of reflecting panels 600 metres in diameter.

zenith The point on the **celestial sphere** vertically above the observer, 90° from the horizon. The point directly opposite, beneath the observer's feet, is the *nadir*.

zenithal hourly rate (ZHR) The number of meteors that would be seen at the maximum of a particular **meteor shower** under good conditions if the radiant were immediately overhead. The ZHR is obtained by applying a number of correction factors to the observed hourly rate of meteors in the shower. The observed hourly rate is always less than the ZHR. There are short-term and long-terms variations in showers' ZHRs as the associated **meteor streams** evolve.

zenith distance The angular distance of a celestial body from the zenith. It is equal to 90° minus the altitude of the body above the horizon.

zero-age main sequence (ZAMS) The **main sequence** on the Hertzsprung–Russell diagram as defined by stars of zero age, i.e. before they have undergone any substantial evolution.

zodiac A belt on the celestial sphere, about 8° either side of the ecliptic, which forms the background for the motions of the Sun, Moon, and planets (except Pluto). The zodiac is divided into twelve *signs*, each 30° long, which are named after the constellations they contained at the time of the ancient Greeks: Aries, Taurus, Gemini, Cancer, Leo, Virgo, Libra, Scorpio, Sagittarius, Capricorn, Aquarius and Pisces. Because of **precession** these signs no longer coincide with the constellations of the same name. The ecliptic also passes through the constellation Ophiuchus, which is not a zodiacal sign.

zodiacal light A cone of faint light, usually fainter than the Milky Way, visible at all seasons in the tropics in the absence of moonlight. It stretches along the ecliptic from the western horizon after evening twilight, or from the eastern horizon before morning twilight. At the point in the sky opposite the Sun, the zodiacal light brightens a little; this is the **gegenschein**. The spectrum of the zodiacal light resembles the Sun's, indicating that the phenomenon results from the scattering of sunlight by particles in the plane of the ecliptic. These particles constitute the *zodiacal dust cloud*, and probably originate partly from matter ejected by the Sun, and partly from the decay of comets and asteroids.

Zond The name of a series of eight Soviet space probes launched between 1964 and 1970, mostly as technology test flights. Zonds 5, 6, 7 and 8 all flew around the Moon and landed back on Earth.

zone of avoidance A region of the sky along the plane of the Milky Way where almost no external galaxies can be seen because of absorption by interstellar gas and dust in our Galaxy. It is between 10° and 40° wide.

Zwicky, Fritz (1898–1974) Swiss–American astronomer, born in Bulgaria. In 1934 he and **Walter Baade** suggested that what is left after a supernova explosion is a neutron star; this was confirmed in 1968 with the discovery of the pulsar in the **Crab Nebula**. Using the newly developed **Schmidt camera**, Zwicky discovered and examined many supernovae in other galaxies, and began a long study of **clusters of galaxies** that culminated with a catalogue of 10,000 galaxies and clusters which he completed shortly before his death.